Special Publication No 31

The Modern Inorganic Chemicals Industry

The Proceedings of a Symposium organised
by the Inorganic Chemicals Group of the
Industrial Division of The Chemical Society

London, March 31st — April 1st, 1977

Edited By
R. Thompson, Borax Consolidated Ltd

The Chemical Society
Burlington House, London, W1V OBN

ISBN 0 85186 158 X

Copyright © 1977
The Chemical Society

First published 1977
Reprinted 1979

PREFACE

The Inorganic Chemicals Group, a subject group within the Industrial Division of The Chemical Society, was established in 1972 to promote the industrially relevant aspects of inorganic chemistry. In pursuit of this objective its committee has organised and run specialist symposia annually in various parts of the country, but decided to take the opportunity of the 1977 Annual Congress in London to arrange a more ambitious meeting which would review the inorganic chemicals industry as a whole. The papers presented at that symposium form the basis of this Special Publication.

It was particularly appropriate that the symposium should coincide with the Centenary of the Royal Institute of Chemistry, an occasion marked by the opening of the new chemistry galleries at the Science Museum. The inorganic sector of the chemical industry experienced sustained and often rapid growth throughout the era, but because many of its basic processes were developed rather early it has a slightly old-fashioned image in the minds of many students and not a few of their teachers. Many areas of industrial inorganic chemistry lack the glamour associated with other branches of the subject; the overall processes can in most cases be described by equations met, and in their simplicity understood, by the pupil in his early years of school chemistry. What is not so readily appreciated by the more mature student, and by practising chemists outside the industry itself, is that even where basic reactions remain unchanged there has been almost beyond recognition an evolutionary development of processes in respect of yield, production rate, scale of operation, environmental control and general efficiency. Raw materials have also changed in some cases. In others, although the basic chemistry has long been known, only in relatively recent times has there been a demand for the product; which perhaps occasioned the search for a totally different production route than that which may have been convenient or economical in the laboratory.

Modern inorganic chemistry textbooks rarely make more than passing reference to current industrial processes. Indeed, this is seldom the intent or, in their authors' views, the function. Information about such processes frequently is available, at least in broad outline, from the open literature and patents by the time the processes are in operation. Various encyclopaedia of chemistry and chemical technology (e.g. Kirk-Othmer,

Mellor's Treatise) provide excellent reference. An up-to-date textbook on industrial inorganic chemistry is, however, hard to find. Several books attempting to cover the general field of industrial chemistry have been published in recent years but most tend to be philosophical, deal with technological economics and make only scant reference to particular inorganic industries. Where there is attention to specific processes the information is frequently out-of-date, often inaccurate in detail and sometimes even misleading overall. The texts contain material derived at second hand by authors who have no experience of the industries concerned.

The symposium was therefore conceived with publication of its papers very much in mind. The intention, entirely fulfilled, was to invite contributions from authors well versed in their subjects through employment in organisations acknowledged to be amongst the leaders in their respective fields. It was believed that only in this way could the book be truly authoritative and accepted as such by its readership. Although the publication is not (and could not be) a fully comprehensive work on the industry, the major tonnage chemicals of the inorganic sector have been included. Reference has also been made in some chapters to the minor tonnage but high value derivatives, the chemistry of which has added interest.

On being invited to contribute, authors were given broad editorial guidance as to scope and format. No attempt was made to impose a rigid, uniform style. Each was left to treat the subject as he, the authority, considered best. Some members of the audience at the symposium expressed the view that technical jargon had crept into the papers. No apology need be made for this; the language of a particular industry is understood by those within and needs to be learnt by persons dealing with or entering it. Comments were also made about the non-uniformity of units from paper to paper, and the relative absence of S.I. units, as it was felt that this might confuse student readers of the book. Perhaps so, but the world outside is as it is; mixed units are still commonplace in the chemical industry, especially in Great Britain and the United States. Conventions also vary in the treatment of thermochemical data. However, it is believed that no unit used is so esoteric as not to be found in tables of conversion factors and swiftly dealt with by the ubiquitous pocket calculator.

The inorganic sector of the chemical industry is truly international and most of the developed nations have within their boundaries manufacturing capabilities for many of the products covered in this book. The U.K. is no less representative than elsewhere, and our chemicals companies are equally alive to overseas technological developments in their fields. Several authors have in fact described in outline their foreign competitors' processes where different. Three of the chapters were written by contributors from companies abroad. They include that which deals with the education and training of chemists for industry, by an author who has spent the past twenty years in U.S. industry but was previously a chemistry professor. Although the educational systems of the two countries differ in detail, the principles are the same and evidently many of the problems are common with the U.K.

Perhaps this Special Publication will help to fill a few gaps in the academic syllabus. It is not aimed at any particular educational level and hopefully the contents will be as interesting and intelligible to college and university students, and to their teachers, as they should be for qualified chemists in industry who require a handy reference work. The Chemical Society has made every effort to reduce production costs so that the purchase price of the book will be within the reach of the average student. For this reason a photographic reproduction process from authors' submitted typescripts was chosen in order to minimise editorial and related expenses. Colour plates and some half-tone illustrations were provided by the authors' organisations, whose generosity is appreciated. The Society is equally grateful for the manufacturing companies' permission for the information to be published and I, as Convener, express my sincere thanks to the authors for their excellent cooperation at all stages of preparation of both symposium and book.

Raymond Thompson
Chairman, Inorganic Chemicals Group.

CONTRIBUTORS

S.P.S. ANDREW. F.R.S. Senior Research Associate, I.C.I. Agricultural Division, Billingham, Teesside.

T.V. ARDEN. Director, Portals Water Treatment Ltd., Brentford, Middlesex.

A.K. BARBOUR. Environmental Scientist, R.T.Z. Services Ltd., Bristol.

D. BARBY. Inorganic Research Manager, Unilever Research, Port Sunlight, Merseyside.
(T. GRIFFITHS, A.R. JACQUES AND D. PAWSON are also with Unilever Research)

A. CAMPBELL. Section Manager. I.C.I. Mond Division, Runcorn, Cheshire.

G.W. CAMPBELL. Process Research Manager, U.S. Borax Research Corporation, Boron, California, U.S.A.

A.F. CHILDS. Central Research & Development Manager, Albright and Wilson Group, Knightsbridge, London.

C.A. CRAMPTON. Technical Coordinator, Interox Group, Brussels, Belgium.
(G. FABER, R. JONES AND J.P. LEAVER are also with the Interox Group)

R.S. DARBY. Research Associate, Laporte Industries Ltd., Grimsby, South Humberside.

J.B. FARMER. Section Leader Research, Borax Consolidated Ltd., Chessington, Surrey.

H.C. FIELDING. Section Head, I.C.I. Mond Division, Runcorn, Cheshire.

W.J. GRANT. Consultant to B.O.C. Gases Division, Merton, London.

B.E. LEE. Information Scientist, I.C.I. Mond Division, Runcorn, Cheshire.

J. LEIGHTON. Senior Process Manager, Laporte Industries Ltd., Grimsby, South Humberside.

R.B. McDONALD. General Manager, Lunevale Products Ltd., Halton, Lancaster.

W.R. MERRIMAN. Manager, Business Development, Great Lakes Chemical Corporation, West Lafayette, Indiana, U.S.A.

A. PHILLIPS. Senior Process Engineer, Sim-Chem Ltd., Cheadle Hulme, Cheshire.

P.E. POTTER. Section Leader, Chemistry Division, A.E.R.E., Harwell.
(J.R. FINDLAY, K.M. GLOVER, I.L. JENKINS, N.R. LARGE, J.A.C. MARPLES AND P.W. SUTCLIFFE are also with A.E.R.E., Harwell)

R.W. PURCELL. Techno-Commercial Manager of Chlor-Alkali Group, I.C.I. Mond Division, Northwich, Cheshire.

S.L. REDFEARN. Manager, Gases Applications, B.O.C. Gases Division, Merton, London.

R. THOMPSON. Research Director, Borax Consolidated Ltd., Chessington, Surrey.

M.E. TROWBRIDGE. Director General, Chemical Industries Association, London.

Lime Burning
Steam and Power
Developments
Trona

By A Campbell

Commercial Position
Chemical Properties of Chlorine
Chlorine Handling Technology
Reaction of Chlorine with Carbon
   Vinyl Chloride
   Ethylene Chlorination
   Dehydrochlorination of 1,2 - Dichloroethane
   Oxychlorination of Ethylene
   Chlorinated Solvents
   Propylene Oxide
Reaction of Chlorine with Oxygen
Reaction of Chlorine with other Non-Metallics
Reaction of Chlorine with Metals and Metal Salts
Future Prospects

By H C Fielding and B E Lee

Introduction
General History
Occurrence of Fluorine
Uses for Fluorspar
Hydrofluoric Acid
Manufacture of Hydrofluoric Acid
Structure of the Industry
Uses for Hydrofluoric Acid
Inorganic Fluorides

By W J Grant and S L Redfearn

By J B Farmer

By J R Findlay, K M Glover, I L Jenkins, N R Large,
J A C Marples, P E Potter and P W Sutcliffe

1 eV (electron volt)	$= 1.602 \times 10^{-19}$ J
bar	$= 10^5$ N m^{-2} (newtons/sq metre)
psi 1 lb/in^2	$= 6.89476 \times 10^3$ N m^{-2}
g mole	= essentially an accepted SI unit. Formally it is defined as the amount of substance which contains as many elementary entities (specified as to molecules ions etc) as there are atoms in 0.012 kilogram of carbon 12.
BTU/ton	$= 1.038$ J kg^{-1}
micron	$= 10^{-6}$ m

Note use of the term micron for μm (micrometre) is not recognized and is to be deprecated.

1 standard atmosphere (760 mmHg)	$= 101.325 \times$ kN m^{-2}
calorie (mean)	$= 4.186$ J
dyn/cm (1 erg)	$= 10^{-7}$ J
ft/sec	$= 0.3048$ m s^{-1}
lb/gallon	$= 99.78$ kg m^{-3}
ton/inch^2	$= 15.444$ MN m^{-2}
1 tonne	$= 1.00 \times 10^3$ kg
1 UK ton	$= 1.01605 \times 10^3$ kg
short ton	$= 9.07185 \times 10^2$ kg

Economic Aspects of the Inorganic Chemicals Sector

By M.E. TROWBRIDGE

Chemical Industries Association

Introduction

In the beginnings of chemistry our predecessors, the alchemists, were fascinated by the transformational aspects of their subject. They sought methods of changing the abundant things they found around them into more valuable materials. In particular, and not unnaturally, they desired the "philosopher's stone" which they believed would change the base metals into gold. Simultaneously they sought the "alkahest" - the universal solvent, and Paracelsus (1493 - 1541), the founder of Iatro Chemistry, led investigations in an attempt to find the Elixir of Life, which, he hoped, would give the secret of eternal youth.

As alchemy began to develop into modern experimental chemistry a generally accepted terminology was developed. Thus those compounds which occurred in the animal and vegetable kingdom were termed "Organic". Those which were of mineral origin were called "Inorganic".

This distinction was first formalised by Lémery in 1675 in his Cours de Chimie, and for a time it was believed that Organic Chemicals - which had in common the fact that they contained carbon - could only be prepared in organisms, depending as they did on the so-called "life force" - the "vis vitalis" of Berzelius. On the other hand, it was realised that inorganic or "lifeless" materials could be prepared in the Laboratory.

In 1828 Wöhler destroyed this life force theory when he prepared urea from ammonia and cyanic acid - two clearly inorganic materials. From that time onwards the concept of organic synthesis - the knitting together of carbon atoms to form increasingly complex molecular patterns - has held a major fascination for chemists. The result is that, today, many chemistry courses at basic level still treat carbon chemistry as a major subject, whilst gathering together the study of all the other elements into a single rag-bag of largely descriptive cook-book alchemy.

This academic value judgement not only formed the basis of a self-fulfilling prophecy as coal tar derivatives revealed their potentialities, but was subsequently given economic support on an industrial scale by the development of a major industry founded on natural gas and, in the USA, on low-cost well head and petroleum refinery gases. In consequence, today, most politicians, trade union leaders and economic commentators are prepared to speak with authority on "the petrochemical industry" and "Britain's North Sea based chemical opportunity". Few, however, would make even passing reference to, say, silicon chemistry in their articles, speeches or television interviews.

It is a defensible view that in the future history of man the chemistry of the non-carbon elements is going to have an importance and an impact at least as great as that of organic chemistry over the last half century. The realisation of these developments may not lie in the short-term future, but such is the lead time required that it is in the near future that the work will have to be done if we are to change the shape of events at the turn of the century.

In this paper it is hoped to set down some of the background economic scenery of the road which has been trodden, and to give a quantitative technicoeconomic snapshot of where we are today. As a by-product, it is hoped to indicate the shape of probable short-term trends, and, albeit tentatively, to

put down some markers to indicate the general scope of opportunities for the
longer range prospects.

COMPARISON OF DATA

Unfortunately there is no generally-accepted definition of the chemical industry and its component parts which is used for all UK and international statistics.   In this paper the figures, unless otherwise stated, are on the basis
on which they were published.   Thus domestic and production figures are based
as the Standard Industrial Classification[14] (SIC).   Trade figures and some
international comparative data are based on the revised Standard International
Trade Classification (SITC(R)).

Minimum list heading 271.1 of SIC Order V (Inorganic Chemicals - including
Inorganic Gases) is defined as follows:-

> "Manufacturing chemical elements (including carbon) and compounds
> (excluding those containing carbon, except for certain simple com-
> pounds such as calcium carbide, carbon disulphide and carbon dioxide).
>
> Notes: 1.   Prepared inorganic pigments are classified in MLH 277.
>
>   2.   Prepared fertilizers are classified in MLH 278."

Since 1968 the SIC classification agrees more closely with the SITC(R) than
formerly, but the differences between them are sufficiently large as to prevent precise comparisons.   Where it is intended to base serious work on the
data, one is referred to the explanations given in the introduction to the UK
Chemical Industry Statistics Handbook 1976.[8]

In summary, it may be generalised that where reference is made to "inorganic
chemicals" in this paper it covers manufactured chemical elements and compounds
excluding those containing carbon with the exception of simple compounds, such
as calcium carbide, carbon disulphide and carbon dioxide.   It does not include,
however, materials containing inorganic elements and compounds such as paints,
nor does it include fertilizers or formulations such as toiletries, soaps and
detergents, synthetic resins and plastics, or other compounded or manufactured
products, which may have an inorganic content.

In tonnage terms the output of the Inorganics Sector is dominated by bulk com-
modity-type chemicals, whose manufacture is founded on mature technology.   In
the case of the UK Inorganics Sector some such commodities are based on indigenous materials such as sea water, air, fluorite, barite, dolomite, salt,
limestone, etc.   Others depend, when there is no economic UK source, on imported minerals such as sulphur, borax, bauxite and various metallic ores.
Thus the basic feedstocks for the Inorganics Sector range widely as to type
and geographic source.

ORIGINS OF THE MODERN INORGANIC CHEMICALS INDUSTRY IN BRITAIN

The pressure of urbanisation which occurred in the early years of the nineteenth century provided a driving force for the development of the industry of
chemistry analogous to the steam pressure which drove the engines which sym-
bolised, and indeed made possible, the Industrial Revolution.   This is not
the occasion to recite the history of growth of the chemical industry.   It is,
however, worth taking a snapshot of the chemical industry in Britain of a century ago (i.e. in 1877), since this gives a key to understanding the present
position of the Inorganic Sector of the total UK chemical industry.

In simplified form this is shown in Fig. 1.   It is interesting to note that,
apart from towns' gas, no product of any significance had a direct domestic
consumer outlet.   All the other products were "enabling" materials required
by the downstream manufacturing industries which grew so vigorously in Britain
during the second half of the Industrial Revolution.

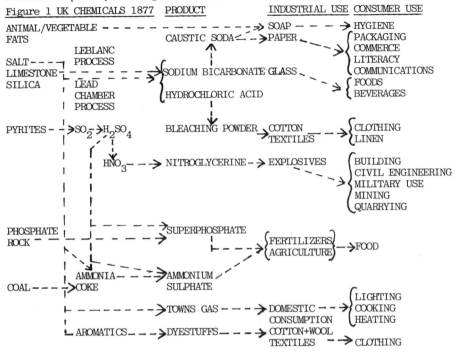

Figure 1 UK CHEMICALS 1877

Figure 2 UK INORGANICS CHEMICAL INDUSTRY 1977 : PRODUCTS

FAMILY	EXAMPLES OF SPECIFIC COMPOUNDS
CHLOR-ALKALIS	BRINE: SODIUM CHLORIDE: SODIUM HYDROXIDE: CHLORINE SODA ASH: SODIUM CARBONATE: HYDROCHLORIC ACID:HYPOCHLORATES
SULPHUR	SULPHUR: PYRITES: ZINC CONCENTRATES: ANHYDRITE: SULPHUR FROM MINERAL OIL: SULPHURIC ACID: ORGANIC SULPHUR
NITROGEN	AMMONIA: AMMONIUM SULPHATE: AMMONIUM NITRATE: UREA NITRIC ACID: CYANIDES: NITROGENOUS FERTILIZERS: (METHANOL)
PHOSPHORUS	ELEMENTAL PHOSPHORUS: PHOSPHORIC ACID: POLYPHOSPHATES: PHOSPHATIC FERTILIZERS: TRIPHENYL TRICRESYL PHOSPHATES
POTASH	CAUSTIC POTASH: POTASSIUM FERTILIZERS: POTASSIUM SULPHATE
ALUMINIUM	ALUMINIUM OXIDE: ALUMINIUM HYDROXIDE: ALUM: ALUMINIUM SULPHATE
BORON	BORIC ACID: BORAX: PERBORATES: BORON CARBIDE
HALOGENS (IODINE:BROMINE)	BROMINE: BROMIDES: IODINE: IODIDES: IODATES
NON-FERROUS METALS	COMPOUNDS OF CHROMIUM: MANGANESE: LEAD: ZINC: MAGNESIUM
FLUORINE	FLUORIDES: ORGANIC FLUORINE COMPOUNDS
SILICON	SILICATES: SILICONES
INDUSTRIAL GASES	NITROGEN: OXYGEN: HYDROGEN: CARBON DIOXIDE: INERT GASES
MISCELLANEOUS	CARBON BLACK: CALCIUM CARBIDE: TITANIUM DIOXIDE: PEROXYGEN CHEMICALS: PRECIOUS AND SEMI-PRECIOUS METALS

It is important also to recall the significance of alkali manufacture in the development of the industry.   Indeed, the term used to refer to what we now call the chemical industry was "the alkali trade".   As Hardie and Davidson Pratt point out[20], the alkali trade was not a one-product industry, but a closely integrated group of processes orchestrated to serve the interests of the textile industry, so important to Victorian prosperity and the links with the developing Empire.   In Britain this alkali trade was originally based on the Leblanc soda process.

Figure 3

RANKING OF COMMONLY TRADED INORGANIC CHEMICALS BY CQM. EXPORT PRICE
BASED ON DEC 1976 CUSTOM STATISTICS. NOTE - THESE REFER MOSTLY TO
BROAD CLASSIFICATIONS AND EXPORT TRADING PRICES.

CLASSIFICATION	£STG/MT	CLASSIFICATION	£STG/MT
GERMANIUM OXIDES	70,000	LEAD OXIDE,RED LEAD,ORANGE LEAD	325
COBALT OXIDES/HYDROXIDES	4,525	BARIUM OXIDES/PEROXIDES/HYDROXIDES	275
TIN OXIDES	3,350	ARTIFICIAL CORUNDUM	250
IODIDES,IODATES,PERIODATES	3,020	IRON OXIDES/HYDROXIDES	250
TUNGSTATES,ANTIMONATES, MOLYBDENATES, ETC.	2,700	SODIUM PERBORATE	240
MOLYBDENUM,TUNGSTEN, VANADIUM OXIDES/HYDROXIDES	2,600	POLYPHOSPHATES	230
IODINE	2,550	CAUSTIC POTASH (FLAKES)	200
CHLORIDES/OXYCHLORIDES	1,850	CARBON BLACK, ETC.	200
ANTIMONY OXIDE (EST)	1,500	MAGNESIUM OXIDES/HYDROXIDES	170
ZIRCONIUM OXIDE	1,135	SILICATES	165
BROMIDES,BROMATES, PER-BROMATES, OXYBROMIDES	940	ALUMINIUM OXIDES/HYDROXIDES	145
HYDRAZINE/HYDROXYLAMINE	890	SODIUM SULPHIDE	135
CHROMIUM OXIDES/HYDROXIDES	880	CHLORITES/HYPOCHLORITES	110
STRONTIUM OXIDES/ HYDROXIDES	460	CAUSTIC SODA 98/99 (DOM)	100
TITANIUM DIOXIDE	450	SODIUM BICARBONATE	90
CYANIDES	450	NITRIC ACID	90
ZINC OXIDES/PEROXIDES	430	CHLORINE (DOM)	69
BROMINE	410	SODA ASH	55
		66° Bé SULPHURIC ACID	30

Warning - These are average
prices derived from customs
returns - they give no guidance
to specific transactions

The far-reaching effects on the British chemical industry of the adoption of the Solvay process, electrochemical developments, the emergence of an organic chemical industry based first on coal tar and then on petroleum, plus the different pressures generated by two world wars, all had their influence on subsequent events.   For the purposes of this paper our attention will be focussed on the inorganic part of the modern chemical industry.   In terms of products and product groups this may be summarised, again in simplified form, as in Fig. 2.   It will be observed how the industry has extended in scope, complexity and fields of application over the last century.   In Fig. 3 some of the principal inorganic chemical materials are listed in approximate ranking of sales price[26]. It should be borne in mind, however, that the ranking is based on 1976 average export price data, so it will vary considerably from particular home price contracts, which will, of course, also depend on quantity, and whether the basis is contract or spot price.

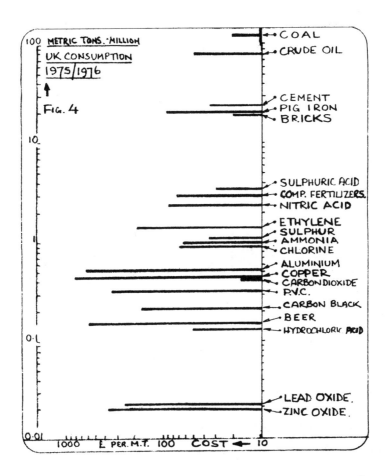

To give a feeling of where inorganic chemicals fit into the commodity market, Fig. 4 compares the price range (1975/76) and UK consumption of various inorganic chemicals with some commonly-encountered commodities[3,7].

Finally, Fig. 5 shows the relationship between raw material feed sources for inorganic chemical manufacture and the final use classification of various groups of inorganic products, either furnished as raw materials to downstream converters, or sold for final use or consumption.

Figure 5: <u>INORGANIC CHEMICALS MANUFACTURE</u>

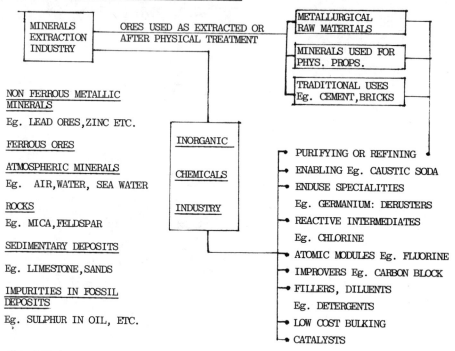

THE BRITISH CHEMICAL INDUSTRY

Before isolating the economic performance of the Inorganic Chemicals segment, it may be helpful to look at the British Chemical Industry as an entity.

According to the provisional result of the 1975 census of production[13], the chemical industry is Britain's third largest sector (after food, drink and tobacco, and Mechanical Engineering) and accounted for 9.4% of the manufacturing industry's net output (value added). In terms of net output per employee it comes second only to coal and petroleum products, and employs about 5% of the labour force whilst deploying in the order of 15% of the invested capital in British industry. Thus, for chemicals as a whole, one gets the picture of a major industry which is labour effective and capital intensive. This picture of the whole industry applies equally to the Inorganic Chemicals segment.

In 1977 the gross output of the total UK chemical industry is expected to reach £13,500 million. The build-up to this sales value is shown in Table 1. However, such figures can be misleading, since they combine sales volume changes with the effects of inflation and the rise of unit sales prices. It is instructive, therefore, to look at UK production in volume terms by correcting to constant money terms. Figures of the index of output of the UK chemical industry are given in Table 2, and compared with the corresponding index of output for all manufacturing.

Table 1   UK Chemical Industry (SIC Order V) Total Sales

Year	Sales	
1963	£ 2,164 million	Note: Sales values
1973	£ 5,579 million	shown are expressed in
1974	£ 8,191 million	"Money of the Day"
1975	£ 9,100 million	(M.O.D.)
1976 (estimated)	£11,500 million	
1977 (estimated)	£13,500 million	

It will be noticed in Table 2 that:

1.   Over the period taken the rate of growth of chemicals production has greatly exceeded that of manufactured products generally in the United Kingdom.

2.   In 1975 production dropped by about 11% in real terms below the 1974 levels - this being the first occasion of a real year-to-year drop in chemicals production over the period studied.

3.   Although in current money terms 1976 production is up about 26% over 1975 levels, in real terms it only reached 98.5% of the 1974 level.

Table 2   UK Chemical Industry Index of Production (1965 = 100)[16]

Year	UK Chemical Industry (SIC Order V)	All UK Manufacturing Industry
1965	100.0	100.0
1966	105.3	101.1
1967	110.7	102.3
1968	120.0	109.1
1969	126.7	113.7
1970	133.3	113.7
1971	136.0	113.7
1972	144.0	115.9
1973	161.3	126.1
1974	170.7	122.7
1975	154.7	114.8
1976	168.3	-

INVESTMENT BY THE CHEMICAL INDUSTRY

As has been said, the chemical industry is largely capital intensive by comparison with British manufacturing industry, generally.   With an output valued at about 1/10th of all manufacturing activity, and a labour force of around 1/20th of the total, the chemical industry deploys around 15% of manufacturing industry's gross fixed capital.   Whilst, historically, fixed capital investment by the industry has been highly cyclical, in recent years a vigorous attempt has been made to even out the level of expenditure.   Indeed, during the 1975 recession the volume of capital expenditure by the chemical industry increased by 22% in contrast with a 14% fall by manufacturing industry, generally.   The pattern of expenditure by the whole chemical industry is shown in Table 3.

Table 3   Gross Fixed Capital Expenditure in UK Chemicals & Manufacturing
           Industry

Year	Investment £ million at 1970 Costs		Chemicals as % of All Manufacturing
	Chemicals Order V	All Manufacturing	
1963	173	1381	12.5
1964	215	1561	13.8
1965	274	1726	15.9
1966	305	1774	17.2
1967	256	1738	14.7
1968	256	1851	13.8
1969	296	1978	15.0
1970	382	2130	17.9
1971	334	1991	16.8
1972	241	1739	13.9
1973	192	1872	10.3
1974	223	2024	11.0
1975	284	1737	15.6
1976 (est.)	289	-	-

It will be noticed that in constant money terms expenditure by the chemical
industry peaked in 1966 and again in 1970.   It then fell away and in 1975
began to build up again to a major investment programme.

According to the most recent CIA survey of investment intentions[25], the UK
Chemical Industry expects to spend in fixed investment during the period
1977 - 1979 more than £3,300 million in current money terms.   This will require in addition an increased commitment of working capital of approximately
£1,800 million, needing a total additional funding over the three-year period
of £5,100 million.

BRITISH CHEMICAL INDUSTRY POSITION INTERNATIONALLY

In international terms the British Chemical Industry is fourth largest of the
Western producing nations (Table 4) and accounts for about 8% of total Western
world output of chemicals.

Whilst on the basis of international comparisons Britain's chemical growth
rate has been relatively slow, it is interesting to compare the ratio of
chemical growth rate to the growth rate of all manufactured products for
Britain and its international competitors (Table 5).   On this basis of comparison, the UK is at the top of the league table, suggesting an outstandingly
good performance in the innovative factors and in export marketing efforts
that contribute to chemical growth.

Table 4   Sales Turnover of Western Chemical Producers

Money values in US $ million equivalent - Money of the Day

Note: * indicates excludes synthetic fibres: for all others fibres included

Year Source	1973 OECD	1974 CEFIC	1975 CEFIC	1975 % of total W.Europe
U.S.A.	65,010	81,676	86,420	73.8%
Japan	27,320	34,541	30,096	25.7%
W. Germany	24,105	32,230	30,295	25.9%
United Kingdom	15,010	20,620	21,600	18.5%
France	14,330	18,575	18,584	15.9%
Italy	9,525	13,995	13,635	11.7%
Spain	5,890	8,290	8,665	7.4%
Netherlands	4,610	7,626	6,722	5.7%
Benelux *	4,303	6,160	5,301	4.5%
Other Europe	8,570	11,946	12,234	10.4%
Total W. Europe	86,343	119,442	117,036	100.0%

Table 5   Comparison of Relative Rates of Chemical Industry Growth
1963 - 1973 [2,7].

Note: * indicates Food, Drink and Tobacco excluded

Country	Average Annual Increase of Output Index 1963-1975 %		Ratio of Average Chemical Industry Growth to Average Growth Rate for All Manufacturing Industry
	Chemical Industry	All Manufacturing Industry	
Europe			
United Kingdom	6.8	3.4 *	2.0
Netherlands	14.6	8.1 *	1.8
West Germany	9.6	5.8 *	1.7
France	10.1	6.4 *	1.6
Italy	8.4	4.6 *	1.8
Comparison			
Japan	13.2	12.4	1.1
U.S.A.	8.0	5.1	1.6

TRADE BY CHEMICALS INDUSTRY

The British Chemical Industry is internationally oriented and an increasing proportion of its output is traded abroad.   Exports as a proportion of chemicals output rose from 21% in 1963 to nearly 30% in 1975.   Despite imports rising from 12% to 20% during this same period, the industry has continued to enlarge its favourable contribution to the balance of payments.
In 1975 exports exceeded imports by £770 million, and the 1976 UK chemicals trading surplus exceeded £1,000 million, representing almost 25% of the favourable balance for all manufactured goods.   Table 6 gives details of the trade balance for Order V chemicals as a whole.

Table 6   Order V Trade Figures   (£1 million at Current Prices)[3,18]

	1963	1975	1976
Exports	372	2,179	3,046
Imports	206	1,409	2,000
Balance	surplus 166	surplus 770	surplus 1,046

Export Growth 1963 - 1973	+	719%	or	+ 17.6% p.a.
Import Growth 1963 - 1973	+	871%	or	+ 19.1% p.a.
Surplus Growth 1963 - 1973	+	530%	or	+ 15.2% p.a.

1976 Chemical Exports as % of Total UK Manufactured Goods Exports	14%
1976 Chemical Trade Surplus as % of Total Trade Surplus of UK Manufactured Goods	24%

INTERNATIONAL INVESTMENT COMPARISON

Finally, in completing the overview of the British chemical industry as a
whole, before proceeding to considerations specific to the Inorganic Sector,
one should review the cumulative fixed investment by the industry over recent
years and compare this with the corresponding figures for the industry's principal international competitors (Table 7).

Table 7   Cumulative Capital Expenditure 1963 - 1974 for UK Chemical Industry
            Industry and Major International Competitors[15]

            (Expenditure M.O.D. in equivalent US $ 1,000 million)

            * indicates including synthetic fibres since 1970

Country	Fixed Investment
U.S.A.	39.2
Japan	15.0
West Germany *	14.0
United Kingdom	8.2
Italy *	7.7
France	6.6

INORGANIC CHEMICALS SECTOR

Materials on all facets of operations dealt with above for the overall chemical industry are unfortunately not available broken down into the individual
sectors.   What has come before has thus been used to illustrate the dimensions of the total canvas.   What follows attempts, where possible quantitatively, to relate the Inorganics Sector to this total picture.   Table 8 gives
a breakdown of chemical industry activity by sectors and it will be seen that
in 1974 inorganic chemicals constituted 8.17% of the gross output, and 8.26%
of the net output.   It should be noted that net output equals gross output
less input cost of fuel and raw materials, payment for work sub-contracted
and duties, subsidies, allowances and levies.   It does not take into account
that a high proportion of the Inorganic Sector output represents inputs to
other parts of the overall chemical industry.

It is also worth noting that at present about 75% by value of the gross Inorganic Chemicals Sector output in the UK is accounted for by four companies[21].

Table 8   Output by Various Sectors of the British Chemicals Industry 1974[10]

	M.L.H.	£ million	
		Gross Output	Net Output
Inorganic Chemicals	271/1	668.9	267.1
Organic Chemicals	271/2	1,624.5	507.4
General Chemicals	271/3	736.5	255.8
Pharmaceuticals	272	873.5	455.0
Toilet Preparations	273	300.7	153.6
Paint	274	419.5	182.1
Soap & Detergents	275	457.1	131.7
Synthetic Resins, etc.	276	1,191.5	437.0
Dyestuffs & Pigments	277	426.2	206.7
Fertilizers	278	532.8	239.3
Polishes	279.1	86.0	38.6
Adhesives, Gelatine, etc.	279.2	130.5	45.4
Explosives, Fireworks, etc.	279.3	163.1	78.4
Pesticides & Disinfectants	279.4	178.6	59.3
Printing Ink	279.5	75.0	39.1
Surgical Bandages	279.6	113.5	58.4
Photographic Chemicals	279.7	212.6	89.8
Total Order V		8,190.5	3,234.7

This breakdown gives a picture for the single year of 1974 of activity across the chemicals sector and shows how inorganic chemicals fit into the total picture. It is interesting also to see how this pattern has changed over the years. This is illustrated in Table 9.

It will be noticed that:-

1.   The rate of growth of inorganic chemicals production has been much slower than that for chemicals, generally - 22.6% growth for inorganics over the decade to 1974 compared with 66.5% for all Order V chemicals.

2.   Whilst the growth rate for all chemicals on an annual compounded basis is about double that for the total manufactured product (see Table 5), inorganic chemicals grow at about the same rate as the manufactured product.

3.   It follows that the proportion of the total chemical industry represented by the Inorganics segment has fallen progressively over the years studied. Taking a volume comparison based on 1970 prices, the proportion of the total chemical industry accounted for by inorganics has fallen from more than 13% in 1965 to barely 9% in 1975.

4.   It is interesting to note by way of comparison that in the USA over the period 1967 to 1976 inorganic chemicals output increased by about 40% compared with an increase in the output of all chemicals of 69%. In the same period in the UK inorganic production increased by 19% compared with an increase in the output of all chemicals of 51% - i.e. a US Inorganics Sector proportionate growth ratio of 1.56 that of the UK Inorganics Sector.

Table 9   Inorganic Chemicals Production Index Comparison (1965 = 100)

Notes: 1. This ratio is based on 1970 prices. It is somewhat higher than
           that derived from Table 8, due to the fact that the price index[4]
           for inorganic chemicals went up slower in the period 1970 - 1975
           than the corresponding index for chemicals generally.

        2. Based on 1970 prices.

        3. Based on 1965 prices

Year	Index of Production (1965 = 100) at Constant Prices			Inorganics Output as Percentage of Total Chemicals Output (1)
	Inorganics HLH 271/1 (2)	All Chemicals (Order V) (2)	All Manufacturing (3)	
1965	100.0	100.0	100.0	13.37%
1966	101.4	104.8	101.1	12.95%
1967	103.4	110.2	102.3	12.55%
1968	107.6	117.6	109.1	12.33%
1969	107.6	124.2	113.7	11.59%
1970	106.8	130.3	113.7	10.96%
1971	107.9	132.8	113.7	10.87%
1972	105.9	140.3	115.9	10.09%
1973	116.7	156.8	126.1	9.95%
1974	122.6	166.5	122.7	9.85%
1975	102.0	150.7	114.8	9.04%

EMPLOYMENT IN THE INORGANICS SECTOR

Out of a total employment by the chemical industry in 1974 of approximately
432,000, the Inorganics Sector is estimated to have employed 45,600 - or, say,
10.6% of the labour force to produce about 8.2% of the output value. This
ratio is partly accounted for by the relatively low unit value of many products
in the Inorganics Sector.

In former times many inorganic chemicals production units, especially those
handling solid/liquid mixtures, were primitive in a process engineering sense.
However, with low unit sales values and increasing unit labour costs the segment has had to become more capital intensive and more labour effective.
Nevertheless, in an overall sense, improvements in output per capita do not
seem to have been as great as in other parts of the chemical industry.

Looking to the future, it is hoped that in new plants, arising from the more
sophisticated technologies employed and especially the development of improved
techniques for solids handling, fewer personnel may be required per unit of
output. However, the need will be for higher quality operators and increasingly qualified staff for environmental control, maintenance and cost
effective operations[17].

DEMAND, PRODUCTION AND TRADE IN INORGANIC CHEMICALS

Tables 8 & 9 gave a picture of the size of the UK production of inorganic
chemicals and how this production grew over the last decade. This study can
now be broadened to encompass the international trading pattern for inorganic
chemicals, and hence the development trends in the home market demand for such
products. Whilst much of this work has been covered by the Chemicals EDC,
it may be fruitful to examine this in further detail. Table 10 shows[1] the
1973 value of inorganic chemicals produced and consumed in the UK, and the
export/import trading patterns. It also shows how these developed in the

decade 1963 to 1973, and compares Inorganic Segment performance with that of
the whole chemical industry, and the performance of the Organics Segment,
which, in the chemical industry, was the pace-setter during that period.

Table 10   Inorganic Segment Data Comparisons 1973[1]

Note: * "Growth" shown is the mean annual compound growth on a volume basis
          over the period 1963 - 1973

| Item | Inorganics MLH 271.1 | For Comparison | |
		Organics MLH 271.2	All Chemicals Order V
UK Consumption Value £ million (1973) Annual Growth 1963-1973	415 4%	813 11.5%	4,342 6.5%
UK Imports Value £ million (1973) Annual Growth 1963-1973	97 15%	237 16%	952 13%
UK Exports Value £ million (1973) Annual Growth 1963-1973	82 7%	244 11.5%	1,365 10%
UK Production Value £ million (1973) Annual Growth 1963-1973	400 3%	820 10%	4,755 6.5%
Trade Surplus/(Deficit) £ million (1973)	(15)	7	413

The following points are worthy of note:-

1.   In 1963 both the UK domestic demand and the UK production of inorganics
      and of organics were comparable in value.   For example, in that year UK
      production of inorganics was £208 million (11.97% of total chemical pro-
      duction), whilst that for organic chemicals was almost the same, at
      £207 million.

      Ten years later, in 1973, organics production was nearly double that of
      inorganic chemicals in current money terms.   Organics, at £813 million,
      had grown to 18.7% of total chemicals production, whilst inorganics, at
      £415 million, had fallen   from approximately 12% to 9.6%.

2.   Production of inorganic chemicals does not correlate particularly well
      with consumer demand.   This is presumably due to the fact that much in-
      organic output is well upstream from the ultimate point of consumption,
      so that there is a considerable flywheel effect when demand changes in
      the marketplace for the final product.

3.   As might be expected, however, the output of inorganic chemicals does cor-
      relate well with the general level of total industrial activity.   This
      was noted previously when considering output index figures.

4.   Despite a relatively high transport cost per unit output value for inor-
      ganic chemicals, exports from the UK have grown significantly more

rapidly than home sales. For example, in the period 1963 to 1973 exports
of inorganic chemicals grew an average of 7% p.a. compared with an average 3% p.a. growth in home sales. Over the same period, total UK manufacturing output grew by an average of 3.3% p.a. and domestic consumption
of inorganic chemicals by 4% p.a. This does suggest that, whilst not in
the same league as the rest of chemicals for new market capture, the UK
Inorganic Chemicals Sector did move ahead marginally in relation to the
general industrial trend.

Figure 6 UK INORGANIC CHEMICALS TRADE LEADERS - 1976 BASED ON
HM CUSTOMS TRADE STATISTICS

IMPORTS TO UK VALUE £K (UNIT PRICE AVERAGE) £MT		GROUPING OF MATERIALS	EXPORTS FROM UK UNIT PRICE AVERAGE) £MT	VALUE £K
3151	2045	IODINE + BROMINE	782	787
3023	3746	COBALT OXIDES + HYDROXIDES	4527	2870
4744	154	MAGNESIUM,BARIUM + STRONTIUM OXIDES + HYDROXIDES	179	1907
208	1465	ANTIMONY OXIDE	N.A.	7927
387	423	CHLORITES + HYPOCHLORITES	111	784
9	6148	IODIDES,IODATES + PERIODATES	3019	625
83	72	SODIUM CARBONATE	55	8174
29	496	SODIUM BICARBONATE	90	2413
7	533	CHLORINE	292	916
820	374	ZINC OXIDE + PEROXIDE	434	4034
127	698	LEAD OXIDE, RED LEAD, ORANGE LEAD	323	2686
NEG	5420	TIN OXIDE	3349	1810
335	42745	GERMANIUM + ZIRCONIUM OXIDE	1581	932
1146	374	TITANIUM DIOXIDE PIGMENTS	450	4722
550	1264	BROMIDES, BROMATES, PERBROMATES, OXYBROMIDES	939	690
114	225	SODIUM SULPHIDE	137	1280
27	47	ALUMINIUM + CHROMIUM SULPHATE	50	605
961	269	POLYPHOSPHATES	231	607
34	626	CYANIDES	454	6549
892	257	SILICATES	169	5154
280	2228	ANTIMONATES, MOLYBDATES + TUNGSTATES	2681	188
151	330	SODIUM PERBORATE	237	1106

5.   Fig. 6 gives details of the Export and Import trade in some of the Inorganic Sector trade leaders in 1976.   The table shows import and export
values and average unit prices at which export and import transactions
were made.   Fig. 7 investigates the import and export volume and value
trade patterns in 1976 for 40 major inorganics, and it will be seen that
there appears to be no distinct price level/volume relationship in UK
inorganics trading.   Elsewhere it has been suggested that in inorganic
chemicals exports from the UK there is less emphasis on the high cost
high added value materials than in exports by other European countries,
especially W. Germany.

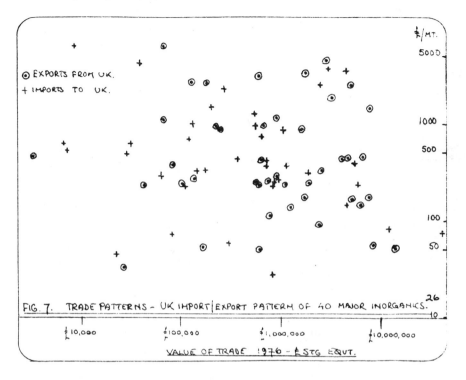

FIG. 7.   TRADE PATTERNS - UK IMPORT|EXPORT PATTERN OF 40 MAJOR INORGANICS.

VALUE OF TRADE 1976 - £ STG EQUT.

6. Acids and alkalis are the largest single identifiable product grouping in
    the Inorganics Sector.   Here the UK consumption pattern has fitted a
    long-term trend line for the last eighty years.   In the case of alkalis,
    the long-term consumption growth trend has been 3.25% p.a. over this
    period.   The view has been expressed[1] that, as a product grouping, inor-
    ganics are probably less sensitive to changes in the environment than any
    other type of chemical product.   Equally, for better or worse, they seem
    less affected by technological changes!

7. The slower growth of demand for, and production of, certain key basic
    inorganics (such as chlor-alkali products, sulphuric acid, etc) in
    Britain, compared with other major chemical-producing countries, is thus
    seen as being mainly due to the slower inherent growth of the general UK
    industrial economy, rather than any underlying lack of chemical capacity
    in the UK.   The spasmodic shortage situations which occurred in 1973 and
    early 1974 seem to have been due largely to international demand/supply
    situations, start-up delays, and equipment failures accentuated by main-
    tenance problems arising from the three-day week.

8. As an exception to those generalisations regarding basic inorganic chemi-
    cals production capacity, it is perhaps worthwhile commenting on ammonia
    as a special case.   About 80% of the consumption of ammonia in the UK
    goes into fertilizer manufacture.   In recent years world price movements
    for ammonia have been much influenced by politico-economic considerations
    influencing feedstock costs.   Thus UK purchasers of ammonia have for
    some time adopted a policy of importing a proportion of their needs to
    take advantage of the low world prices.   For a period in 1974, arising

from the oil crisis, the world surplus of ammonia disappeared - but this
subsequently appears to have been re-established. One may expect that,
although the emphasis may change from time to time and from place to
place with changes in natural gas prices, this basic pattern regarding
ammonia supply will continue.

FUTURE TRENDS IN DEMAND, PRODUCTION, AND TRADE PATTERNS FOR INORGANIC CHEMICALS

At the basic end, bulk inorganic chemicals represent a unique and generally
irreplaceable raw material for downstream industries. In addition, they
generally represent a relatively small proportion of such industries' input
costs. Thus, so far as the markets for basic inorganics are concerned, the
short and middle term future demand for inorganic products is likely to continue to be closely related to the movements of industrial activity generally.
In this part of the market it may also be expected, on the same time-scale,
that inorganic demand will be less sensitive to changes in the world economy
or technology change than other types of chemical products.

There are undoubtedly entrepreneurial sections of the inorganic market where
substitution, product innovation or improvement, process development, and increased market penetration through enhanced technical sales service are possible. For the companies concerned, these will represent exciting opportunities. However, the total market is so dominated at present by the flywheel
effect of the basic products that these glamour opportunity sectors will have
little effect on the overall figures.

By their nature and maturity, many inorganics have commodity status with the
output being controlled by downstream demand. Thus, the total market demand
for inorganics has been thought of as being relatively price inelastic. It
has been felt in the UK that greater sales demand would not be created by investing in larger plant to obtain cost reductions through the economies of
scale. In this sense, much of the Inorganics Sector has not been subjected to
the "bigger and bigger" temptations which producers have been subjected to at
the basic end of the Organic Chemicals Sector over the last fifteen years.

It must be said, however, that whilst the larger unit may not represent any
attraction on an industry-wide basis, where unit sales values of particular
products are high enough to justify the transportation costs, it may offer
benefits to individual companies. In such cases the larger unit may increase
market share or enable international markets to be captured. Even with existing markets, so long as the benefits are not swallowed up by higher transportation costs, the larger plant, given a high degree of automated operation, may
improve profitability in a world of increasing unit labour costs.

In looking to the future, one also needs to give credit to the fact that some
user industries for inorganic products may grow faster than the general manufacturing industry trend. Thus there is the possibility for UK producers
that Britain may gain a greater share of the growth of Europe's petrochemical
output than in the past. So to the extent that inorganic products, as
catalysts, reactants or components, are raw materials for such organic processes, this should represent an additional opportunity area for the inorganics
industry in the UK.

The most recent authoritative forecast of future growth opportunity for the
Inorganics Sector is that prepared by the Chemicals EDC[1] which covers the
period up to 1980. This is summarised in Table 11.

Table 11   Chemicals EDC Forecast of Inorganic Sector Growth to 1980[1]

Notes: (1) "Expected Growth" is the forecast mean annual compound growth on a volume basis over the period 1973-1980

(2) Value figures are in constant money terms (1973 prices) and expressed as £ million

| Item | Inorganics MLH 271.1 | For Comparison | |
		Organics MLH 271.2	All Chemicals Order V
Expected Growth 1973-1980. (Note 1)			
UK Consumption	5%	6½%	4½%
UK Imports	7%	8½%	6½%
UK Exports	4½%	8½%	7½%
UK Production	4½%	6½%	5%
Value (Note 2)	1973 - 1980	1973 - 1980	1973 - 1980
UK Consumption	415 - 588	813 - 1260	4342 - 5910
UK Imports	97 - 156	237 - 427	952 - 1490
UK Exports	82 - 112	244 - 427	1365 - 2280
UK Production	400 - 544	820 - 1260	4755 - 6700
Trade Surplus (Deficit) (Note 2)	(15)- (44)	7 - Nil	413 - 790

The following points are worth noting in comparing these figures with those given in Table 10 for the period 1963 - 1973:-

1. Whilst overall chemical growth rates and growth rates for Organic Sector products are expected to fall, the demand for inorganics is expected to increase faster than it has over the last decade. (Average 5% p.a. versus an historical 4% p.a. growth in UK consumption).

2. To meet this increased growth rate in consumption demand, production capacity is expected to rise faster (4½% p.a. versus an historical 3% p.a. growth in UK inorganic chemicals output).

3. However, to make good the deficit, the adverse trade balance is also expected to grow - although remaining at a modest level compared with the overall chemical surplus.

4. Despite the increased domestic growth rate, the growth rate for inorganic exports is expected to fall from an average of 7% over the last decade to 4½% p.a. in the period up to 1980. This reflects a less optimistic view of the general world economic climate over the years ahead.

INORGANIC CHEMICALS - PRICE MOVEMENTS

Table 12 shows the movement of the index of home selling prices of inorganic chemicals since 1963, and compares this with the comparable movements for all Order V chemicals. The third and fourth columns give comparable home price index figures for West Germany. Since the UK figures are based on domestic prices in pounds sterling, and the German figures are based on DM prices in Germany, they are not comparable in any absolute sense due to the change in rate of exchange (£ Stg to DM) over the period. They are, however, of some interest in giving a comparative customer's-eye view of price movements of inorganic chemicals in the two countries.

Regarding the increment of prices in the UK, it is interesting to note that, despite the dramatic jump in the price of naphtha feedstock to the Organics Sector since 1973, from 1963 to date the index of inorganic prices has risen faster than the index for the total chemical industry.   This reflects, in part, the lesser volatility of the inorganic market and, in part, arises from the absence of the "economy of scale" savings that had such a profound effect, for better or worse, on the price of basic organics over the decade 1963 to 1973.

Table 12   Home Sales Price Indices (1963 = 100) for Britain[18]
            (in £ Stg basis) and for W. Germany[19] (in DM basis)

Note: * indicates change in basis of calculation

Year	UK		W. Germany	
	Inorganics	All Chemicals	Inorganics	All Chemicals
1963	100.0	100.0	100.0	100.0
1964	101.3	100.1	104.6	100.1
1965	101.9	101.0	134.3	103.5
1966	105.6	101.8	127.3	103.0
1967	105.9	102.1	123.9	101.8
1968	110.1	105.5	* 123.3	* 95.8
1969	113.3	106.5	122.7	94.2
1970	118.8	111.2	125.2	94.4
1971	128.3	120.3	128.2	94.7
1972	133.4	126.5	130.3	94.6
1973	138.7	132.2	132.7	97.5
1974	167.1	170.0	149.6	122.4
1975	234.1	209.9	172.2	124.6
1976 prov.	287.8	245.8	173.5	126.1

COMPARISON OF INORGANIC PRICE MOVEMENTS WITH THOSE OF COMMODITIES

Since inorganic chemicals are frequently sold to other commodity-using indus-tries, either as raw materials or as enabling materials, it is interesting to compare the price movement for inorganics in recent years with the corres-ponding domestic price movements in the customer and related industries. This is done in Table 13, and it will be seen that the price index of inorganic chemicals rose faster than chemicals generally, but slower than most commodi-ties.

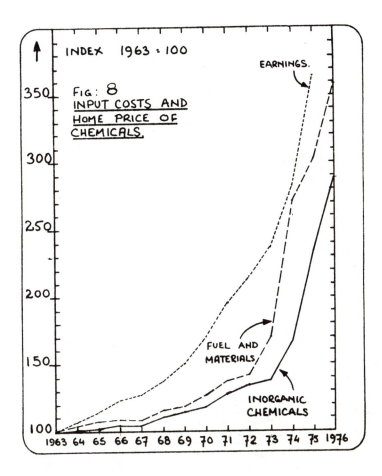

INDEX   1963 = 100

FIG: 8
INPUT COSTS AND
HOME PRICE OF
CHEMICALS.

EARNINGS.

FUEL AND
MATERIALS

INORGANIC
CHEMICALS

Fig. 8 shows the movement since 1963 of input costs (index of employee
earnings, and cost of materials and fuels) and compares these indices with the
home sales price of inorganic chemicals.

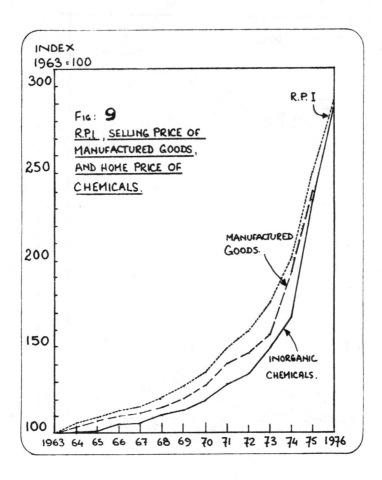

Fig. 9 compares the index of inorganic chemical prices with the index of prices of all manufactured products, and with the retail price index for the period 1963 to 1976.

Table 13   Relative Cost Increases of Inorganic Chemicals and Other Materials
               1963 to 1976

	% Increase 1963 - 1973	% Increase 1963 - Dec 1975	% Increase 1963 - Dec 1976
Copper	275	193	291
Softwood	171	226	428
Tin Ingot	120	241	454
Steel Tube	79	251	337
Wood Pulp	77	328	417
Coal	76	304	369
Iron Castings	69	222	275
Crude Oil	66	554	715
Tinplate	64	177	249
Paper & Board	54	190	250
Natural Rubber	49	70	159
Glass Containers	47	163	208
Steel Sheet	40	120	227
Aluminium Ingot	37	132	224
All Chemicals (Order V)	32	122	157
Inorganic Chemicals (MLH 271.1)	39	147	201

RAW MATERIALS SOURCES

In any consideration of the economic background to the inorganic chemical
industry, one must touch, however briefly, on the special considerations regarding the cost and continuing availability of raw materials sources.   So
far as UK production is concerned, raw materials fall into three main categories:-

a)   Basic raw materials which are indigenous (such as salt, limestone,
     dolomite, air, sea water, natural gas).

b)   Basic raw materials which have to be imported, since they are not in-
     digenous, or at least not so in suitable form (such as bauxite, borax,
     sulphur, phosphate rock, metallic minerals, etc).

c)   Derived raw materials required for further processing which, for economic
     reasons (eg, high energy dependence, etc), are imported (such as elemen-
     tal phosphorus, calcium carbide, etc).

Some of the raw materials on which the industry is dependent are shown in
Fig. 10.

At the time of writing the "storm or catastrophe" dilemma, which was triggered
by the oil crisis which followed the Yom-Kippur War and the subsequent hike
in energy and raw materials prices, remains unresolved.   Although four years
have passed, traditional economic theories are under such doubt that it is a
fairly barren speculation to try to decide whether the current economic environment is a pause between periods of rapidly growing living standards, or
whether it is a phase of transition to a slower growth era.

Figure 10   MINERALS USED IN INORGANIC CHEMICAL PRODUCTION

MINERAL OR ORE	WORLD CHEMICAL CONSUMPTION[27] ('000 TONS)	TYPICAL USE IN CHEMICALS
BARITE	750	BARIUM COMPOUNDS, PIGMENTS, FILLERS
BAUXITE	8000	ALUMINIUM OXIDE, HYDROXIDE, SULPHATE
BORATES	2400	BORAX, BORIC ACID, GLAZES, DETERGENTS
BROMINE COMPS.	300	BROMIDES, AGRICULTURAL CHEMICALS, ETC
CHROMITE	1200	CHROMATES, BICHROMATE, CERAMICS, PIGMENTS
DOLOMITE	–	MAGNESIUM AND REFRACTORY USES
FLUORITE	1500	ALUMINIUM FLUORIDE, ORGANIC FLUORINES
IODINE COMPS.	12	IODATES, IODIDES ETC.
LIMESTONE	60,000	SODA ASH, LIME, CALCIUM CARBIDE
LITHIUM COMPS	23	LITHIUM CARBONATE, HYDROXIDE ETC.
MAGNESITE	3000	MAGNESIUM COMPOUNDS
MANGANESE COMPS	1000	MANGANESE DIOXIDE, PAINT DRIERS, ETC.
PHOSPHATE ROCK	120,000	FERTILIZERS, DETERGENTS, WATER TREATING
POTASSIUM COMPS	25,000	FERTILIZERS, CAUSTIC POTASH, ETC.
RARE EARTHS	30	COMPOUNDS OF RARE EARTH ELEMENTS
SAND/SILICA	1000	SILICATES, DETERGENTS, SILICONES
SODIUM CARBONATE	8000	CLEANING FORMULATIONS, CAUSTIC SODA
SODIUM CHLORIDE	120,000	CHLOR-ALKALI PRODUCTION
SODIUM NITRATE	550	FERTILIZERS, EXPLOSIVES
SODIUM SULPHATE	1500	DETERGENTS, PAPER + GLASS CHEMICALS
SULPHUR	50,000	SULPHURIC ACID MANUFACTURE
TITANIUM COMPS.	4350	TiO_2 PIGMENTS

Although recently the unity of the OPEC countries has begun to show fatigue cracks, their initial successes in obtaining major increases in crude oil prices became both the example for, and the cause of, similar attempts in other mineral exploitation areas. Such producers, apart from wanting to benefit from the example of the OPEC countries, needed to finance their own increased costs of oil imports. This was especially the case for the producers of bauxite and phosphate rock.

From current considerations of demand/supply balance, and of increased cost exploitation, mineral prices are expected to continue to rise. However, cost considerations apart, the widely-held view is that supplies of major mineral raw materials will remain adequate beyond the turn of the century, although there may be short-term supply bottlenecks due to lack of investment in mineral exploitation during the low price era of the 1960's and to the induction time required to bring new mines into production.

It is interesting to note, however, that increasing market prices, increasing unit transport cost, and increasing cost of mining development will probably have a significant effect on the pattern of raw materials supplies. Such effects may include:-

1. Increasing recovery and re-use of higher value materials.

2. The development of higher quality sources, even though they may be more distant from the centres of use.

3. More concentration and upgrading plants at the point of production prior to transportation.

4. The re-introduction of processing of inferior ores located close to the point of use.

5. Greater attention to the differential costs of meeting varying environmental standards and other social costs at producing sites in different countries.

EXAMPLE OF THE CHANGING RAW MATERIAL PATTERN - SULPHUR

Sulphur is a good example of the changing raw material supply pattern (Fig. 11)

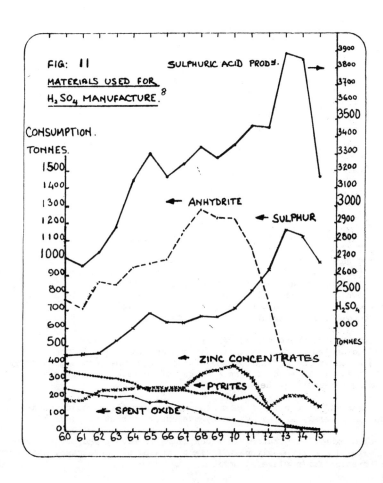

As the basic raw material for sulphuric acid, sulphur is one of the key raw
materials for the inorganic chemicals industry.   Currently the consumption
rate in the UK for sulphuric acid production is around 1.25 million tons per
annum at a cost of about £45 million.   A decade ago sulphur accounted for
only half of the sulphuric acid produced in Britain.   Now it is the domina-
ting raw material with a movement away from the use of pyrites, indigenous
anhydrite, and zinc concentrate.

The energy crisis has emphasised the advantages of sulphur, since the exothermic reaction on burning sulphur yields a significant heat output which can
be used for steam generation.   World sulphur supplies seem to be adequate to
meet chemical industry needs and will be augmented by increasing quantities of
sulphur recovered from the purification of sour oil and associated gases.

ANOTHER VIEW OF RAW MATERIAL VULNERABILITY

In the last few years Britain has been given special hope for its economic
future by the good fortune of North Sea Oil.   Other countries less fortunate
have been more preoccupied by the problems of availability of hydrocarbon
feedstocks.   Perhaps because it is less easy to popularise, the broader issue
of inorganic raw materials availability has received less general attention.
It did, however, receive attention from Meadows, and others, in the study
sponsored by the Club of Rome[12], which was subsequently commented on by
Lord Ashby[11] in the 21st Fawley Foundation Lecture in 1975.

Meadows and his colleagues published data on how long it seems that known
reserves of certain non-renewable resources would last.   However, since
"hunger makes the best cook", as has been previously suggested need will
stimulate exploration and exploitation.   Thus there is some justification
to a view exactly opposite to that of the Club of Rome.   A view that there
is no practical limit to growth and:-

1.   New mineral deposits will be discovered.

2.   Lower grade ores will be worked.

3.   Recycling will become more economic and more readily practised.

4.   Substitutes will be found for scarce raw materials.

5.   Alternative processes and products will be developed.

However, Lord Ashby points out that in the future the problems are not only
economic or technological - but they are likely to be geopolitical.   He
argues that long before new reserves of ores are physically exhausted they
are likely to become politically inaccessible except on terms dictated by the
producing country.   Thus in the supply of such materials he forecasts a
series of geopolitical confrontations of which OPEC is but the forerunner.

Even so, although producers of mineral ores may be encouraged by the successes
of the OPEC countries to believe that they can substantially increase the
world price of their products, from the standpoint of the chemical industry
they are less likely to enjoy the same degree of success for a number of
reasons.   Such reasons include:-

a)   The withdrawal of supply of such mineral raw materials would not have the
     amount of leverage as that given by the energy impact of crude oil.

b)   For solid mineral raw materials stockpiling is easier and more practical
     than with oil.   Thus, given sufficient warning of the danger, the users
     could prepare for a siege situation.

c)   The possibility of substitution is greater, and with it the danger of
     losing a market for the material being withheld.

For these reasons, it seems unlikely that cartels could be established for
raw materials for inorganic manufacture with the strength and unity of purpose
of OPEC.   Nevertheless, one may expect that in the future high quality
sources of minerals will command higher prices than in the past for the practical economic reasons given above.   It is also to be expected that there will
be increasing international interaction between the supply of raw materials
and the distribution or marketing of other exports from the primary producing
countries to the developed world - the conditional supply situation!

INVESTMENT BY THE INORGANIC SECTOR

As was stated in reviewing the performance of the chemical industry generally,
the present total fixed capital expenditure after peaking at £380 million in
1970 dropped to less than £200 million in 1973, but is now back up the £300
million p.a. level (all in 1970 money terms).   As a broad generalisation,
investment in the Inorganic Sector has for a number of years run at 10% - 15%
of total chemical industry investment.   In the period 1968 - 1973 inorganics
investment fell from a peak of £35 million to £20 million in 1973 (again all
in 1970 prices).   Now, however, the investment in inorganics seems to be
turning up again with the upturn of investment in the chemical industry
generally.   Some general factors to note relating to investment in the sector
include the following:-

1.   Lead time in planning major new extensions is now 5 years or more[23,24].

2.   In consequence, there is a substantial commitment of financial resources
     at least 3 years ahead of production.

3.   The minimum economic size of plant for many products is such that an
     assured market is required - or risks are high.   A single unit may
     represent several years' growth of domestic demand.   Hence, new plants
     are difficult to justify in a slow growing economy.   This is a factor
     which continues to make investment decisions in the Inorganic Sector
     more difficult in Britain than in an economy with a more vigorous indus-
     trial growth rate.

4.   Small companies in the industry cannot afford to take swings and rounda-
     bouts decisions[21].   Thus, phasing of capital spending may be conditioned
     by available cash flow or an inability to earn sufficient surplus to
     increase financial gearing with safety.

5.   A further general factor which also influences the Inorganic Sector is
     that low capacity utilisation following a period of price restraint has
     reduced the cash flow available to fulfil planned programmes.   This,
     coupled with effects of rapid cash inflation[22], causes anxiety about the
     likelihood of sufficient investment taking place to fulfil the basic
     projections required to meet the expected demand to the end of the decade
     (see Table 11).

THE FUTURE

Fortunately, when this paper was invited, emphasis was laid on its role as
providing a backcloth to the specialist papers making up the remainder of the
session.   Thus the prognostication of new developments and the forecasting of
specific future trends can safely be left to those who follow.   There are,
however, a few general comments which can be made in conclusion on probable
trends and influences which may have a significant impact on the future development of the UK inorganic chemicals industry.   In no particular order of
importance, these include the factors listed below:-

1. Whilst inorganic chemicals have been, and will remain, a relatively slow-growing segment of the chemical industry, and whilst inorganic demand is relatively price inelastic, it is the only segment of the chemical industry which is expected to grow faster in the period 1975 - 1980 than it has over the past decade.   During this period growth of inorganic chemicals in the UK is expected to run at about 2% p.a. above the growth of manufacturing industry generally, with an extra 0.5% for export stretching possibilities.

2. The reason for this higher growth rate is that the sector contains some products with special market and new application opportunities.

3. Apart from such special situations and some specialty products, the growth of basic inorganics will remain crucially dependent on the growth of industrial activity in the UK, although due to the flywheel effect inorganics as a whole will be less sensitive to changes in the economic environment than other parts of the chemical industry.

4. In view of this dependent growth rate, and trends in the export markets, it is probable that more significant developments or dramatic increases in market penetrations cannot be achieved without a radical change in UK industrial electric power costs in relation to competing countries[5,6]. Since this action would require decisions with crucial and far-reaching political impact such a fundamental change is improbable.

5. With this backcloth, the growth of the UK Inorganic Sector relative to the continental European Inorganic Sector will depend on the relative rates of growth of industrial activity generally, subject to two caveats:-

    a) The more rapid growth of the petrochemical industry in the UK arising from North Sea Oil and other considerations may give "above trend-line" growth opportunities for some of the enabling and reactant materials produced by the Inorganic Sector.

    b) The high natural purification capacity of estuarial and coastal sites in Britain, and the more rapid development of a favourable ecological infrastructure to support the petrochemicals developments, may make the UK a particularly attractive location for growth of high unit value inorganics.   Nevertheless, in the UK, as elsewhere, there will be increasing pressure for yet more stringent limits on the discharge levels of solids, liquids and gaseous effluents.

6. The other side of the environmental coin is that requirements of water and waste treatment by other industries, and improved sewage treatment standards, should generate additional markets for inorganic specialities.

7. The Inorganic Sector is relatively energy dependent and to a significant extent relies on raw materials which have to be imported into the UK. Many of these imported raw materials have a disproportionately high transport cost content.   In consequence, one may expect R + D emphasis on new process development and modification to improve yields and to reduce specific energy demands.

8. Despite the favourable environmental considerations referred to above, there will be continuing and increasing pressure to eliminate or consume waste products.   The incremental cost of this naturally becomes less punitive if additional values can be extracted to compensate for additional costs.   However, general experience indicates that the additional cost per unit of increased yield may be disproportionately high.

9. As regards indigenous raw materials exploitation will become more costly - as calculated in constant money terms.   The activities of lobbies and pressure groups will make planning permission for mineral extraction operations more difficult to obtain.   Conditions of use, requiring costly restoration of land and landscaping, will become more exacting, and hence more expensive per unit of output.

10. Social action at the consumer level may work backwards to influence the market for some products forming raw materials to consumer supply industries.

11. In developing business in the UK and in formulating export market strategies it may be expected that there will be a relative shift in emphasis from the present production preoccupations to total business planning considerations.   This could be especially important in moving export effort towards the more sophisticated and higher added value end of the international trade spectrum.   However, for companies other than those of the first magnitude of size, this will necessitate the evolution of radically new international marketing structures, which may require significant moves towards consolidations or cooperative marketing programmes.

Note: Whilst gratefully acknowledging the material, ideas, help and suggestions provided by colleagues, the author stresses that the opinions expressed in this paper are offered in a personal capacity.   They may or may not coincide with the views of the Chemical Industry Association and/or of its member companies.

BIBLIOGRAPHY

1.   Chemicals Economic Development Committee, UK Chemicals 1975 - 1985,
     (NEDO, London, March 1976)

2.   Chemicals Economic Development Committee, Industrial Review to 1977 -
     Chemicals, (NEDO, London, 1973)

3.   L. S. Adler, Private Communication, (CIA, London, February 1977)

4.   B. G. Reuben and M. L. Burstall, The Chemical Economy, (Longman, London,
     1973)

5.   Anon, Energy Trends: January 1977, (Department of Energy, London, 1977)

6.   Anon, Energy Price Statistics: February 1977, (CEFIC, Brussels, 1977)

7.   H. H. Pinzenthal, Private Communication - 1975 Study, (London, 1975)

8.   Anon, UK Chemical Industry Statistics, (Chemical Industries Association,
     London), Volumes 1969 to 1976

9.   Anon, European Chemical News, 1977, February 18, 18

10. Anon, Census of Production, (London, 1968)

11. The Lord Ashby of Brandon, A Second Look at Doom: 21st Fawley Lecture,
     (University of Southampton, Southampton, 1975)

12. Meadows, et al, The Limits to Growth

13. Anon, Census of Production 1975 - Provisional Results, (Trade and Industry
     London, 28-1-1977)

14. Anon, Standard Industrial Classification of Central Statistical Office,
     (HMSO, London, 1968)

15. K. J. Wey, Private Communication, (CIA, London, February 1976)

16. Anon, Activity in the Chemical Industry, (Trade & Industry, London,
     7-1-1977)

17. Chemicals Economic Development Committee, Chemicals Manpower in Europe,
     (NEDO, London, 1973)

18. K. J. Wey, Private Communication, (CIA, London, February 1977)

19. V.C.I., Private Communication, (V.C.I., Frankfurt, February 1977)

20. D. W. F. Hardie and J. Davidson Pratt, A History of the Modern Chemical
     Industry, (Pergamon Press, London, 1966)

21. Chemicals Economic Development Committee, Financial Results of UK
     Chemical Companies, 1971/72 to 1974/75, (NEDO, London, 1976)

22. Anon, Some Aspects of the Changing Structure of the Chemical Industry,
     (CIA, London, 1976)

23. Process Plant Working Party, A Necessary Partnership Reconciling Conflic-
     ting Objectives in the Supply of Process Plant, (NEDO, London, 1975)

24. Mechanical and Electrical Engineering Construction EDC, <u>Engineering Construction Performance</u>, (NEDO, London, 1976)

25. Anon, <u>Chemical Industry Investment Intentions Survey, 1977 - 1979</u>, (CIA, London, 1977)

26. Anon, <u>General Overseas Trade Statistics - Imports and Exports, 1966</u>, (H.M. Customs & Excise Statistical Office, London, 1977)

27. B. M. Coope, <u>Industrial Minerals for Inorganic Chemicals</u>, (E.C.N. Large Plants Supplement, London, 1976)

---ooo0ooo---

Environmental Aspects of Inorganic Chemicals Production

By A.K.BARBOUR, (Environmental Scientist)
RTZ Services Ltd.

        All who are concerned with the production and sale of
inorganic chemicals - and in this context I will include metals
as well - are familiar with the bad publicity accorded to the
chemical industry by the media during recent years, and the
concern which this has generated in the minds of both the lay
public and some scientists.   Some of the issues which have
aroused public concern are:

ENVIRONMENTAL   CONCERNS

. Exhaust fumes from cars
. Marine and river pollution
. Indiscriminate cyanide dumping
. World-wide distribution of persistent insecticides,
        e.g. D.D.T.
. Mercury and cadmium poisoning incidents in Japan
. Spoil dump hazards, e.g. Aberfan
. Hazards of large chemical plants, e.g. Flixborough, Seveso
. Safety of nuclear plants and nuclear waste disposal.

        There are those who would argue that media publicity
is ephemeral and that the environmental "scare" publicity on
the above and related topics is merely a "nine-day-wonder".
In this talk I will try to illustrate the dangers of this
viewpoint, concentrating on the following:

. Legislative developments;
. Scientific problems in standard setting;
. Some areas where new technology assists environmental pro-
        tection;
. Areas where new basic information and new technology is
        required.

Regulatory Systems

        In organisational terms, UK regulatory systems seem
complex, especially in comparison with superficially simpler
organisations overseas:

UK Regulatory Bodies

Atmospheric Emissions

Alkali Act Registered Works - Alkali and Clean Air Inspectorate (Central - Health & Safety Executive - Secretary of State for Employment)

Non-Registered Works - Environmental Health Officer (Local)

In-plant Conditions

Factory Inspectorate (Central - Health & Safety Executive)

Liquid Effluents

Regional Water Authorities (Autonomous Regional - Minister for the Environment)

Solid Waste Disposal

Local Authorities designated as Waste Disposal Authorities (+ RWA's) (Minister for the Environment)

The Mining, Quarrying and Nuclear Industries are regulated by separate Inspectorates.

Although this organisation may seem cumbersome, and from an industrial standpoint necessitates most companies dealing with several regulatory inspectors for each plant, this situation is not in fact much different from overseas systems. In practice, it works effectively because by and large, industry respects the professionalism of the inspectors and the systems permit rational technical interchange before standards are defined. The other characteristic of the British system is that it deals with individual plants or point sources rather than imposing national or state-wide fixed standards. The philosophy of deciding upon individual criteria differs, however, between atmospheric emissions and aqueous effluents, for in the latter the basic UK criterion is the receiving body's capability to absorb contaminants without damage to so-called target organisms. The Alkali Inspectorate's "best practicable means" approach only indirectly concerns itself with the national ambient atmosphere; in the UK we have to date avoided the pitfalls of some recent overseas systems of setting ambient air standard targets first and then deriving point source emission standards.

The Alkali Inspectorate system is based on the following:

Best Practicable Means.

Practicable - "Reasonably practicable having regard, amongst
                other things, to local conditions and circum-
                stances, to the financial implications and to
                the current state of technical knowledge".

Means   - "Design, installation, maintenance, manner and
                periods of operation of plant and machinery, and
                the design, construction and maintenance of
                buildings".

        In general, emission standards are not specified in
UK legislation; "Best Practicable Means" for each individual
emission are ultimately expressed by the Alkali Inspectorate as
so-called "Presumptive Limits" which are based upon the following
criteria:

Alkali Inspectorate Criteria for Presumptive Limits

I   No emission can be tolerated which constitutes a <u>demon-
      strable health hazard, either short or long-term</u>.

II   <u>Emissions</u>, in terms of both concentration and mass, <u>must
      be reduced to the lowest practicable amount</u>.

III   Having secured the <u>minimum practicable emission</u>, the height
      of discharge must be arranged <u>so that the residual emission
      is rendered harmless and inoffensive</u>. (For highly toxic
      metals, the concentration of each source to the existing
      background concentration shall not exceed one-fortieth of
      the Threshold Limit Value for a factory atmosphere on a
      three-minute mean basis. In deciding on the most impor-
      tant parameter, the effects on vegetation, animals and
      amenity are also considered.)

        It will be seen, therefore, that implementation of the
above criteria provides a fully satisfactory level of health
protection for persons living in the proximity of major point
sources of pollution, consistent with the best current medical
knowledge at the time. Two points need to be emphasised. One
is that TLV's or Threshold Limit Values (that concentration which
can be tolerated by an average person exposed eight hours per
day forty hours per week for prolonged periods) are currently
set by an independent group of professional specialists (the
American Conference of Governmental Industrial Hygienists) and

adopted without modification by the UK Factory Inspectorate. The ACGIH is, of course, totally independent of either the Alkali Inspectorate or the Factory Inspectorate. The second point concerns definition of the phrase "demonstrable health hazard, either short or long-term". It must be emphasised that this is a moving target, reflected usually in lower (or more stringent) TLV's adopted by the ACGIH in the light of increasing toxicological and biochemical knowledge. Scientists can usually accept without difficulty that increasing knowledge can result in changing standards; the general public often finds difficulty in grasping the idea that current knowledge is far from complete at any given time and hence that standards are not fixed for all time.

In the area of liquid effluents, the UK philosophy is to categorise receiving bodies of water into end-uses and then to fix pollutant concentrations in the receiving bodies at levels appropriate for the natural survival of "critical organisms" consistent with that end-use. Individual effluent standards are then derived so as to be consistent with these receiving body criteria. This "Environmental Quality Objective" (EQO) approach contrasts with some overseas legislation (e.g. EEC) which seeks to impose national fixed standards for the concentration of pollutants in effluents regardless of the end-use category of the receiving waters. In the extreme, effluents discharging into estuaries would have to be purified to the same level as those discharging into fresh water streams used for potable waters. For the so-called "black list" pollutants (including mercury and cadmium) specified in the Dangerous Substances in the Aquatic Environment Directive, the UK is currently setting up quantitative EQO-based schemes designed to be at least as effective environmentally as the EEC fixed standards approach (which, for "black list" substances approaches the zero-emission concept) without incurring the expense and misallocation of resources which it implies.

Thus, for aqueous effluents, UK philosophy embraces something of the ambient air philosophy which to date has been absent from Alkali Inspectorate control methods. The reason for this apparent paradox is that river systems are much more separated than the ambient atmosphere. This omission has been recognised recently in the Fifth Report of the Royal Commission on the Environment which suggested that Ambient Air Quality

Guidelines might be considered for adoption in the UK, supplementing the best practicable means approach. In my view, this suggestion is worthy of detailed analysis for it would provide national air purity targets which are difficult to assemble by aggregating hundreds of individual "b.p.m." consents. National guidelines will become increasingly sought as local authorities and the general public become increasingly interested in environmental matters. Such guidelines would, however, have to be consistent with b.p.m. philosophy and hence the Alkali Inspectorate should play a major part in defining them. We must guard against the major error made in some overseas countries who set Ambient Air Standards first, probably at an arbitrarily low level, and then attempted to derive point emission source standards from them. It is, indeed, fortunate that the existence of the proven b.p.m. system provides a safeguard which should prevent this error from being repeated in the UK.

The Fifth Report of the Royal Commission also made the complementary suggestion that the best practicable means approach might be considered for application to liquid effluents. Although this suggestion appears to have received a less than enthusiastic reception from the RWA's, I feel that it could have merit in providing solutions to aqueous effluent problems on the basis of less complex analysis than the EQO approach outlined above. The Report also proposed the establishment of a combined pollution Inspectorate (HMPI) but I propose not to go into the detail of this question, which remains under active current discussion.

Standards Setting

To the scientist, I believe that this remains the most fundamental, and probably the most challenging, aspect of the environmental scene. I suspect that, together with improved arrestment techniques, it is also the area which is receiving less than its due share of research resources.

The chemical industry must face the fact that major pressures are inexorably pushing in the direction of ever-tighter standards, in some countries regardless of the availability of technology to meet them. These factors are both sociological and scientific.

Sociologically, we all feel that the chemical industry

should be, and should be seen to be, safe both to the people who operate its plants and to those in the outside world, particularly near-neighbours. But what is safe? We have long since passed the time when exposure of work forces to the atmospheres of chemical plants caused acute, clinical poisoning - there has been no UK case of clinical lead poisoning, for example, during the last fifty years. Increasingly, attention is being turned to the possible effects of chronic, long-term, low-level exposure which may ultimately lead to degenerative disease of various types. Some scientists take the extreme position that exposure to a level of toxicant which gives rise to <u>any detectable bio-chemical change</u>, whether or not it is reflected by changes in a person's feeling of well-being, or clinical diagnosis, cannot be tolerated and that exposure standards should be set at levels which avoid these subtle biochemical changes. Scientists who take this view sometimes express it by saying that work force exposure must not be greater than general population exposure - the obverse of the more general view that TLV concentrations safeguard operator safety and hence general population safety, since the latter is generally several orders of magnitude lower.

Continuing developments in analytical chemistry in all its manifestations provide something of a scientific base for the above sociological concerns. Modern embodiments of gas-liquid chromatography, thin-layer chromatography, atomic absorption spectroscopy, mass spectroscopy and so on permit the detection of a very wide range of chemicals in miniscule quantities in a wide range of substrates. We can now talk confidently of con-centrations in the p.p.m. and even p.p.b. range for many pollu-tants - concentrations which would have been quite beyond the limits of detection until recently. There is no sign that this revolution in analytical capability has yet spent itself.

Complementing these developments in strictly chemical analysis are those in biochemical testing - the measurement of the effect of minute concentrations of chemicals on biochemical systems. The main gap in this area naturally continues to be the precise relevance of such "in vitro" testing to the human situation, particularly the effects of very long-term, very low-level exposures. Consequently, much attention has been paid, and will continue to be paid, to epidemiological surveys, i.e. judging the effects of pollutants on the basis of statistical

analyses of large populations (sometimes not particularly large)
of relatively highly exposed persons.   Such epidemiological
approaches have widely-appreciated defects, particularly the
separation of the effects of one pollutant from another, personal
living factors, synergistic effects and so on, but in many cases
the epidemiological approach is the only tool currently available
to judge the effects of pollutants on human populations - and
naturally, judgments tend to be weighted towards the "safe" side.

This is particularly the case where epidemiological or
other testing suggests a link between exposure and the development
of cancer.   An increasing number of regulatory bodies now exists
to sponsor such studies and to develop new criteria levels for
exposure on the basis of the results.   In the United States such
bodies tend to propose levels approaching the limits of analytical
detection where a link with cancer is suggested.   But such
effects, even if they can be associated with exposure to a
specific and identifiable substance, are usually the result of
historical exposures over many years at concentrations very much
higher than now.   So there is a tendency for some current
criteria to represent enormous reductions in control levels for
which technology does not at present exist.   Examples in the
subject area of this paper are inorganic arsenic and cadmium compounds, but several other metals are under current study from
this standpoint.   I would certainly not wish to minimise the
possible hazard of carcinogenicity but I do feel that some overseas regulatory agencies interpret epidemiological data in an
unjustifiably restrictive way.   Having said that, the difficulty
of balancing such data with one's responsibility to the work
force has to be admitted.

I hope that the foregoing will also have demonstrated
the close inter-relationship between toxicology and environmental
protection or ecology as it is sometimes termed.   The integra-
tion between these two scientific areas is often not as close as
it should be and I urge that all interested chemists should do
what they can to improve scientific communication between these
areas.

Modern Process Technology and Environmental Protection

During the course of this symposium individual authors
will deal with modern technology for producing virtually the

whole range of major inorganic chemicals, and there is a paper specifically devoted to water purification. In this section, therefore, I will review some of the problems encountered in non-ferrous metals production and try to point out areas where modern process technology substantially reduces environmental impact or makes it easier to control. I will also mention a few points about the environmental aspects of some inorganic products because these may contain some lessons of general applicability, and conclude with some areas which, in my view, require new scientific information or greater concentration of research effort.

1. Non-ferrous metals, particularly zinc/lead/cadmium, copper and aluminium.

The production of non-ferrous metals starts of course with mining, often in modern practice by open-pit methods and usually followed by enrichment of the ore by milling and flotation. Since very little metal mining is now operational in the UK, I will not go into this aspect in detail, except to mention the very large scale of the solid waste disposal problem which often surprises those outside the industry.

This is particularly so for copper which does not occur naturally in high concentrations. Currently, deposits in the range 0.4 to 0.7% copper are being worked world-wide. At our operations in South Africa and Papua New Guinea which, in this sense, are typical, something around 30 million tons of earth has to be excavated, ground, milled and separated to produce the equivalent of 100,000 tons of copper. And all the many millions of tons of finely-divided tailings have to be disposed of in an environmentally acceptable manner.

(i) Zinc, lead, copper - atmospheric emissions - the sulphur dioxide problem

Zinc, lead and copper - but not aluminium - occur most widely as the sulphides and so roasting, usually to the oxide, is the first stage in most commercial zinc, lead and copper processes to provide a reducible or dissolvable feed for subsequent processing. Thus, a major environmental concern in the production of these three metals is capture of sulphur dioxide down to very low levels.

This is due to the fact that, although non-ferrous

smelting represents a relatively small percentage of the total
burden of sulphur dioxide in the atmosphere, as indicated below
for the United States, uncontrolled emissions from smelters can
result in very high local concentrations with resultant damage to
vegetation and amenity.

U.S. SULPHUR DIOXIDE EMISSIONS (1966)

Source	Millions of tons SO_2	%
Power generation (coal)	11.93	41.6
Power generation (oil)	1.22	4.3
Other coal consumption	4.70	16.4
Other oil consumption	4.39	15.3
Ore smelting*	3.50	12.2
Oil refinery operations	1.58	5.5
Sulphuric acid manufacture	0.55	1.9
Coke processing	0.50	1.7
Miscellaneous	0.28	1.1
Total	28.65	

*Copper 2.83m   :   Lead 0.15m   :   Zinc 0.51m

Fortunately, improvements in basic technology for all
three metals, initiated originally for reasons of economies in
labour, capital cost and/or purity of product, now permit the
off-gases from the roasting stage to be converted to sulphuric
acid in double contact acid plants with routine efficiencies of
better than 99.5%, corresponding to a stack emission of below
500 ppm.   Such acid .plants, which will be discussed in detail
elsewhere in this symposium, became commercially available as the
result of developments in Germany about fifteen years ago.

From the smelting point of view, success in the capture
of sulphur dioxide depends on the cleaning or dedusting of
roaster off-gases to a very high level by the use of both wet-
scrubbing (packed towers, venturi scrubbers, etc.) and dry
(electrostatic precipitators) methods and, critically, upon the
ability of the roasting process to deliver gases containing a
fairly constant concentration of sulphur dioxide in the 5-7%
range.

For zinc production, the electrolytic process, which
invariably treats high-grade zinc concentrates, now achieves this

by fluid-bed roasting, another development which has become fully
established over the last fifteen or twenty years. The Imperial
Smelting blast-furnace process, developed at Avonmouth by the
former Imperial Smelting Corporation Ltd. is unique in that it
treats simultaneously a mixture of zinc and lead concentrates.
For this reason, fluid-bed roasting is inapplicable since the
lead content causes the charge to agglomerate and "hang-up"; but
updraught sintering on a conventional moving grate is readily
operated to provide a gas of the required strength. With both
processes, conventional practice world-wide is to convert the
sulphur dioxide produced from roasting to sulphuric acid via a
double-contact acid plant. Similar developments are in train at
plants which produce lead only by means of lead blast furnaces.
Thus a combination of effective gas-cleaning allied to modern
basic process technology provides a complete solution to sulphur
dioxide pollution from zinc and lead production on presently
envisaged emission levels.

The same is essentially true for copper production,
though the time scale is a little more recent. Traditional
reverberatory copper smelting, still widely practised in all
parts of the world, finds great difficulty in meeting modern
emission standards because the strength of the sulphur dioxide
produced is very low (ca. 1%) from the matte or reverberatory
stage and cyclic from the second or converter stage. Fortunately,
more modern continuous or semi-continuous processes (electric
smelting, flash smelting and the Noranda and Mitsubishi continuous smelting processes) are designed to provide a constant-
concentration sulphur dioxide which can be converted to sulphuric
acid in a double-contact acid plant.

There may well be situations where either process factors or capital cost requirements may militate against the replacement of single- by double-contact acid plants. In such
cases, a very wide range of sulphur dioxide scrubbing technologies
is now available many of which were devised in the hope of cap-
turing emissions from power stations. Time does not permit a
review of such methods.

However, sulphuric acid manufacture is not always an
economic proposition, particularly, as in the United States,
where most copper smelters are located very long distances from
acid markets and transportation costs are prohibitively expensive

for a low value commodity such as sulphuric acid.   Looking to
the longer term, the ultimate solution for such operations will
probably be to convert their sulphur dioxide through to elemental
sulphur which is easy to store and, on a contained sulphur basis,
much cheaper to transport. Two direct reduction processes to achieve
this goal have been developed to the full commercial scale, one by
Allied Chemical and one by Outokumpo.

I   High Temp. Reduction: Catalytic (Allied) or Non-Catalytic
          (Outokumpo) $SO_2 \longrightarrow S + (COS, H_2S, CO, H_2$, etc.)

II   Lower Temp. Reduction to remove impurities

III Claus Reaction: $2H_2S + SO_2 \longrightarrow 3S + 2H_2O$

IV   Tail Gas Scrubbing

Outokumpo : Natural Gas, Naphtha, Pulverised Coal
Allied   : Natural Gas

          Increasingly in the future, I believe that copper
smelters will turn to processes of this type.   Their economics
are not yet fully competitive but the product attraction is
obvious, particularly as it permits the full treatment of concentrates at the mine site - a concept of considerable attraction in developing countries who desire increasingly to move
"down stream" in minerals as well as in oil production.

(ii)   Zinc, lead, copper - solid by-products

          The basic steps in electrolytic zinc production are
outlined below:

1.   Concentrate roasting $\longrightarrow$ zinc oxide calcine plus sulphur
          dioxide for conversion to sulphuric acid

2.   Calcine leaching with sulphuric acid to give impure zinc
          sulphate

3.   Solution purification

4.   Electrolysis to zinc metal plus sulphuric acid for
          recirculation to leaching stage.

          The electrolytic process depends critically on refining
the zinc sulphate fed to the cells to very high levels and

so much technology has had to be developed to remove impurity
metals, particularly lead, iron, copper and cadmium. From the
calcine leaching stage, lead remains as insoluble lead sulphate.
During the past decade, technology has been developed to precipitate iron as a highly crystalline jarosite $(NH_4,Na)Fe_3(SO_4)_2(OH)_6$
The flowsheets for copper and cadmium removal vary considerably at
different plants but invariably involve multi-stage precipitation
and thickening/filtration of copper and cadmium which valuable
constituents are subsequently recovered. The environmental consequence of this is that electrolytic zinc plants accumulate two
kinds of solid residues:

(i)   A relatively high-lead dump which is eligible as a feed for
      a lead blast-furnace, although its finely-divided nature
      requires that it be carefully handled.

(ii) A dump of jarosite, $(NH_4,Na)Fe_3(SO_4)_2(OH)_6$, which has no
      known use at present and is, in practice, ineligible as a
      feed for an iron blast furnace because its iron content is
      so low.

Both materials will have other heavy metals associated with them
and hence the run-off from them has to be collected and treated
before disposal to a water course.

On the other hand, the only solid by-product from the
zinc-lead blast furnace is a fused, granulated slag which, because
of its siliceous glass-like nature is inert and innocuous. It
may have possible application as a grit-blasting material. Much
the same is true for pyrometallurgical copper processes - the
by-product slag is inert and harmless environmentally.

(iii) Aluminium

In contrast to zinc, lead and copper, aluminium arrives
at the smelter or reduction plant as a high purity alumina, produced by the Bayer process. Only one alumina plant remains in
the UK, so disposal of the irony "red-muds" which are a by-
product of the Bayer process is not a significant environmental
problem in this country.

In the classical Hall-Herault process, alumina is
dissolved in a cryolite melt contained in carbon-lined steel
cells, or "pots", prior to electrolytic reduction.

Hence, the main emission to be dealt with consists

mainly of fluorides and these can be captured by either wet or
dry scrubbing, both of which are currently practised in the
recent generation of primary aluminium plants built in this
country.   Wet-scrubbing techniques have the disadvantage of
leaving a fluoride-containing effluent for disposal but they can
be designed to deal with the problem of recovering cryolite including the cryolite component of used cell linings.

    The dry-scrubbing approach, developed over the last
decade and illustrated schematically below, diverts a proportion
of the alumina feed to the cell through a specially designed bag
filter plant where it reacts with the off-gases from the cells,
effectively neutralising their fluoride content.

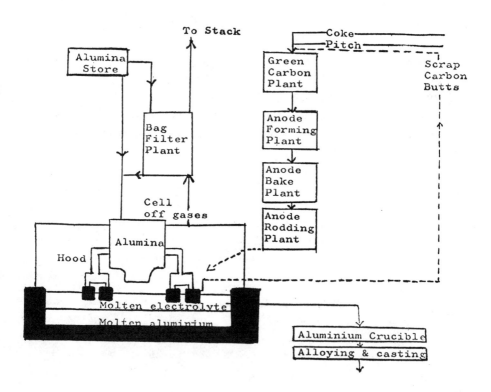

    This approach has the advantage of providing "instant
recycle" of the fluoride content of the cell emission with no
aqueous effluent problems but great care has to be taken that

alumina used for scrubbing fluorides is carefully controlled.
Since it now contains traces of fluoride, cell hoods have to be
tightly fitting and well maintained, otherwise traces of fluoride
can be emitted to the external atmosphere at a low level.

The major solids disposal aspect of primary aluminium
production, particularly where dry-scrubbing of the cell gases is
practised, concerns disposal of the carbon cell or "pot" linings.
After a service life of eighteen months or two years, these have
become anthracitic in nature but they may well have a content of
cryolite in the range 40-50%. Whilst they do not present a
major hazard from the disposal standpoint, a practicable method
of recovering the cryolite would be advantageous, both environmentally and from the economic point of view. Several technologies for achieving this objective have appeared in the
patent literature but none yet seems to be practised on the
commercial scale; this remains a desirable development goal.

The original Hall-Herault technology for aluminium production has remained basically unchanged for over eighty years,
although modified by a variety of engineering improvements.
However, in recent years several chloride-based processes have
been patented which claim to offer better energy utilisation and,
of course, to avoid emissions of fluorides. Alcoa has announced
the construction of a semi-commercial plant based on this type of
technology and its operating results are awaited with keen
interest.

(iv) Product Aspects

Whilst this Symposium is primarily concerned with
processes, the impact of environmental regulation on products and
their markets cannot be overlooked. The most obvious example in
the industries with which this paper is concerned is the long-
running debate as to whether the addition of lead tetra-alkyls is
or is not sufficiently dangerous as to be a general environmental
or health hazard. But other illustrations are under serious
discussion. Arsenic is an unavoidable by-product of many copper
operations and its markets have been seriously circumscribed so
that essentially only its use as an excellent timber preservative
now remains as a tonnage outlet. In this application, the
copper-chrome-arsenite compound is tightly bonded to the timber
substrate and I believe its economic value as a timber preservative enormously outweighs any conceivable environmental effect

which may result from either the treatment itself or the ultimate disposal of the treated timber.

Cadmium is an unavoidable by-product of the smelting of zinc concentrates in which it typically occurs in concentrations of up to 0.5%. Unlike many chemicals, production of cadmium cannot be discontinued without at the same time stopping the production of zinc.

It disseminates widely through the general environment through its market outlets in:

- Pigments
- Stabilisers for plastics
- Plating anodes and salts
- Batteries
- Solders and other alloys

and cannot easily be replaced in many of these uses.

From the environmental point of view its acute toxicity is well appreciated (the current TLV, a ceiling value, for cadmium oxide fume is 0.05 mg/m^3) and both producers and processors of the metal employ stringent safety precautions, usually backed up by medical monitoring of the work force. Increasing concern has, however, concentrated on its chronic effects, for it tends to accumulate over the long-term in the kidney and liver. It is the rate at which this accumulation occurs, and a more precise evaluation of the risks of kidney damage, which is the subject of much current speculation and concern.

These long-term effects are not easy to replicate in animal experiments which can sometimes be interpreted to indicate that no serious problem exists.

We urgently need more research to quantify these long-term effects for long-term, low-level exposure to human beings, and this implies better analytical diagnoses for cadmium concentrations in various human substrates. More detailed thought will also have to be given to the best routes for disposing of cadmium-containing end-products. In the UK at present, analytical surveys of various water courses reveal no disturbing concentration of cadmium but concern is beginning to be expressed that its concentration in sewage works sludge could, in one or two areas, be sufficiently high to prevent disposal of the sludge as an agricultural fertiliser.

2. Emission and Effluent Technology

This subject is dealt with in detail elsewhere in the Symposium and so I will only draw attention to some of the major issues which seem to me to justify continuing research attention.

On the emission side, much progress has been made recently in arrestment equipment of the conventional type, i.e. venturi scrubbers, electrostatic precipitators and bag-filters, particularly of the mechanical type often provided with synthetic fibre bags. However, the problem of fine fume in the sub-micron range remains incompletely solved and in parallel with this is the toxicological aspect of the finer fractions in the dust-fume particulate range.

In the area of effluent treatment, producers of heavy metals generally rely upon lime treatment followed by thickening and filtration to various degrees of sophistication to purify their aqueous effluents, and such treatment processes are usually effective in meeting most current UK standards. However, if much more onerous point emission standards become necessary, either through EEC legislation or otherwise, then other technologies may be called for. These might include ion-exchange, reverse osmosis, or solvent extraction - technologies which may require substantial development to make them applicable to the large volumes and complex ionic and suspended solids content of primary and secondary smelter effluents.

In conclusion, it is my view that the chemical and metallurgical industries can and should develop and grow in the UK in a manner which is environmentally acceptable by the standards of an industrial nation. For this to happen, it is necessary that our legislators should continue to provide a regulatory framework which can be interpreted in a scientifically sound manner by professional inspectorates.

For their part the chemical and metallurgical industries must:

. continue to take positive steps to develop more, sound basic knowledge about the environmental effects of its processes and products;

. insist upon the highest standards of production and maintenance efficiency;

. maintain and sustain relationships with Government so that regulatory officials continue to be fully briefed on the overall impact of environmental legislation, particularly from the EEC.

Education and Training of Chemists for the Inorganic Chemicals Industry

By G.W. CAMPBELL

U.S. Borax Research Corporation

I.   Introduction

The rate of growth of the science of chemistry in recent years is unparalleled in history. These advances have been accompanied by an increasingly sophisticated and intellectually challenging educational program for the chemist. That the universities are successful in training capable scientists is attested to by the continuing growth of chemical knowledge. There is a problem, however. While university graduates are proving to be increasingly capable researchers, they are becoming progressively less acceptable to the chemical industry. One line of evidence of this is in the job market for chemists. In 1976 the starting salaries of chemists in the United States were 70 percent, 75 percent, and 80 percent of the starting salaries for B.S., M.S., and Ph.D. chemical engineers, respectively.[1]

The educational processes by which chemists are prepared to function as professional chemists in industry differ in details from one country to the next and even from one university to the next. There are the basic fundamentals, however, that are common to all such programs. In the United States there is a program of the American Chemical Society whereby the chemistry curriculum of the university is evaluated by professional chemists functioning as members of the Committee on Professional Training. This Committee was established in 1936 by a resolution of the Council of the American Chemical Society. The Committee maintains extensive records on institutions offering baccalaureate degrees in chemistry. Upon invitation, it consults and cooperates with institutions desiring assistance and evaluation. It publishes a list of colleges and universities offering programs that meet the Committee criteria for professional baccalaureate degrees in chemistry. It is, therefore, important to the institution to be on the list of schools that are approved by the American Chemical Society.

Due to the background of the author, this paper is necessarily slanted toward the situation that exists in the United States. Discussions with chemists from other countries, both from industry and the universities, have indicated that the situation in the United States is not unique but in many respects the programs of the universities and the attitudes within the universities are common to most countries.

In 1974 the late Richard L. Kenyon wrote an editorial in which he stated: "A large part of the graduates will seek their places in industry. Yet past history tells us that most will go out to their careers with very little sense of how industry works. Anecdotal evidence tells us that there are today many research chemists frustrated in a career in industry that is much different from what they expected. Adjustment is often very hard."[2]

Many in recent years have recognized a divergence of interests between industry and academia, and this separation certainly plays a part in the educational deficiencies of the new chemistry graduate. At the recent "ACS Ninth Biennial Education Conference," the first recommendation made by the participants specifically dealt with this issue.[3] This recommendation stated "in recognition of the special importance of industrial chemistry to our economy, that ACS undertake activities leading to better training of chemists for the practice of applied chemistry, with immediate action in industrially significant but academically underemphasized fields, such as polymer chemistry."

If the problem is to be dealt with effectively and positively, we should give attention to the causes. It appears that one of the most important is the fact that very few university professors have ever been industrial chemists. The timetable that must be met by the prospective young professor does not allow for such employment. In the United States, the most common route to a university professorship includes a B.S. at about age 21 or 22 followed directly by a Ph.D. at about age 26. At this point he must acquire a post doctoral appointment for a period of about two years. He is then ready for his first academic appointment as an instructor or assistant professor at about age 28. Most schools allow from five to seven years for new staff members to earn the tenure rank of Associate Professor, which corresponds to Reader in the United Kingdom. To gain tenure he must produce a multitude of research publications. Therefore, the new staff member gathers graduate students or postgraduates and searches for research support. This leaves little time for him to concentrate on educational processes or to prepare lectures for the courses he is required to teach. At the age of about 35, he receives the appointment to Associate Professor. He then has a secure position and can plan his future academic activities. At this point in his career he has been successful. He has concentrated on research and he has an active research staff of graduate students under his direction. The strong tendency is to continue doing those things that have brought him this success. And with justifiable pride he turns out more chemists in his own mold.

There are good things to be said about this pattern. It is this academic process that has developed the brilliant research professors and has resulted

in the outstanding advances in chemical theory in recent decades. These contributions deserve the recognition they have received. However, the university professor in this educational program is seldom, if ever, exposed to the
methods, the operations, and the problems of chemical industry. As an
example of the consequences, the chemistry department chairman of a well-
known university, a man of international repute as a chemist, recently admitted that he and his students had been working with a particular class of
compounds for more than twelve years, yet he had no idea how they were made
commercially.

This separation of academia and industry is promoted by the employment
policies of the chemistry departments of many universities. Departmental
reputations are built by the research publications of their faculty, and the
reputation attracts more and better graduate students in a progressive cycle.
This leaves industry on the outside. Some university administrators frankly
recognize this situation. The university professor that leaves the university to accept employment in industry has taken a "one-way street" and, with
very few exceptions, there is no return route.

An unfortunate consequence of the gulf between the universities and industry
is the development of attitudes and perspectives in the student that are incompatible with the nature of industrial science. The idea develops that the
industrial chemist has sold his soul for the monetary returns. Ideals and
attitudes are persistent and the university years are the formative years.
The proper time to achieve a philosophically balanced orientation is then.

The academia-industry cleavage that is so evident in chemistry has not developed in chemical engineering, at least not to the same degree. In chemical
engineering it is recognized from the beginning that the student is preparing
for a career in industry. It is also quite common for professors of chemical
engineering to have experience in chemical industry. It might be expected in
universities that include a chemical engineering program that training in
chemistry would include the benefits of contact with those experienced in the
industry. Unfortunately, this has not been realized. It was recently observed[4] that "chemistry and chemical engineering educators at American universities have not really talked to each other in 30 years." In this same
article it was observed that "nonacademic opportunities (in chemistry) are
. . . limited because the training of the young chemist has been so thoroughly academically oriented."

A comparable development can be observed in the United Kingdom where most of
the older universities tend to adhere to programs in pure chemistry only,

although a few also offer an Applied Chemistry degree course. The larger proportion, by far, of chemistry graduates at the United Kingdom universities is in the "pure" area.*

The growth pattern of the chemistry curriculum has followed a course based upon the growth of chemical theory. This has been a path of least resistance and it has ignored the primary responsibility of the university—the student and what he needs to become a chemist! If it were not for the student, the university could not exist. Since about 82 percent of the scientists trained in the university ultimately find employment outside of the university,[5] it is the responsibility of the university to prepare them realistically for their careers.

In this paper we will look at the chemistry curriculum to see where it can be improved from the industrial chemist's viewpoint. In addition, it is recommended that the universities consider the addition of a faculty position of "Professor of Industrial Chemistry," and the development of a course in Industrial Chemistry is proposed. We will also examine some of the efforts that are now being made to better prepare students for a career in industry.

The development of the present academic situation in chemistry has taken many decades. Educational philosophies are deeply entrenched and rapid changes in attitudes will not take place. It might be hoped, however, that with concerted effort the educational process for chemical industry might be significantly improved in the coming years.

II. The Chemistry Curriculum

In the United States, the four-year undergraduate program for students whose principal subject is Chemistry generally consists of one year of General Chemistry, one year of Organic Chemistry, one year of Analytical Chemistry, and one year of Physical Chemistry — not necessarily in this sequence. At the third or fourth year, several advanced courses are commonly included. This general program is consistent with the criteria of the American Chemical Society Committee on Professional Training and it has not changed significantly for many years. However, the content of the courses has undergone a gradual change to keep pace with the advances in chemical theory and understanding. Historically, General Chemistry contained an introduction to chemical principles and methodology and the mathematical problems related to chemical processes followed by or intermeshed with the descriptive chemistry

---

*Personal communication, R. Thompson, Research Director, Borax Consolidated Ltd., Honorary Professor, University of Warwick, Special Professor, University of Nottingham.

of the elements. All of the General Chemistry textbooks now in use include
treatments of the descriptive chemistry of the elements. The second half of
General Chemistry commonly included a detailed treatment of Qualitative
Chemical Analysis both in lecture and laboratory. In recent years, however,
many instructors have found that there is not sufficient time to present all
of the material that should be included. Hence, much of the descriptive
chemistry has been squeezed out of the General Chemistry course to make room
for modern concepts of atomic structure, nuclear chemistry, molecular structure, coordination chemistry, etc. The trend has been to make General
Chemistry into a "watered-down" physical chemistry.

There is validity to the argument that an improved understanding of the basic
principles is important to understanding descriptive chemistry. Unfortunately,
the displacement process to make room for more theory eliminates rather than
postpones the treatment of those parts of the course that are displaced.

The objectives of the chemistry curriculum, especially at the undergraduate
level, should include the commitment to provide instruction, in classroom and
laboratory, to prepare the chemistry student to function as a professional
chemist upon graduation. To so function he must have an understanding of
basic chemical theory, he should know a considerable amount of descriptive
chemistry, and he must be able to use the chemical literature. There is
another requirement, however, which is perhaps even more important. The late
Professor James B. Ramsey* referred to the development of a Chemische Gefühl,
or chemical intuition, as the most important single achievement of a chemist.
This is especially true for the industrial chemist who usually cannot afford
the luxury of concentrating on a small branch of his profession but must deal
with various problems.

It is beyond the scope of this paper to go into detail on what should and
should not be included in the undergraduate program of a chemistry student.
However, it is appropriate to point out some of the areas that are of importance to industry which have been given relatively low priorities in many of
the universities.

Descriptive Chemistry of the Elements. The changes in the introductory course
in General Chemistry to include discussions of the more recent developments
of chemical theory have usually been made with a reduction of the descriptive
chemistry of the elements. Yet it is here that chemistry can be made to have
real meaning to the students. The discussion of operating industrial processes which touch the lives of every individual provides a stimulation to learn

---

*Professor of Chemistry, University of California at Los Angeles, 1923-1959.

more about the science. In contrast, a highly theoretical treatment of chemical theory at this stage will frequently cause the student to lose interest in chemistry and turn him away from this profession.

Another important related point is that a presentation of chemical theory without discussing the applications to industry implies the absence of such applications or that such applications are unimportant. The result is the formation of attitudes that are oriented against industry and the applications of the science.

There is a point of balance which must be reached where both chemical theory and application are dealt with appropriately. The evidence is that most of the weights are now on the theory pan.

Qualitative Chemical Analysis. Qualitative analysis provides an exceptional opportunity to teach a wealth of useful descriptive chemistry in a logical and comprehensive manner that interrelates laboratory observations with pertinent theory. This background for the industrial chemist is of critical importance. Yet, Qualitative Chemical Analysis is now frequently relegated entirely to the laboratory where it is presented over a period of about six weeks.

Quantitative Chemical Analysis. The Quantitative Chemical Analysis program has changed drastically with the virtual elimination of classical quantitative analysis in some cases. Instrumental methods of analysis have become widely used and their importance to industry cannot be overlooked. Nevertheless, it is a rare industrial laboratory that does not make some analyses by the classical gravimetric or volumetric procedures. The fact that the classical methods are still used, however, is only part of the value of the study of this branch of chemistry, both in the classroom and in the laboratory. Some of the more valuable concepts that are fundamental to this study include ionic equilibria, oxidation-reduction equilibria, electrolysis, and adsorption phenomena. The adsorption processes that can influence the purity of a precipitate in a quantitative gravimetric determination are also going to degrade product purity in an industrial manufacturing operation.

Thermochemistry. Thermodynamics, which is presented in the course of Physical Chemistry, deals also with thermochemistry and the heats of reaction, crystallization, fusion, etc. The calculation of a heat balance for a chemical process usually involves nothing more than simple algebra. Yet most chemists coming from the university are unaware that they have the capabilities to make these calculations. This is one of the areas that chemists have turned over to the chemical engineers. Perhaps this is one of the reasons that new chemical engineers are more valuable to industry.

The course content of Physical Chemistry is now such that it is not appropriate to displace any of it to devote more time to applied thermochemistry, but it is suggested that a course dealing with industrial chemical problems would be an appropriate addition to the undergraduate curriculum.

Phase Equilibria. The course in Physical Chemistry provides an introduction to phase equilibria. As in the discussion of thermochemistry, this is a topic that is of great interest to the chemical industry, yet it gets too little attention. Selective crystallization processes, solvent extraction processes, and techniques to shift phase equilibria by changing the chemistry of the system all serve to emphasize the importance of this subject. This too could be included in a course on industrial chemistry both at the undergraduate and the graduate level.

Other Chemical Areas of Concern. Discussions comparable to those preceding could also be developed for other topics including Industrial Electrochemical Processes, Fundamentals of Corrosion, and Practical Reaction Kinetics. Obviously, due to time limitations, some of these topics could best be dealt with at the postgraduate level, although the only significant prerequisite would be Physical Chemistry.

Corrosion is commonly ignored in the undergraduate chemistry program, yet it is the most universal problem faced by the industry. For this reason it should be given a high priority in a course dealing with industrial chemical problems.

Technical Writing in Industry. The most common and usually the most obvious weakness of the new graduate chemist is the inability to write an acceptable technical report. The requirements of the technical report in industry are quite different from what is expected of the academic technical report. The university professor submits a technical article that is written primarily and in practice almost exclusively for his peers. In industry the report must be written for those who need the information that is contained. These include both technical and nontechnical personnel, the latter sometimes being the most important. Hence, the writing job is more difficult. This problem is not uniquely a problem for chemists, but applies to all fields of technical endeavor that must also relate to other disciplines. Contrary to this recognized deficiency in the educational program, many of the universities are actually reducing the English composition requirements of their students. It thus appears that much more attention to this form of English composition is called for in the universities. It is also pertinent at this point to note that the American Chemical Society has developed an audiovisual course

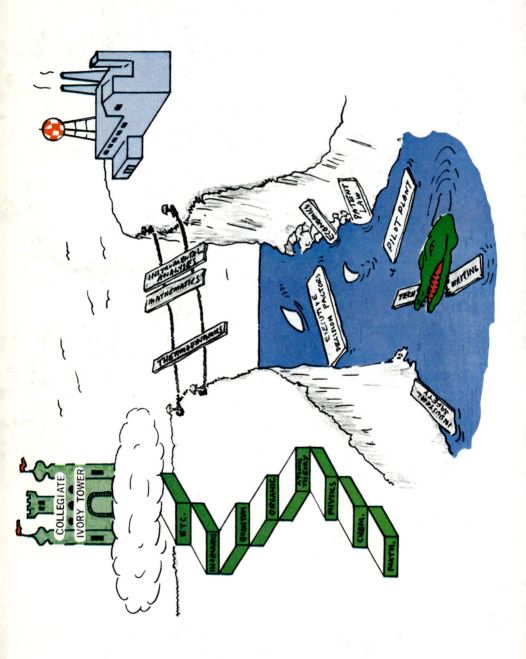

on "Practical Technical Writing."* At the recent "ACS Ninth Biennial Education Conference," the need for improved communication skills was given a very high priority among the recommendations developed at the Conference.[3]

Evaluation and Consequences of Curriculum Recommendations. At the undergraduate level these recommendations will represent primarily a change in emphasis more than an addition to subject matter. However, some additions to the undergraduate program are needed and therefore they must displace some of the current course content. It is suggested that consideration be given to returning some of the following subject areas to the postgraduate level, as was once the case: atomic and molecular structure theory, coordination chemistry and the relevant bonding theories, and the theory of rate processes.

Many universities have condensed the postgraduate course offering to a one-year program. This has advantages in that it promotes earlier specialization. However, this has been one of the factors that has forced more of the sophisticated chemical theory into the undergraduate program and displaced much of the chemistry that is needed by a versatile industrial chemist. This trend should be carefully reevaluated in view of the needs of 82 percent of the undergraduate students and certainly the majority of the graduate students. This trend was also a topic of discussion at the ACS Ninth Biennial Education Conference,[3] and it was recommended that "where applicable, graduate chemistry departments restore a better balance between research and course work."

III. A Course of Study in Industrial Chemistry

The discussion of the chemistry curriculum dealt specifically with the content of the courses that have virtually become standards for the B.S. degree in chemistry. Advances in the science of chemistry have caused changes in the content of these courses, most of which were made to keep pace with the growth; but some of these changes have made the program less relevant to the ultimate career of the chemistry student. Chemical industry has also been changing and growing through the years, and it is appropriate to provide an orderly study of the responsibilities of the industrial chemist in his professional training.

Many of the things that a competent industrial chemist can and should do have been given over to the chemical engineer by default. The determination of material and energy balances for a chemical process is at least as appropriate for a chemist as it is for a chemical engineer, but because the chemists are reluctant or unable to make these calculations, which are needed for the

---

*Available through the American Chemical Society, Department of Educational Activities, 1155 Sixteenth St., N.W., Washington, D.C. 20036.

design of a plant, the chemical engineer by necessity steps into the breach. Courses in industrial chemical calculations, including the material and energy balances, are standard in all chemical engineering programs, and excellent texts are available.[6,7] This course should also be available to chemistry students and presented from the chemist's point of view rather than the engineer's. It should be emphasized that this is an area in which a chemist can provide a valuable service to the chemical industry. There are a number of other topics that should be added to the education of the industrial chemist.

<u>Chemical Economics</u>. There is probably no area of concern to the industrial chemist that is as important to him as the very broad field of chemical economics. The academic chemist seems to take special pride in disregarding the interaction of chemistry with the financial aspects of chemical processes and the marketing function. As a result, the chemist entering industry commonly has no knowledge of how his contribution to his company's activities can and must provide profit for the company. Speaking for the chemist in industry, Herbert and Bisio[8] observed that "our practical knowledge of industry economics and our skill in using economic techniques frequently have been self-learned, often in bits and pieces." Unfortunately, this self-education approach is uncertain and slow to reveal the broad and complex picture of the interactions of economics with chemical industry. An orderly and systematic introduction to the economics of chemistry as a part of the training of a chemist for industry is needed.

If a chemical manufacturing company is to make a profit on its operations, the income from the sale of its products must exceed the costs that are associated with the operations of the company. The chemist should know the sources of these costs. For example, for a prospective new plant, he should have an understanding of the methods of predicting the cost of construction and total capital required, the operating costs including personnel, raw materials, fuel and energy, maintenance, etc. Each of these costs has complicating factors. The location of the plant will influence worker availability, raw material costs, fuel costs, and marketing costs.

The price that is charged the customer for the product is the most sensitive single factor to profitability. But price is not fixed by production costs. The price is determined by customer demand or need for the product, competition in the marketplace, and other factors that include production costs, company image or reputation, and advertising arts. Price has been defined as "the interface at which the dynamic interaction of the forces of demand and supply come together.[8]

Profitability and the return on investment (ROI), or the discounted cash flow
(DCF), is usually the most important basis for investing in new manufacturing
facilities. The industrial chemist should appreciate the significance of the
ROI and the DCF. The DCF is generally more useful because it provides a
schedule for the investment returns. The final decision to invest, however,
involves much more than the predicted ROI or DCF of a new venture. One of
the factors is the reliability of market predictions and the risk involved if
the predictions are too optimistic or if competitors take a larger than anticipated share of the market. Commonly, the larger the scale of operation
the higher the ROI, provided the consequent higher availability of the product
does not adversely affect the marketplace. The dependability of the raw
material source or sources is a very important factor and it can be critical
to the ultimate decision.

Because of the risks involved in a new venture, it is frequently expedient to
find another company to share the risk, either by joint venture or by a contract to produce. In this case the profits are not as large, but if the
project goes bad the losses are not as large either.

These are all part of the economics of any industrial chemical operation.
The chemist in industry is not and in most cases should not be an economist.
However, he should have an understanding of the basic economic processes and
problems that relate to the business of his company and have a very direct
bearing on what he does in the laboratory or the manufacturing plant. A
course of study for the industrial chemist should include an introduction to
this important field.

Patents and the Industrial Chemist. The entire subject of chemical patents
is commonly ignored in the university program. Yet the patent position of a
company is frequently one of its most valuable assets, and the chemists
employed by the company are largely responsible for the establishment of that
patent position. In the United States, the patents that have been acquired
by a company constitute its "exclusive right"[9] to inventions and innovations.
These inventions and innovations take various forms, but they are bought at
a price to the company through the efforts of the technical staff, and this
price is sometimes very high. They are frequently the basis for constructing
new plants and marketing new products. The stated purpose of establishing a
patent system in the United States was to "promote the progress of science
and the useful arts."[9] The protection provided by the patent ensures that
the inventor has the right to exclude others from making, using or marketing
the invention for a period of seventeen years. In exchange for this protection the inventor must provide a full and complete disclosure of the invention

and it will be published by the U. S. Patent Office. Hence, the published
patent adds to the body of scientific and technical knowledge that is available to all.

The chemistry student should learn the importance of chemical patents and how
they are obtained. "An invention, to be patentable, must be novel, useful
and unobvious."[10] He should learn the legal definition of these terms.
Novelty in the patent sense has some unusual legal interpretations, but to
deny novelty, according to the law, prior use or knowledge of the invention
must be described in a patent or other publication. The invention must be
useful. The fact that a new compound or composition has been developed is
not enough, but it must have some beneficial use. The requirement that it be
"unobvious" is probably the most subjective of the three requirements. However, if the patent meets the first two requirements and the utility value is
high, the fact that it solves a long recognized problem is contributory evidence that it is not obvious. There is now available an audiovisual course,
"Introduction to Patents," offered by the American Chemical Society.*

There is often a question whether an invention should be patented or kept as
a "trade secret." There are instances where the "full disclosure" required
in a patent would provide useful information to the competition. Indeed,
disclosure of an improved process might, in some cases, be readily adopted by
competitors under circumstances where it would be virtually impossible to
detect or to police the infringement.

The question of who is the inventor of the patent should also be discussed.
To the laboratory chemist who does all of the experimental work in developing
an idea, it sometimes comes as a shock when his name is left off of the
patent. Unless the initial idea was his, however, he may not have made any
creative or inventive contribution.

The further question of patent ownership will depend upon the employment terms
of the chemist. Most industrial chemists have sold their services to their
employer and the employer has the right to ownership of any patentable inventions he makes.

Finally, the patent system of the United States should be compared with that
of the United Kingdom and with the patent requirements of some of the other
countries. Specifically this should deal with what is required to obtain and
retain a patent in a given country. The rules for filing patents established
by the International Patent Convention should be discussed. In those

---

*Available through the American Chemical Society, Department of Educational
Activities, 1155 Sixteenth St., N.W., Washington, D.C. 20036.

participating countries a patent must be filed within one year of the first filing date for the United States Patent.

At the present time the E.E.C. or "Common Market" countries are progressing toward a European patent system that is designed to be uniform for all member countries.

<u>Chemical Process Design</u>. The industrial chemist is inevitably involved with chemical processing. He should gain an understanding of how a chemical plant functions, contrasting the laboratory techniques with methods used on a large scale. This should include an introduction to changes necessary to make a process continuous rather than batchwise. It is appropriate therefore for him to learn the basics of process or plant design. This is particularly useful in the early stages of the development of a process since it provides a means for an early process evaluation before chemical engineers are brought into the program. An example of what could be included in the training of an industrial chemist is presented in a recent series of articles by J. Peter Clark.[11, 12, 13] In these articles he provided some simple guidelines for the preliminary design of a plant based upon laboratory data. A nine-step procedure is presented that includes:

1. Draw a flow sheet
2. Calculate a material balance
3. Calculate an energy balance
4. Estimate equipment sizes
5. Estimate the capital cost of the project
6. Estimate operating costs
7. Evaluate the project
8. Optimize the process
9. Present the results

At the preliminary stage, none of these steps is particularly difficult since high accuracy at this point is not necessary. Some of these steps are more specifically chemical engineering than chemistry. This is especially true of Steps 4, 5, and 6. However, guidelines are readily available in the literature[14] for these estimates. A greater appreciation of the process will be gained by the chemist that goes through this procedure, enhancing his value as a chemist.

<u>The Transition — Laboratory to Plant</u>. The most important event in the work of an industrial research chemist takes place when the new process in which he has had a part in developing moves from the laboratory to production. However, this is not a simple process and it takes place in several steps, each of which must provide satisfactory answers to critical questions.

Assuming the laboratory studies are successful and the preliminary design and
economic evaluation are favorable, the process is ready for piloting.
Arthur L. Conn[15] has observed that piloting is the key step in the development program because it represents the first major commitment of manpower and
money. The pilot plant is a collection of small units in which all major
process steps that require experimental investigation can be studied. In
most cases it is not necessary to study every major step since some are wholly
and reliably predictable from previous studies or by calculation.

During the design and operation of the pilot plant, communication between the
bench chemists and the pilot plant engineers is essential. If this communication is to be effective, the chemist must have an appreciation for the kinds
of information that the engineer needs. Specifically, this involves reaction
kinetics to help size the reactor, thermochemistry to determine whether heating or cooling coils or other types of heat exchangers will be needed, the
nature of solvents to be used and their volatilities, flammabilities and
toxicities, etc. It is important for the chemist to recognize that the pilot
plant is a small or scaled down plant, not a scaled up laboratory experiment.

The objectives of the pilot study are: (1) to confirm the technical feasibil-
ity reported from the laboratory, (2) to provide data for definitive engineering design, (3) to optimize the operating parameters, and (4) to provide the
data for a more precise economic evaluation.

In the pilot plant program there will be repeated need for laboratory support.
Design changes sometimes mean changed chemical process steps. Different solvents will need evaluation. The bench chemist is not bothered by a few extra
steps, but every step is costly in an industrial operation, both in equipment
and in product recovery.

It is sometimes desirable to extend the pilot plant operation to obtain more
specific information for equipment selection or design or to test the market.
In this case it may be necessary to build a larger, fully integrated pilot
plant which is sometimes called a demonstration plant. This latter step is
expensive and the decision to construct and operate such a plant must consider the ultimate scale of investment, the degree of risk involved if the
design is based upon small pilot plant data, and the economic evaluation. If
the market is uncertain, the sustained small scale production may be essential
to testing or developing a demand for the new product.

The transition from laboratory to plant requires a close relationship between
the chemist and the chemical engineers; yet is is in this process that the
two fields are most readily distinguished. The treatment of this topic in

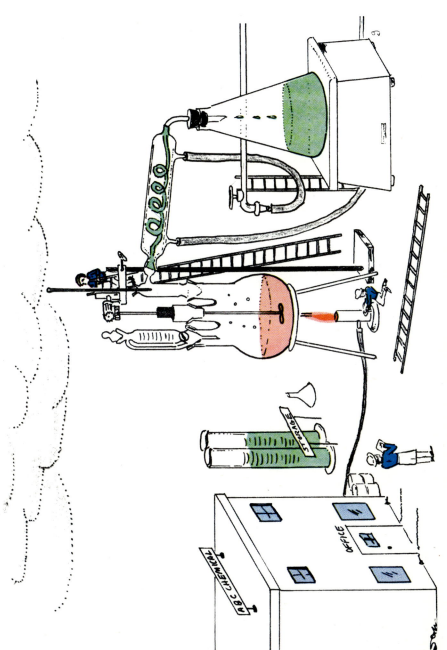

THE BENCH SCALE RESULTS WERE SO GOOD WE BYPASSED THE PILOT PLANT

the classroom is very effectively achieved by means of the "case study" approach where specific chemical principles or sets of principles are transformed into viable industrial processes or products. This approach also provides a convenient means of inspecting a proposed process to determine which steps are in need of pilot plant study and specifically what studies are appropriate.

Safety in the Chemical Industry. The hazards to workers in chemical industry include essentially all of the hazards that are common to manufacturing in general, but it also has many additional safety problems. Chemical manufacturing commonly utilizes volatile and flammable solvents that constitute fire and explosion hazards. An additional concern of nearly all chemical systems is the toxicity of the components. The hazards include all of the possible means of contacting the toxic materials; breathing, touching, ingesting, etc. Many chemical processes require pressurized equipment. For a variety of causes the pressures can get out of control and exceed the bursting strength of the vessel.

The chemist has a responsibility to recognize the hazards inherent to the process and to see that steps are taken to protect the workers. In the United States, the Occupational Safety and Health Administration (OSHA) has made an intensive study of the hazards of industry and provided detailed procedures to protect the workers. The chemist should be familiar with OSHA publications. However, his responsibility goes beyond the OSHA guidelines since the specific process may not be anticipated by the general safety standards that were prepared by OSHA.[16]

The toxicity problems are only now beginning to be dealt with effectively, and much more effort is needed in this area. The fire and explosion hazards have been recognized for many years. The Dow Chemical Company has developed a Hazard Classification Guide[17] which has worked well for them. The emphasis is to recognize the hazard prior to the design of the plant. Then "design safety in and the plant will be safe."

Environmental Protection. The problems relating to the protection of the environment are largely chemical problems. The industrial chemist, in his training, should be exposed to the types of problem involved and the more common methods of controlling refinery emissions, containing and disposing of waste, etc. The various agencies in the United States are progressively restricting the allowed discharges with a goal of zero effluent. The chemical challenge of attaining this objective will require an increasing amount of effort by the chemical industry.

<u>Public Responsibility and Liability</u>. The responsibility of the employer to
provide a safe working environment for the worker (OSHA) is now widely recognized. It is now being realized that it is inconsistent to provide a safe
environment for the worker while endangering the environment or health of the
consumer. It is inevitable that in the near future the legal obligations of
a manufacturer to market a "safe" product will be made more stringent and
demanding. In these matters, not only is the "company" going to be liable,
but responsible individuals within the organization are likely to have crimi-
nal liability on the basis of "superior knowledge."

The new chemist in industry should be aware of the role he is to play in this
relatively new aspect of the chemical industry, and his training should prepare him for it.

IV.   <u>The Role of Executive Decision Factors in Chemical Industry</u>

It is not uncommon to hear a technical person complaining because his new
process or new product was turned down by top management even though all of
the essential requirements (from his point of view) had been satisfied; the
process is technically feasible, the economics are favorable, a marketing
study was favorable, etc. The training of a chemist for industry should
emphasize that these obvious criteria for evaluating a new venture or a new
process are only a part of the requirements that must be met if the company
is to invest a substantial amount of capital. Some time should be devoted to
the "Executive Decision Factors." These are the factors that identify the
"good" research programs in the eye of management. They are often very sensitive factors that relate to the long range objectives of the company, sometimes so confidential in nature that only the top members of management are
privy to them. Some of these factors are described below.

<u>The Nature of the Company</u>. The nature of the company relates to the kind of
product the company markets. Hence, a company that is primarily a producer
of detergents is not likely to be willing to invest in a process to manufacture a new selective herbicide. This is an extreme example. However, much
closer product relationships might still be unacceptable to the company executives for factors such as company size or capital resources compared to the
companies that are already marketing in the potential new product area. The
entry of a new producer could cause the competition to cut prices, which
would drastically change the economic evaluation and wipe out expected profits. This is a risk factor that must be considered in any venture involving
a new product.

Established Company Expertise. The history of a chemical company is reflected in the technical reports that have accumulated through the years, and in the special capabilities of the technical and production staff. To a large degree these constitute the expertise of the company. New ventures which are consistent with these special capabilities involve less risk. To a degree, this is related to the factor discussed above under The Nature of the Company, but in the present context it is based upon technical and manufacturing process relationships rather than end use or product relationships. A company which uses aqueous extraction process steps to recover the valuable components from an ore may have no expertise in flotation processes or electrolytic processes. Hence, even with the discovery of a new and valuable ore body, the executive decision may be to sell the ore body rather than to develop it.

Timing. This factor is probably responsible for the rejection of more "good" projects than any other. The new process or new product is ahead of its time and the necessary market development problem appears too formidable, or it is too late and the competition in the marketplace is too intensive.

Another example of unfavorable timing is when two (or more) projects come at essentially the same time and the company has the resources to develop only one. It is particularly unfortunate for a new process or new product development to coincide in time with a mandatory big expense associated with the primary operations of the company. The latter commitments of capital are necessary for the company to stay in business, and they must be made. By the time the company has recovered from those expenses the "new process" is likely to be too late. In situations where the capital resources are limited, it is understandable that corporate management might be reluctant to discuss their reasons for rejecting a project, and this, in turn, would likely lead to frustration of the chemists involved in the development.

V. A Professorship of Industrial Chemistry

The chemical industry is dependent upon the universities to provide the education and the technical training of its chemists. It was pointed out earlier in this paper that very few of the chemistry faculty members have ever had experience in industry and that the present system in the front line universities virtually legislates against bringing faculty members in from industry. At the same time, 75 to 85 percent of the chemistry students are being trained for industry, but by industries' judgment the training is quite inadequate.

As a first step toward improving this situation, it is recommended that the universities consider establishing a faculty position of Professor of Industrial Chemistry. This should be a full-time appointment with a salary commensurate with the other senior professors in the department. Under no circumstances can the Professor of Industrial Chemistry be a second class member of the chemistry department.

The Professor of Industrial Chemistry should have the scholarly qualifications expected of a university faculty, but the basis for the evaluation must recognize achievements which are meaningful to industry. His credentials would include success within the industry, papers published, and patents granted. Likely candidates would include research directors, managers or supervisors, senior research scientists, etc. Members of top management would be less likely candidates because they are usually too far removed from the chemical process problems and from the laboratory.

The responsibilities of the Professor of Industrial Chemistry would include the development and presentation of a course in industrial chemistry. The contents of the course should include topics such as are discussed in Section III of this paper. It would also be appropriate for him to develop a program leading to a graduate degree in industrial chemistry. This has been done at Florida Technological University at Orlando, Florida.[18] Their program will be discussed briefly in the next section of this paper.

In the United Kingdom, full-time professorships in industrial chemistry or applied chemistry do exist in some of the universities.

VI. Industrial Chemistry in American Universities

Some institutions have initiated programs to provide training for the industrial chemist. These programs have a wide variation in approach and in level of effort. Several of these programs have been selected for review.

Florida Technological University (Orlando, Florida). It appears that the most comprehensive and extensive program directed specifically toward the education of industrial chemists is being offered at Florida Tech. This is a relatively new school of about seven years where innovations in the curriculum are not facing long-standing traditions within the institution. At the undergraduate level, a course in industrial chemistry has been developed and offered as an elective in the senior year.[19] This course is designed to supplement the chemistry curriculum that is accredited by the American Chemical Society. Divided into two parts, Part 1 of the course involves introduction to selected engineering and business concepts, and Part 2 consists of selected case studies. This is a pioneering effort at Florida Tech, and the authors of the course state that they "are trying to turn out chemists that can communicate and interact effectively with engineers."[20] They are also planning to publish a textbook for the course entitled "Principles of Industrial Chemistry."

In addition to their undergraduate course in Industrial Chemistry, Florida
Tech is now offering an M.S. degree in Industrial Chemistry.[21] The program
will emphasize the application of chemical principles to the development of
products and processes.

The course of study calls for advanced courses in chemical structure, chemical
dynamics, and chemical synthesis. In addition, it includes courses in chemical processes, process kinetics and control, and chemical process economics.
An industrial research project and report are also required. Florida Tech
has an unusual advantage in this program since all of the involved faculty
members have had some industrial experience.

The program at Florida Tech is a fine program that has been well planned and
could serve as a model for other institutions.

The University of Akron (Akron, Ohio). An excellent course has been developed
at the University of Akron by Dr. Joseph P. Kennedy that deals specifically
with the professional aspects of the work of the industrial chemist.[22] The
course deals with those areas usually overlooked in the training of the
chemist, including the business, legal, economic, and societal factors. The
course therefore is not an industrial chemistry course but rather a course
for the industrial chemist. It is based upon the principle that "a chemist
who understands only chemistry cannot understand chemistry."

It is offered at the graduate level, although senior undergraduates are permitted to attend. There is no special textbook for the course, but several
books are used for specific topics.

The early part of the course deals with chemistry in industry and the types
of professional jobs that are available in the chemical and allied industries.
It is observed that nearly 40 percent of the approximately 100,000 chemists
in United States private industry are in work assignments other than research
and development.[5] The differences between academic and industrial jobs are
contrasted.

Considerable time is spent on patent law and the role of patents in industry.
This is followed by the roles of marketing and management and finally by a
discussion of the interface of research and development with the other
functions of industry.

This course, therefore, is complementary to the program now offered at Florida
Technological University, and it appears to be the only one of its kind now
offered in the United States.

Northeastern University (Boston, Massachusetts). A quite different approach

to education for the chemical industry is taken at Northeastern University.
At Northeastern the curriculum is essentially the standard program approved
by the American Chemical Society. However, a cooperative work-study program
with local industry has been developed to provide on-the-job experience to
the chemistry students as they earn their degrees.[23, 24, 25]

In this program the course work is interspersed with work periods on a schedule that leads to the B.S. degree in five years. The industrial assignments
are carefully selected to provide a realistic perspective of chemistry in
industry. In a typical program pattern, the first three quarters are devoted
to class work; the second, third, and fourth years alternate quarter by
quarter between work periods and the classroom; and in the fifth year there
are two quarters in the classroom and one quarter at work. Hence, there are
seven quarters of industrial experience, not all of which are with the same
employer.

Programs very similar to this have also been developed in the United Kingdom
and in Canada, where they are referred to as "sandwich courses." There is a
practical benefit to this program since the students are paid by the employer
at the same rate as other employees at the same level of capability. The
development of attitudes that are in tune with industrial objectives is of
greater benefit to their future careers.

Northeastern University recently extended the work-study approach to the
graduate program. Considerable flexibility is provided in this program.
Usually 12 to 15 months are spent in industry and the research for the doctoral thesis is frequently based upon or related to the industrial assignment.
During the time in industry the student takes one graduate course each quarter
in the evenings and attends seminars and conferences at the university, thus
maintaining close contact with his academic advisors.

More than 50 employers in the Boston area, industrial, governmental, and
clinical laboratories, are cooperating with Northeastern University in this
program.

The University of Wyoming (Laramie, Wyoming). At the University of Wyoming,
a new B.S. degree program in Chemistry and Chemical Technology[26] has been
developed. This program differs from the Florida Technological University
program in several ways although both are directed toward preparing a student
to enter an industrial chemical environment immediately upon graduation. The
most significant difference is that at Wyoming, the program immediately emphasizes the basic laboratory techniques and practice of chemistry with specific
orientation to industry. Much of the design of this "program was based upon

the premise that the education of B.S. chemists should include significant exposure to the nonacademic aspects and attitudes of chemistry." Attention is also given to related areas such as engineering, economics, and management. In the laboratory courses, report writing, problem solving, and the use of chemical literature are stressed.

Although the program is designed primarily for the student entering an applied area of the chemical profession with the B.S. degree, it includes the necessary preparation for most graduate work. The University emphasizes that "there is a solid core of experience in the first two years . . . and considerable flexibility in the last two years for a student to specialize in an area of interest. Students should have 'marketable' skills at any time after the first year.

Another factor in the Wyoming program is their cooperation with the development of associate degree (two-year) programs in chemical technology with seven Wyoming community colleges wherein the associate degree programs are fully transferable toward the B.S. degree at the University of Wyoming.

Programs in the United Kingdom. Although the older universities are still dealing primarily with "pure chemistry," the development of programs designed to provide relevance to the chemical industry appear to be significantly ahead of comparable programs in the United States. The work-study approach, which was described for Northeastern University, has been in operation for many years in several British universities and polytechnics under the term "sandwich courses." A number of universities have established professorships of Industrial Chemistry, commonly termed Applied Chemistry.

There is now a growing trend in the United Kingdom in which industrial research executives are brought into the universities on a part-time basis as "Associate Professors" or "Visiting Professors" or "Honorary Professors." In this relationship the part-time professor gives his time to advising on courses, lecturing to undergraduates, and in some cases supervising some postgraduate research. The lecture presentations are commonly an integral part of the chemistry program of the university. By drawing upon industry to supplement their faculty at this level, the student is exposed to the attitudes and the philosophies of industry at a relatively early stage of their development.

It is evident that there is interaction between industry and academia and that the British universities are in the process of responding to the needs of industry and their students. In the United States the response appears to be getting started. In both countries the front line universities or the older universities have been slow to act.

VII.   Conclusions and Recommendations

In most cases the professional chemist, upon graduation from the university, has not been prepared to function as an industrial chemist. His education in chemistry has been completed, but only after joining industry does his education for industry begin. Since statistics show that the great majority will end up in industrial jobs, it must be concluded that the universities are not fulfilling their responsibilities to their students.

There are many routes to providing more relevant training for industry, several of which have been discussed in this paper. A shift in emphasis in the undergraduate program to include more descriptive chemistry and to include more of the fundamentals that are important to industry is highly recommended. In no way need this reduce the quality of the undergraduate program, and it should certainly meet all of the guidelines established by the American Chemical Society for curriculum approval. Most important of all at the undergraduate level would be a shift in attitude which would be more favorable to the application of chemistry as an honorable and scholarly endeavor — an attitude that recognizes that it is the responsibility of the university to provide the necessary training for those who will seek their places in industry.

It is recommended that the universities change their employment practices and policies such that qualified chemists that have experience in industry can be employed as university professors. It is further recommended that universities consider establishing the position of Professor of Industrial Chemistry with the purpose of ensuring that the university program provide appropriate options to the student who specifically desires training for industry.

The development of a course dealing with the professional aspects of the responsibilities of a chemist in industry is especially recommended. The course given at Akron University is an excellent example, and it might well be expanded to include topics such as safety in the chemical industry, environmental protection, and public responsibility and liability, as discussed in Section III of this paper.

The program adopted by a given university will certainly vary based upon the judgment of those responsible. There are several approaches described in this paper that have been reduced to practice. Each of the programs described is completely different from the others. There are strong points in each program, and all of them contain the vital factor of a favorable attitude within the university. The corollary to this is that if there were such a change in attitude, programs comparable to those that have been described above would certainly be developed.

Finally, the fact that industry finds that the end products of the university program, the chemistry students, have not been prepared to function effectively as industrial chemists calls for action in the university. The quality of the chemistry student is at least as high as the quality of the chemical engineering student, and the educational program is at least as demanding. Yet the starting salary of the chemical engineering graduate at the B.S. level is 42 percent higher than his chemistry counterpart. Even at the Ph.D. level the salary of the chemical engineer is 25 percent higher than that of the chemist. It is incumbent upon the universities to work toward the removal of this penalty to their chemistry students.

References

[1] Chemical and Engineering News, 25 October 1976, 36.

[2] Chemical and Engineering News, 1 April 1974, 2.

[3] J. Chemical Education, 1976, 53, 672.

[4] H. A. McGee, Jr., CHEMTECH, 1976, 355.

[5] Chemical and Engineering News, 28 February 1977, 19.

[6] O. A. Hougen, K. M. Watson and R. A. Ragitz, Chemical Process Principles, Part I, John Wiley and Sons, Inc., New York [2nd Ed. 1954].

[7] W. K. Lewis, A. H. Radasch and H. C. Lewis, Industrial Stoichiometry, McGraw-Hill Book Co., Inc., New York [1954].

[8] V. D. Herbert, Jr. and A. Bisio, CHEMTECH, 1976, 174.

[9] United States Constitution, Section 8, Article I.

[10] J. Schimmel, Patents for Chemical Inventions, E. J. Lawson and E. A. Godula, editors, Advances in Chemistry Series, Vol. 46, American Chemical Society, Washington, D.C. [1964].

[11] J. Peter Clark, CHEMTECH, 1975, 664.

[12] J. Peter Clark, CHEMTECH, 1976, 23.

[13] J. Peter Clark, CHEMTECH, 1976, 235.

[14] R. H. Perry and C. H. Chilton, <u>Chemical Engineers Handbook</u>, McGraw-Hill Book Co., Inc., New York [5th Ed. 1973].

[15] Arthur L. Conn, CHEMTECH, 1975, 154.

[16] Nicholas B. Ashford, CHEMTECH, 1976, 676.

[17] A. Richard Albrecht, CHEMTECH, 1974, 690.

[18] Chemical and Engineering News, 21 July 1975, 18.

[19] C. A. Clausen and G. C. Mattson, CHEMTECH, 1975, 535.

[20] G. C. Mattson, private communication.

[21] Chemical and Engineering News, 21 July 1975, 18.

[22] J. P. Kennedy, CHEMTECH, 1974, 156.

[23] K. Weiss, R. N. Wiener and B. L. Karger, J. Chemical Education, 1973, <u>50</u>, 408.

[24] B. C. Giessen, G. Davies and K. Weiss, CHEMTECH, 1976, 106.

[25] B. C. Giessen, G. Davies, P. W. LeQuesne and K. Weiss, J. Chemical Education, 1976, <u>53</u>, 149.

[26] D. A. Nelson, S. L. Holt, V. S. Archer, R. J. Hurtubise and R. E. Barden, J. Chemical Education, 1976, <u>53</u>, 148.

Water Purification and Recycling

By T.V. ARDEN

Portals Water Treatment Ltd.

1. Introduction

In contrast with all other raw materials used by industry, water is produced naturally in a relatively pure form, and in quantities greatly in excess of the daily requirements for domestic and industrial uses. Average daily precipitation over Gt. Britain is 400 million tons, of which about 50 million tons are used for all purposes. Even this figure has no absolute meaning, since with the exception of totally negligible quantities which enter chemical reactions (e.g. the hydration of cement to form concrete) or are exported in products (e.g. whisky) no water is consumed. All other water used, whether for irrigation, in industry, or for human consumption, eventually re-enters the water cycle by direct evaporation from the soil and vegetation, or by returning to the sea for re-evaporation. Net consumption is nil.

With a large excess of supply over demand, and a substantially pure source, it would initially seem that the only problem is distribution, no process of a manufacturing or extractive nature being required. Historically, this was largely true until the beginning of this century. Water for human consumption was given no more processing than simple filtration. Even today, although more elaborate clarification and disinfection procedures are used, they represent only a minor proportion of the total delivered cost.

However, as rain must fall on the ground, and being distilled water, is a fairly aggressive solvent, the liquid distributed is not water but an aqueous solution containing 50-500 mg/l dissolved solids. The human body is highly tolerant of changes in drinking water composition, but industry is less flexible. The most exacting industrial processes impose a maximum impurity level of 0.02 mg/l for their input water, and the impurity levels in water supplies may therefore be up to 25,000 times the maximum level which can be tolerated for a wide range of industrial uses. The extraction of pure water from these solutions is in practice just as much a manufacturing process as is the production of phosphoric acid from phosphate rock. It requires in fact the same reagents, in similar quantities, and plant designed on an equal scale.

In the case, for example, of power producing boilers, used by the electricity generating authorities, improvements in fuel consumption efficiency, which themselves are related to boiler pressures, have been governed largely by the availability of methods for producing boiler feedwater of increasing purity, as shown in Table 1.

Table 1.   Changes in water quality and treatment between 1925-75

Year	Boiler pressure (atmospheres)	Treatment	Water TDS (ppm)	Quality (μS/cm)
1925	12	None	350	700
1935	25	Softening	350	800
1945	35	Evaporation	5	15
1950	65	Deionisation	1.0	3
1955	100	"	0.3	1
1965	130	"	0.1	0.5
1970	155	"	0.03	0.1
1975	155	"	0.02	0.08

The purity level achieved for boiler feedwater, and many other industrial water supplies, is compared with the best obtainable results for other chemicals in Table 2. Water purity standards, at 99.999998% are far higher than those of analytical grades of other reagents.

Table 2.   COMPARATIVE IMPURITY LEVELS

	Impurities mg/l Fe,Cu,Ni, etc.	Total
HCl (analytical grade)	1.2	15
H_2SO_4 " "	1.5	25
NaCl " "	5.0	250
NaOH " "	30.0	500 (+1%Na_2CO_3)
H_2O (Boiler Feed)	0.01	0.02

This purity standard is not however the real problem. Since 1955 there has been no fundamental difficulty in producing water of this standard, even though it was totally impossible in 1945. One problem is that of scale. Other pure chemicals are produced on the scale of grams, kilograms, and occasionally tons per day. Water of greater purity is manufactured on the scale of millions of tons per day.

Because of the scale factor the final key problem arises. The manufacture of pure water must be achieved at costs which are lower

by several orders of magnitude than those of any other comparable
manufacturing process. Table 3 shows the purification factors and
the corresponding costs for the industrial and pure grades of
three common chemicals.

Table 3.   Comparative costs of purifying industrial chemicals

Material	Total impurities			Price		
	Industrial ppm	Purified grade ppm	Improvement factor	Ind. £/ton	Purified grade £/ton	Price Ratio
HCl	2400	15	160	65	1320	20
NaOH	20,000	500	40	44	1275	39
Water	(Mains)	(Distilled)				
	300	1.0 (Boiler) Feed	300	0.1	1-10	10-100
	300	0.02	15,000	0.1	0.5	5

It is the combination of three incompatible requirements - the
ultimate in purity, the maximum in scale, and minimum in cost,
which gives to the water manufacturing industry the great complexity
and interest in what at first might seem a simple and almost
mundane subject. This paper reviews the processes used, section 2
describing an overall diagram of the routes by which various
starting point waters are converted to different finished products.
In each case, the initial water is one which is a standard natural
supply, not significantly contaminated with industrial effluent.
Certain special cases are considered separately.

2.   Water Sources and Treatment Summary.

Table 4 categorises in col.1, five main types of water (A) -(E)
which may be regarded as normal starting points for supply.[1]   In
addition there are included two subsidiary types Seawater (F) and
Brackish water (G), which are potential usable sources, but must
be regarded as abnormal. In the last column 15 types of use are
classified, broadly in order of increasing purity standards.

TABLE 4 WATER TYPES, PROCESSES AND USES

WATER TYPE	CHARACTERISTICS* SS	DS	H	O	MAIN PROCESSES	RESULTS* SS	DS	H	O	USES	ADDITIONS	EXTRA USES
Treatment 1: Physical Purification												
(A) Silted River	5000	300	200	30	nil	5000	300	200	30	A	nil	–
(B) General Surface Water	50	300	200	5	sedimentation	50	300	200	5	AC	nil	–
(C) Deep Well Water (Limestone)	1	350	250	1	coagulation, filtration / nil	1	300	200	1	ACD	disinfectants	E
						1	350	250	1	ACD	antiscalants & anticorrosives	G
(D) High Mountain Water	1	50	30	1	nil	1	50	30	1	ACDFH	disinfectants	E
(E) Peaty Hill Water	15	50	30	15	coagulation, filtration / nil	15	50	30	15	ACD	anticorrosives	G
Treatment 2: Partial Chemical Purification (Softening, Dealkalisation)												
(A,B) Treated as in 1.	1	300	200	1	softening	0	300–350	0	1	DF	anticorrosives	G
(C) Untreated	1	350	250	1	dealkalising, degassing plus softening	0	50–100	0	1	DHF	"	G⁺
(D,E) Treated as in 1.	1	50	30	1	softening	1	50	0	1	DHF	"	G⁺
Treatment 3: Deionising												
All waters					cation, degassing, anion	0	5	0	0.5	I		J
					above plus mixed bed	0	0.02	0	0.2	I⁺	"	J⁺
					above plus total organic removal	0	0.02	0	0.1	I⁺⁺	"	J⁺⁺

Substandard Sources

Source	SS	DS	H	O	Treatment	SS	DS	H	O		Usage
(F) Seawater	20	35000	4000	50	50 ← filtration →	1	35000	4000	50	none → [disinfectants, anticorrosives, antiscalants]	B
					← distillation →	0	20	2	0.5	DFI → [disinfectants, anticorrosives]	E
					above plus mixed bed 1↑ →	0	0.02	0	0.1	I⁺(I⁺⁺) "	G⁺J
					filtration plus reverse osmosis or electrodialysis →						
					above plus mixed bed →	0	0.02	0	0	I⁺⁺ "	J⁺(J⁺⁺)
(G) Brackish Well Water	1	3000	300	1		0	300	10	0	DEF "	G
						0	0.02	0	0	I⁺⁺ "	J⁺⁺
(H) Recirculated effluents – see text											

* SS = Suspended solids (mg/l)
  DS = Dissolved Inorganic Solids (mg/l)
  H = Hardness as $CaCO_3$ (mg/l)
  O = Organic matter by 4 hour permanganate test, as O_2 (mg/l)

USAGE CODE

A   Irrigation only (coastal power stations)
B   Condensor cooling (coastal power stations)
C   Crude industrial use: mining, coal and gravel washing, tank cooling, coke quenching
D   General purpose domestic and industrial use (cold water)
E   Drinking
F   Laundering, bottle washing, woolscouring, etc;
G   Cooling (low temperature differences), water heating boilers, boiler make-up (15 atmospheres)
G⁺ Cooling water (high temperature differences), boiler make-up (25 atmospheres), boiler make-up (all packaged boilers)
H   Industrial process water, low quality (general chemical manufacture, general photographic processing, etc.)
I   Process water, distilled quality (pure chemical manufacture, metal finishing, air humidification, etc.)
I⁺ Process water, ultrapure quality (photographic emulsion manufacture, special photographic processing, etc.)
I⁺⁺ Process water, ultimate quality (electronics industry)
J   Boiler make-up (35 atmospheres)
J⁺ Boiler make-up (70 atmospheres)
J⁺⁺ Boiler make-up (possible future requirement, 150 atmospheres)

<u>Water Types and Characteristics (Cols. 1 and 2)</u>. The five
headings typify various key water characteristics, but do not form
a comprehensive list. Thus there are, for example, surface waters
which have suspended solids and hardness levels as in (E). The
types chosen have been taken to illustrate the occurrence of the
four main undesirable factors, at different input levels as
follows:

<u>Suspended Solids (SS)</u>.   Very high levels, (A) do not normally
occur in Great Britain, except for short periods during highly
abnormal weather. They are encountered in rivers such as the Nile.
In general, when suspended solid levels are high, a predominating
proportion is relatively coarse and settles rapidly. The more
normal levels of about 50 mg/1 are typical of most British rivers.
The suspended matter is normally extremely fine, and cannot be
settled out in any practicable time without the use of coagulation
procedures.

<u>Dissolved Solids (DS) and Hardness (H)</u>. In Britain, waters fall
largely into two types. Those in hard rock areas have low
dissolved solids, and therefore low hardness. Those in soft rock
areas have high dissolved solids and hardness, as the soft rocks
are mainly limestone ($CaCO_3$) or dolomite ($CaCO_3/MgCO_3$).

In other countries intermediate cases exist, and waters are
found which are high in dissolved solids in the form of sodium
salts, but low in hardness. Although such sources are not
included in Table 4, the treatment logic contained in the table
can still be applied. Thus a high NaCl water would not require
softening for use F, but would require de-ionising for use I.
Similarly, a high $NaHCO_3$ water could be treated by dealkalisation
and degassing, without softening, for use H.

Seawater and Brackish waters are special cases in which the
hardness is high as an absolute value, but low as a proportion of
dissolved solids.

<u>Organic Matter (O)</u>. The organic content of water supplies is low
in relation to TDS, but can be extremely important from the
viewpoints of taste and colour of potable supplies, and of
interference with industrial use.

<u>First Stage Treatment Process and Results (Cols. 3 & 4)</u>.

<u>Waters (A)</u> - Silted - are suitable without treatment only for
irrigation A. After sedimentation, they are closely similar to
waters (B) and are considered together.

Waters (B) - Cloudy - are directly usable for all industrial processes which are essentially dirty in nature, B. Examples are coal and gravel washing, coke quenching, mineral classification, etc. Crude cooling, as on the outside of solvent storage tanks comes into this category, but piped cooling circuits do not.

After coagulation and filtration, waters (B) become identical with (C) and they are considered together.

Waters (C) - Deep Well - are clear, but hard. They are normally directly suitable for all general domestic and industrial purposes D, including drinking E, and are so used in much of Europe. In Great Britain, however, disinfection of municipal drinking supplies is standard, and has been included in column 6. While softening of such water is desirable for all industrial processes involving heat, they can nevertheless be directly employed for use G with the addition of antiscalants and anticorrosives.

Waters (D) - High Mountain - are of course uncommon in Britain, but widely found in Scandinavia and Central Europe. They are normally clear, low in dissolved solids, including organic matter, and are therefore soft. They are used without treatment for all purposes up to laundering and other uses F, although again in Great Britain they would be disinfected before inclusion in potable supply. These waters, having high levels of dissolved CO_2, are highly corrosive, and must be dosed with anticorrosive chemicals before use in cooling circuits, low pressure boilers etc., G.

Waters (E) - Peaty Hill. This category is intended to include the waters of Scotland and Wales which, passing over fairly hard rock areas, do not pick up major concentrations of inorganic salts. They do however run through highly vegetated or peaty areas before reaching the rivers and are thus coloured, with fairly high organic content. They are normally coagulated and filtered before use and are then suitable after appropriate additive treatment for all purposes up to G.

Note: In column 6, disinfection and anticorrosion treatments are shown as alternatives. In practice, disinfection is compatible with all other processes, and is widely used even when not strictly necessary. Anticorrosive and antiscalant chemicals are however, normally incompatible with potable use E.

Second Stage Treatment Processes and Results. For all uses more
complex than G, some form of chemical purification is required in
addition to the physical steps of stage 1.

Those waters (A,B,C) which are hard at this point, require
softening, to allow their use to be extended to laundering etc. F,
and to light duty cooling and low pressure boiler feed G. Partial
deionising by means of carboxylic ion exchange resins, combined
with softening, produces a soft water with fairly low dissolved
solids.

It is suitable for a wide range of chemical and other industrial
process water H, in cases where very high purity is not needed.
After anticorrosive additives, it is used for heavy duty cooling,
and for the feed to medium pressure or packaged boilers, both of
which require to be fed with water of reasonably low dissolved
solid content. Waters (D) and (E), although relatively soft, may
on occasion require to be totally softened for the purpose. The
carboxylic process cannot, however, be applied in these cases,
as it operates only with waters containing appreciable alkalinity.

Third Stage Treatment and Results. For all purposes beyond H, all
water supplies require to be deionised. For feedwater to
boilers operating below 35 atmospheres pressure, and for all
purposes in which distilled quality water is needed, two stage
deionising is used. The intermediate degassing stage shown in
the table is sometimes omitted. For boilers operating over 35
atmospheres at all pressures up to super critical J*, mixed bed
deionising is used as an extra stage. The combined process gives
virtually the ultimate level of removal of inorganic matter, down
to 0.02 mg/1, together with virtually all ionised organic matter
which has escaped earlier stages.

Finally, there is the case of the electronics industry, which
uses water for washing transistors, T.V. tubes and similar items.
For these purposes, the ultimate possible purity is required, I**,
economics being a secondary consideration. Traces of neutral or
slightly polar organic matter which occur in many waters can pass
through all the systems so far listed. In the past, elaborate
trains of extra equipment have been used to remove these materials.
It is now standard practice to replace all such systems with
reverse osmosis (q.v. below) which gives virtually total removal
of organic matter, while contributing considerably to the
reduction of the inorganic load.

Reverse osmosis is a relatively expensive process, and its use
is accordingly restricted. It may however, be applied to the

feedwater (J**) to the highest pressure boilers of over 150
atmospheres. Evidence is increasing that trace organic matter can
be a source of operational difficulty in this field.

<u>Sub Standard Water Supplies.</u>

<u>Seawater</u>, In spite of its excessively high hardness and TDS,
seawater is extensively used for condensor cooling in coastal
power stations, with no treatment other than chlorination (to
minimise the growth of mussels etc. in pipework) plus
anticorrosive and antiscaling additives. For all other purposes,
distillation is required, which takes the usage potential up to
all purposes from A to I, after which standard processes will
purify further to any required level. Distillation costs are high
however, and seawater cannot be considered a suitable source of
supply unless none other is available.

<u>Brackish Water</u> - Waters containing 2-5000 mg/l dissolved salts are
often distilled, particularly in Middle East regions where fuel is
plentiful, but can equally be desalinated down to about 200 mg/l
TDS   by electrodialysis or reverse osmosis. They then correspond
roughly with waters C or D and can be further treated accordingly.

<u>Recirculated Effluents</u> - Sewage and industrial effluents form a
potential large source of industrial water, to which increasing
attention has been paid, as a result of the succession of drought
years in the mid 1970's. Possible treatment systems are
considered below.

<u>3. Clarification</u>

For all purposes other than irrigation surface water must be free of
suspended solids. If the water source contains heavy silt, as in
the case of the River Nile, preliminary sedimentation is carried
out either in elongated basins with longitudinal scraper gear or
in circular thickeners in which the water flow is radial from a
central feedwell to a peripheral take-off launder. The silt is
scraped back to the centre for desludging. Such systems are not
used in Great Britain. For the removal of light suspended solids
at input levels of up to 50 mg/l, together with dissolved or
collodial organic matter which causes discolouration of water
supply, the universally used procedure is clarification by means
of metallic hydroxides.

Aluminium or ferric hydroxides precipitates are formed by one of
the following reactions:

$$Al_2(SO_4)_3 + 3H_2O \longrightarrow 2 Al(OH)_3 + 3H_2SO_4$$
$$Fe_2(SO_4)_3 + 3H_2O \longrightarrow 2 Fe(OH)_3 + 3H_2SO_4$$
$$4 FeSO_4 + O_2 + 10H_2O \longrightarrow 4 Fe(OH)_3 + 4H_2SO_4$$
$$2 FeSO_4 + Cl_2 + 6H_2O \longrightarrow 2 Fe(OH)_3 + 2HCl$$

The quantity of coagulation reagent used is 20-30 mg/1, and the
by-product acid produced is normally neutralised by the natural
bicarbonate alkalinity of the raw water. In practice, according
to water analysis, it may be necessary to add either acid or
alkali to the water to obtain the optimum coagulation pH of about
6.5 for alum coagulation or 8.5 for ferric sulphate. The
precipitated metallic hydroxides form positively charged flocs
which attract and carry down the negatively charged colloidal
clay particles and also dissolved or colloidal organic acids.

Alum is the most widely used coagulation reagent, followed by
chlorinated ferrous sulphate. It is possible in some cases to use
ferrous sulphate alone, the dissolved oxygen in the water being
sufficient to produce the ferric hydroxide precipitate, but the
use of chlorine gives more positive control. In recent years
there has been an increasing use of ferric sulphate as the primary
reagent.

In all cases, the initial precipitates are in the form of
extremely fine floc, with a large surface area to give maximum
absorptive power. To separate the floc from the water, the
principle of the "sludge blanket" is used. In its simplest form,
used in relatively small installations, an inverted conical or
pyramidical tank is employed, the water rising at a steadily
decreasing linear rate. Collision of the floc particles causes
coagulation and growth so that a dense blanket is formed at the
level at which the upflow rate equals the falling velocity of the
heaviest particles. The blanket then acts as a filter entrapping
all finer rising particles. Continuous desludging is carried out
through perforated pipes at the blanket level.

For larger scale applications, many different types of unit have
been developed, of which four have been widely used in Britain.

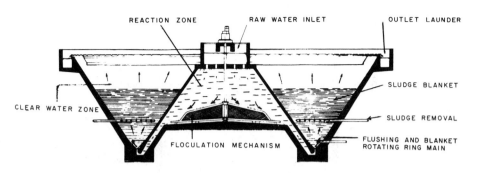

FIG I PRECIPITATOR

The Precipitator, (Fig. 1) is in effect a direct extension of the conical upflow tank. Water enters the centre zone of a circular unit, coagulation chemicals being added at the same point and gently mixed by a slow speed stirrer. The water flows through an annular orifice, so sized that at the operating flow of the unit, there is an upflow of about 0.2 m/sec, sufficient to prevent any settling of the sludge in this region. The annulus is fitted with swirl plates which impart slow rotation to the whole of the outer region, and just above this orifice there are fitted a series of high velocity tangential jets which are used when necessary to introduce poly electrolyte flocculation aids, the most successful use of which requires them to be added after the formation of the initial fine floc. The overall upflow rate in the blanket zone is 3-4 m/hour.

The fine sludge then rises in the outer region, forming a sludge blanket exactly as in a simple upflow tank, sludge removal being from perforated pipes at blanket level as before.

The Pulsator, (Fig. 2) is also a vertical flow clarifier with a system of water flow pulsing which ensures by the alternate expansion and contraction of the blanket, a homogeneous sludge. The concentrated sludge flows into hoppers from which it is periodically extracted.

The Pulsator can be fitted with an inclined plate system, generally similar to the floc barrier (see over page), being then known as the Superpulsator. The plates are situated in the blanket zone, resulting in a higher sludge concentration, and permitting rising velocities of 6-10 m/h.

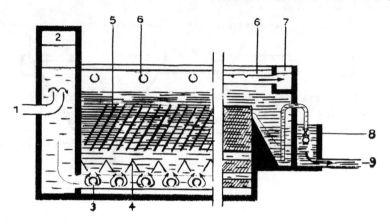

1. RAW WATER INLET

2. VACUUM CHAMBER

3. PERFORATED RAW WATER DISTRIBUTION PIPES

4. STILLING PLATES

5. PLATES

6. PERFORATED CLARIFIED WATER OUTLET PIPES

7. CLARIFIED WATER OUTLET

8. SLUDGE EXTRACTION VALVE

9. SLUDGE OUTLET

FIG. 2   PULSATOR

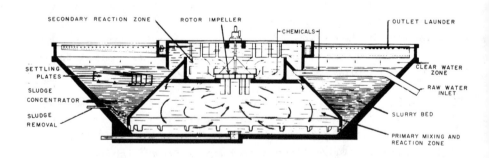

FIG. 3   ACCENTRIFLOC

The Accentrifloc, (Fig. 3) uses the principle of sludge
recirculation to promote maximum absorption of the newly formed
floc. Primary mixing of old sludge with new water occurs in the
central section of the lower zone, and flow continues upwards to
a secondary reaction zone in which new coagulation chemicals are
added. The water then spills over into the outer annulus at about
its centre level, where the blanket forms, filtration of floc
occuring in the upper part of the blanket and sludge concentration
in the lower. Sludge take off is from the bottom of the
concentrator section with subsidiary removal as required from the
bottom of the mixing zone.

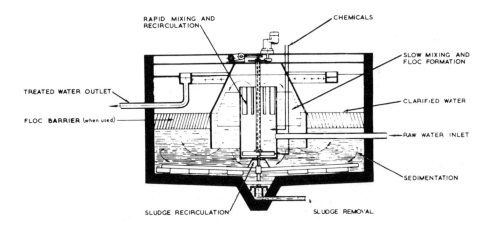

FIG.4 REACTIVATOR

The Reactivator, (Fig. 4) also uses a sludge recirculation
principle. It is vertical sided, with concentric zones.
Upflowing raw water with its added alum or ferric salts is mixed
fairly rapidly in a central zone with recirculated sludge picked
up from the bottom of the zone. The flow is then diverted
downwards through the intermediate annulus where mixing is slow
and floc growth is encouraged. Finally, the flow is again
reversed in the outer annulus to give the normal upflow conditions
with reducing velocity towards the top of the tank. The blanket
is maintained at low level, and sludge is removed from the bottom
by scraping to the centre as in a standard thickener. Because
of the necessity of keeping the blanket low, this unit does not

operate at such high linear flow rates as the others, but it has
the advantage of acting as a combined thickener and clarifier in
one unit. It is thus particularly for waters containing small
amounts of silt which do not require a seperate thickener unit.

Floc Barriers. An interesting addition to all types of upflow
tank is shown in the diagram of the Reactivator (Fig. 4). Blocks
of rectangular section tubes at about 60 degrees to the horizontal
are mounted in the upflow tank just above normal blanket level.
The upflow rate can then be increased until some floc escapes from
the blanket, and must then rise through the floc barrier. Further
settling then occurs in the barrier, the settled sludge sliding
down the lower surface of the tubes to be returned to the blanket.
The output of existing plants can frequently be increased by as
much as 50% using this system. The blocks of inclined tubes are
fairly expensive to manufacture and install, so that for the
design of a new unit, there is an economic balance between a large
standard unit without floc barrier, or a smaller one containing
the barrier. For the extension of existing plants, floc barriers
are normally more economical than the installation of completely
new units.

Dissolved Air Flotation.[14] This technique, which has been widely
used for many years for the treatment of fat-bearing effluents, is
now finding increasing use for the removal of floc from solution.
The chemistry of coagulation is unchanged, and a proportion,
normally 10-30% of the raw water is saturated with air under
pressure, by means of a pressure vessel connected to a compressed
air supply. The air saturated water is now mixed with the main
flow at the bottom of a simple open tank. With the release of
pressure, air bubbles form, picking up the fine floc particles,
and carrying them to the surface where they form a thick sludge.
The outgoing water passes below a submerged weir to emerge in
clarified form while the sludge is scraped from the top for
disposal.

Dissolved air flotation is not yet in widespread use for water
treatment. It shows considerable promise for the future, as the
operating units are smaller in size than sludge blanket tanks for
the same flow rate. Moreover, the waste water consumption and
sludge disposal problems can be considerably less than in a sludge
blanket tank, as the floating sludge has a lower water content
than that pumped from a normal blanket region.

Filtration. Water emerging from coagulation units requires final polishing by filtration through sand beds about 1m deep. Classically, rapid gravity filters or pressure filters have been used, operating downflow at rates of up to 3mm/sec. Filtration takes place mainly at the top surface of the sand, and according to the clarity of the feed, filter runs may be 8-24 hours, after which the filter is backwashed to remove the solids which are choking the top surface. Backwashing regrades the sand so that the finest particles always remain at the top. The backwash water represents a significant loss of output and a number of modified filter types are now in use, to increase the filter loadings.

(A) Deep Bed Filtration - coarse sand (1-2mm) is used in beds about 2m deep. Filtration takes place in depth at flow rates up to 5 mm/sec with considerably longer filter runs. The water clarity at the beginning of each cycle may be imperfect.

(B) Dual Bed Filtration - a 50cm layer of fairly coarse anthracite (ca. 2mm) is placed on a similar layer of standard filter sand (ca. 1mm). Filtration takes place in depth in the upper layer while polishing is achieved at the interface. During backwash for filter cleaning, the anthracite, due to its lower S.G. remains above the sand in spite of its larger particle size.

(C) Upflow Filtration - fairly deep beds (up to 2m) of standard filter sand are operated upflow, the sand bed being prevented from rising by means of a grid system just under the top of the sand bed. Increase in upflow rate breaks the arches and permits normal backwash.

(D) Varivoid Filtration - on a main bed of coarse (2-3mm) sand is placed 10% of fine (0.5-0.8mm) sand, which is then mixed with air so that it disappears into the voids, the bed depth decreasing correspondingly. This combination produces both large and small voids uniformly distributed throughout the depth of the filter, permitting filtration in depth at high flow rate while maintaining a high standard of filtrate quality. Backwashing reseparates the two sand types, opening up the voids and permitting easy cleaning of the filter.

4. Disinfection

In Great Britain sterilisation of the public water supplies, using chlorine, is standard, and most industrial abstractors of water carry out the same process. The practice is by no means universal elsewhere, many EEC water authorities supplying water without any form of disinfection. Some countries, particularly

France, use ozone in preference to chlorine, and there are some
installations in Britain. Ozone has a more powerful   bactericidal and anti-viral action, but has the great disadvantage of
instability which makes it impossible to maintain a residual
bactericidal power in the treated water. It is thus impossible
to protect against possible re-contamination of piped supplies
resulting from cracks or leaking joints in the piping. Ozone
has advantages in all cases where contamination of water sources
with sewage occurs, but increasingly in Europe it is being found
necessary to provide back-up chlorination to ensure sterility in
the pipelines. Even this system can provide problems with
seriously contaminated supplies, as organic compounds in the
water can become chlorinated to produce possible carcinogens.
For the treatment of the seriously polluted Rhine, ozone, followed
by carbon, and final chlorination is now under consideration.

5. Softening and Dealkalisation

Softening - The removal of magnesium and calcium ions from water,
is practised in all cases where the presence of these metals
would give rise to precipitation on heating, or by reaction with
chemicals such as soap. The ions may be replaced by sodium, or
totally removed, together with an equivalent quantity of
bicarbonate. The second process is now generally known as
de-alkalising, although in the past it was also called softening,
no distinction being drawn between the two procedures.

Chemical Methods

Dealkalising with Lime ("Lime Softening") - This process was
widely used in Great Britain, and plants are still in operation.
It is rare however for new plants to be installed, ion exchange
methods now being preferred. Lime softening only applies to
waters containing bicarbonate alkalinity, the reaction being:

$$Ca(HCO_3)_2 \ + \ Ca(OH)_2 \ \longrightarrow \ 2CaCO_3\!\downarrow + \ H_2O$$

By operation at pH 10.5 almost complete removal of bicarbonate
hardness can be achieved. The process is carried out in plants
similar to those described for coagulation.

Lime-Soda Softening - In most waters, the total hardness exceeds
the total alkalinity, the difference being "Permanent hardness",
that is salts such as $CaSO_4$ and $MgCl_2$ which do not give rise to
precipitation on heating, and are not removed by lime softening.

In the classical lime-soda process, sodium carbonate was used to precipitate this remaining hardness.

$$CaSO_4 \ + \ Na_2CO_3 \ \longrightarrow \ CaCO_3 \downarrow \ + \ Na_2SO_4$$

Both these processes have mainly been replaced by ion exchange.

Ion Exchange Softening[3] - Ion exchange resins are solid, insoluble, water permeable, reactive polymers manufactured in bead form. The cation exchange materials used for softening consist of a cross-linked polystyrene network to which sulphonic groups are attached. They are thus able to absorb and exchange cations, while rejecting the anions. The resin, which is contained in cylindrical vessels, is converted to the sodium form before first use, and subsequently after every exhaustion, by passing through it an excess of sodium chloride, after which passage of hard water gives rise to replacement of calcium in solution by sodium.

$$\text{Softening} \quad 2RSO_3Na \ + \ Ca(HCO_3)_2 \ \rightleftharpoons \ (RSO_3)_2Ca \ + \ 2NaHCO_3$$

$$2RSO_3Na \ + \ MgSO_4 \ \rightleftharpoons \ (RSO_3)_2Mg \ + \ Na_2SO_4$$

$$\text{Regeneration} \ (RSO_3)_2Ca \ + \ 2NaCl \ \rightleftharpoons \ 2RSO_3Na \ + \ CaCl_2$$

Both forward and reverse reactions are equilibria, and accordingly the degree of completion can be controlled by the quantity of regenerant used. When this quantity is about 2g.eq/litre of resin the efficiency of usage is approximately 70% and the operating capacity to the breakthrough point is about 1.4g.eq/l. Under these circumstances the treated water hardness is maintained below 1mg/l until breakthrough. Reduction of the regenerant quality to 0.7g.eq/l gives an efficiency of usage of over 99% at the expense of a hardness leakage of up to 10 mg/l. This figure is acceptable for almost all purposes for which soft water is required, and accordingly, there are considerable cost savings to be made by reducing regenerant levels to the optimum figure. The leakage problem can in any case be eliminated by the use of counterflow regeneration.

Ion Exchange Dealkalisation - Carboxylic cation exchange resins are cross-linked acrylic acid copolymers containing -COOH active groups in place of -SO₃H. They are accordingly weakly acidic in character, and are virtually non-ionised below neutral pH. They are therefore of no value on a sodium cycle, but have useful properties on a hydrogen cycle, in which they remove from solution

calcium and magnesium equivalent to the bicarbonate content of the
water.

$$2 \ RCOOH \ + \ Ca(HCO_3)_2 \ \xrightleftharpoons \ R(COO)_2Ca \ + \ 2CO_2 \ + \ H_2O$$

(Virtually total forward action)

$$2 \ RCOOH \ + \ CaCl_2 \ \xrightleftharpoons \ R(COO)_2Ca \ + \ 2HCl$$

(Virtually no forward action)

After removal of CO_2 in an atmospheric degassing tower, the result
of the carboxylic acid resin process is thus chemically identical
with lime softening, but without the problem of sludge disposal.
The nature of the carboxylic resins results in their giving 99%
efficiency of regenerant usage under all practical conditions.
They are thus economical and give the minimum of effluent disposal
problems. If the ion exchange dealkalisation process is followed
by ion exchange softening, the result is identical with the old
lime soda process, giving a soft water with substantially reduced
total dissolved solids.

6. De-ionising

The use of sulphonic acid resins on the hydrogen cycle (regeneration with sulphuric or hydrochloric acid) causes conversion of
all metal salts present to the free acids:

$$RSO_3H \ + \ NaCl \rightleftharpoons RSO_3Na \ + \ HCl$$

Passage of the resultant solution through a basic ion exchange
resin, that is one containing amine groups, then causes
absorption of the free acid.

$$R-N(CH_3)_2 \ + \ HCl \ \xrightleftharpoons \ RN(CH_3)HCl \ \text{(weakly basic resin)}$$

$$\underline{or} \ R-N(CH_3)_3OH \ + \ HCl \ \xrightleftharpoons \ RN(CH_3)_3Cl \ + \ H_2O \ \text{(strongly}$$
basic resin)

Weakly basic resins are the anion exchange equivalents of the
carboxylic acid resins, and are accordingly readily regenerated at
high efficiency, by means of caustic soda. They do not however,
remove from solution weak acids such as silica and carbon dioxide.
The latter is taken out of the system by atmospheric degassing,
the resultant water being equivalent in quality to distilled water,
except for its silica content, which is normally less than 10 mg/l
and does not present any problem for a wide range of process uses.

Two step de-ionising with weakly based resins is accordingly very widely practised throughout Europe. In Great Britain it has become the custom, sometimes without full technical justification, to use a somewhat better quality water, obtained by using a strongly basic resin in the same system. This then gives removal of silica, together with any residual CO_2 at the expense of a higher running cost, due to the less efficient consumption of regenerant. As the cation unit permits the passage of traces of sodium ions, which are converted to caustic soda in the anion unit, the treated water contains a few mg/l of caustic alkalinity. This factor is desirable in boiler feed circuits, but may be obnoxious in process water.

Accordingly the use of mixed bed units to remove final traces of ionised matter has become almost universal. When hydrogen form cation resins are intimately mixed with hydroxide form anion resins, each disturbs the equilibrium of the other. with the result that the exchange reactions proceed to the ultimate extent:

$$RSO_3H \ + \ RN(CH_3)_3OH \ + \ NaCl \longrightarrow RSO_3Na \ + \ RN(CH_3)_3Cl \ + \ H_2O$$

The resins are regenerated by taking advantage of the greater density of cation resins, in comparison with anion materials. They are separated by means of an upflow of water, and are then seperately regenerated, in the same unit, using a pipe network at the interface to achieve the appropriate flows.

The mixed bed system, which was first introduced in 1950, represents a major advance in water quality. All earlier purification systems, including two bed systems and distillation, had given treated water containing 1-2 mg/l of dissolved inorganic solids. Accordingly, the development of the modern power industry, using high pressure boilers, could not have taken place without the prior invention of the mixed bed process, which enabled dissolved inorganic solids to be reduced to 0.02 mg/l, two orders of improvement.

<u>Ion Exchange Equipment</u>. Classical ion exchange equipment was developed directly from pressure sand filters, and is substantially unchanged. Resins are operated downflow, in beds 1-2m deep, contained in cylindrical pressure vessels 3-3.5m tall. Water collection below the bed is by means of a pipe network equipped with strainer buttons or, in the case of simple softeners, a sand or gravel underbed. Good design technique gives linear flow through the bed and the operating results of large industrial

units differ only marginally from those obtained in laboratory
diameter units (25mm) of the same bed depth.

Downflow regeneration was standard until about 1965, and is still
frequently used because of the simplicity of plant design
resulting from a common path for water and regenerant.
Increasingly however, counterflow regeneration procedures are used.
A typical system is shown in Fig. 5.[4]
A collector system of pipes and strainer buttons is buried just
below the top of the bed. Service water flow is downwards from
the top of the unit past this pipe system, to the bottom
collector. Backwash for bed cleaning takes place upwards, the
water leaving from the top of the unit. Regenerant is also upflow,
and out through the buried collector. Simultaneously a downflow
of untreated water applies a pressure to the top of the bed and
prevents it from lifting and channelling. Compressed air rather
than water may be used to hold down the bed.[5]

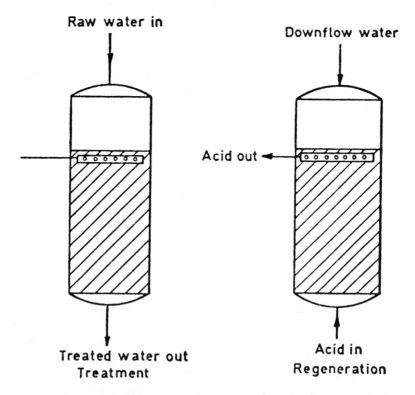

Fig 5. Buried Collector System

Counterflow regeneration causes major improvement in treated
water quality and regeneration efficiency. Fig. 6 shows results
with a water containing 7mg eq/l (i.e. 350 mg/l as $CaCO_3$) total
cations, of which 50% are Na, 40% Ca and 10% Mg. The anions are
60% HCO_3, 20% SO_4 and 20% Cl. A cation column, regenerated with
50 g/l HCl, maintains a leakage of less than 1 ppm until 1.2
g.eq/l of cations have passed through the column. A co-flow
column gives no water of this quality and is exhausted after the
passage of 1 g.eq/l.

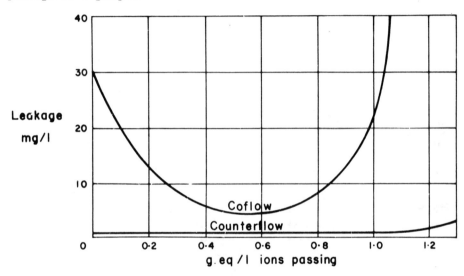

Fig 6. Counterflow operation

Continuous Counterflow Systems.

   A number of different systems have been developed for the
continuous treatment of water by ion exchange. All depend on the
same basic principle of moving the resin through a column, or
series of columns, in which the processes of service flow,
backwashing and regeneration are carried out simultaneously in
seperate regions, all liquids flowing counter to resin movement.
Owing to the technical problems of pumping solid resin particles
continuously, and in a uniform manner, none of the industrial
systems is fully continuous. The resin is moved in slugs at
intervals varying from 3-20 minutes, liquid flows being stopped
or reversed during the resin movement. The continuous units give
chemical advantages similar to those of fixed bed counterflow
units, together with a lower resin inventory. In general, however,
they are complex, and many designs cause damage to the resin
resulting in high replacement costs.

Although many continuous units have been installed, particularly
in the U.S.A., they have not found general acceptance, and fixed
bed counterflow operation remains the preferred system in Great
Britain.

The Higgins Contractor[6] (Fig. 7) The resin moves in closed
circuit through 4 chambers, under the force of hydraulic pulses
applied on one section, simultaneously with the opening and
closing of the appropriate valves to reverse the normal flows.
Service water flow is upwards and regenerant flow downwards.

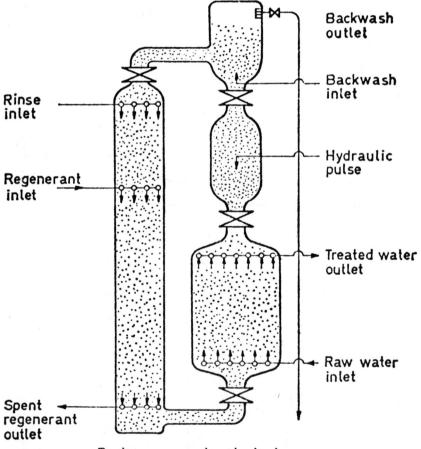

Resin movement — clockwise
All liquid flows — counter-clockwise

Fig 7. Higgins Contactor

The Asahi Process[7] (Fig. 8) The three processes are in seperate
vessels, each with a feed chamber. The units are all identical
in principle, but shaped differently to allow for their various
duties. Water is upflow and resin downflow in each case. Fig. 8a
shows the service flow, combined with resin transfer. Fig. 8b
shows resin movement into the operating unit, service flow being
stopped.

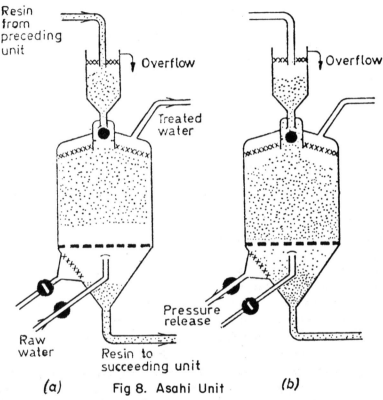

Resin from preceding unit

Overflow

Overflow

Treated water

Pressure release

Raw water

Resin to succeeding unit

(a)   Fig 8. Asahi Unit   (b)

The Cloete-Streat Process[8] (Fig. 9) This system uses fluidised
resin beds, as against the packed beds of the other two described.
Water and regenerant travel upflow, through a series of shallow
resin beds (25-30 cm settled depth), held between perforated plates.
At intervals, water flow is briefly reversed and controlled
quantities of resin travel downflow through the plates, and from
the lowest compartment into the top of the opposite unit. This
equipment has the advantage of very mild action, causing minimal
damage to resin, but is restricted to rather low linear flowrates
governed by the free fall rate of the resin beads. The units are
thus larger than packed bed units having the same duty.

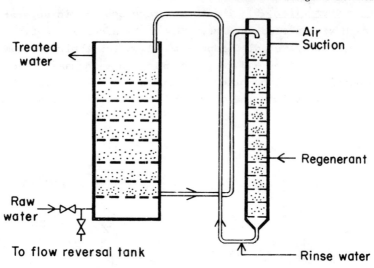

Fig 9.   Cloete-Streat Process

7. Seawater Distillation[9] (Fig. 10)

Distillation is a major subject, and cannot adequately be
reviewed here. As applied to the recovery of freshwater from
seawater, simple distillation systems would be prohibitively
expensive and some form of multi-stage system is universally used.
A multi-stage flash evaporation system consists of a number of
chambers operating at a series of temperatures from 90°-50°C, and
the corresponding steam equilibrium pressures which are, of course
all below atmospheric pressure. There may be as many as 50 stages
with temperature differences of under 1°C. Heat enters the system
on the left in Fig. 10, through an external heat exchanger, while
cold seawater enters at the right, as final cooling water for the
last stage. Brine is recirculated through coils from right to
left, condensation occuring on each coil, the distillate
collecting in trays, and passing to the right (lowest pressure)
section, from which it is pumped as product. The brine is
accordingly heated in each stage, receives its extra heat input
from the heat exchanger, and then re-enters the first stage when
it flashes off as vapour, cooling correspondingly. The process is
then repeated in each section until the most concentrated brine,
at the lowest operating temperature, receives its admixture of new

seawater in the last chamber, and recirculates through the system, with rejection of a proportion to maintain the balance of dissolved salts. A substantial proportion of the latent heat of evaporation is recovered in the system, the heat input being balanced by the energy consumed in separating salt from water, together with the heat losses in cooling water and blowdown. The overall process which is highly efficient, is nevertheless expensive, and its use is confined to regions of extreme water shortage.

Scale formation in seawater evaporators can be a major problem, which is minimised by destroying the alkalinity present with sulphuric acid or by the use of antiscalant chemicals such as polyphosphate.

In recent years, soluble acrylate or other carboxylic polymers have been used with considerable success as antiscalants. They permit higher temperature operation, and therefore higher efficiency, than the polyphosphates, without the corrosion problems associated with the use of mineral acids. External softening of the input seawater by ion exchange has been used, but the process does not operate with high efficiency, because of the sodium salts present, and internal treatment is preferable.

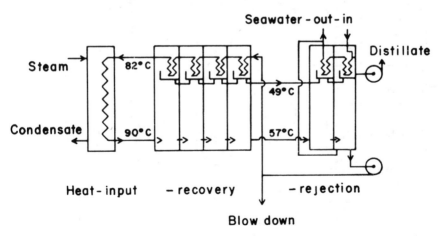

Fig 10. Seawater Distillation

8. Brackish Water Treatment

While distillation is equally suitable for seawater and brackish waters, the lower salinity of the latter permits a number of extra processes which are uneconomic although theoretically possible for the former.

Electrodialysis[10]   Ion exchange resins can be manufactured in
thin sheet form by a variety of methods. Heterogeneous membranes
consist of finely ground standard resins in an inert polymer
matrix, while homogeneous membranes are produced from sulphonated
polythene, aminated P.V.C. etc.

An electrodialysis cell is shown schematically in Fig. 11. It
consists essentially of a series of compartments, containing the
solution to be desalted by alternate cation and anion membranes.
The two end compartments contain electrodes and in them the normal
electrolytic processes take place. These actions are, however, of
minor importance. An electrodialysis stack normally contains
several hundred unit cells, and in all except the end ones, the
process occurring is not electrolysis, but simple transfer. The
sulphate ions in each compartment travel to the right towards the
anode, until they are halted by a cation exchange membrane through
which they cannot pass. Similarly, the sodium ions travel to

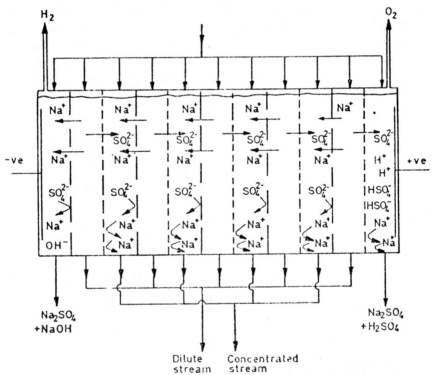

Fig II. Electrodialysis Cell

the left, until stopped by an anion exchange membrane. The net
result is that the solution in one set of alternate compartments
becomes denuded of salts, while the other alternate set becomes
enriched. If, now, the saline solution is fed into all
compartments, while the effluents from the two sets of alternate
compartments are combined into two streams by means of a manifold
system, the result is a continuous supply of desalinated water in
one stream, and concentrated reject solution of sodium sulphate in
the other. By suitably adjusting the flow rates of the two streams,
up to 90% of the input water may be recovered in purified form.
The process is normally operated to take the dissolved solids
concentration down to about 100 mg/1, below which the electrical
resistance of the cell pack becomes too high. The whole unit is
constructed in a similar way to a plate and frame filter press,
with special design features to minimise electrical resistance and
eliminate short circuits.

Reverse Osmosis[11] If water and a saline solution are contained in
two branches of a U-tube, separated by a membrane of cellulose
acetate or other permeable material, water will pass into the salt
solution, whose level will rise until sufficient head is developed
to arrest the process. This phenomenon is osmosis, and the head
developed is the osmotic pressure of the solution. Application of
a higher pressure to the saline side of the unit reverses the flow,
water leaving the salt solution, and passing into the pure water
section. This process is known as reverse osmosis, and is now
applied industrially on a wide scale. In order to achieve reasonable water flow, the membranes must be extremely thin, and as the
pressures required range from 5 to 20 atmospheres, the membranes
require support. One method of achieving this result is to form
them on the inside of paper or fabric tubes of 1-2 cm diameter,
these tubes being themselves fitted into perforated stainless steel
supporting cylinders. Alternatively, the membranes, in sheet form
are formed into spiral rolls with interlayers of supporting guaze
and impervious plastic. In either case, pressure end seals are
required so designed as to permit the raw water entering the system
to pass through the tubes or spirals, and emerge as a concentrated
stream, while the permeate is collected separately. The
engineering problems involved are quite complex, and reverse
osmosis equipment is accordingly expensive. A widely used third
method is the manufacture of the membranes as fine tubular
filaments, of hair like dimensions, polyamide resins being used as
the membrane materials. The filaments are coiled into skeins,

which are packed tightly into a pressure cylinder about 1m long by
15 cm diameter. The open end of the vessel is then sealed with
epoxy resin, the resultant block finally being sawn through, so as
to expose millions of unobstructed hollow filament ends. Feedwater
is led into the pressure chamber, the pressure being on the
outside of the hollow fibres, which being extremely fine, and
tightly packed, are self supporting. The permeate emerges through
the cut ends into a non-pressurised end chamber from which it
flows to service.

Reverse osmosis membranes do not give complete retention of
inorganic salts, the proportion which passes increasing with input
pressure. Output, and treated water quality, from a given unit are
thus governed by opposite factors, and a compromise is normally
made at about 90% desalination, giving a treated water of 300 mg/l
from a typical brackish input of 3000 mg/l. The proportion of
purified water recovered, as compared with feed, is of course, a
simple relationship of the reject and feedwater flows, but the
wastewater flow setting must be governed by the input analysis, to
avoid saturation of any salt in the concentrated stream. In
practice, acid dosing of input water is widely practised to avoid
carbonate scale.

Ion exchange softening is also advisable with many waters, and
filtration to a high clarity standard is essential to avoid blockage
of membranes. The pretreatment stages represent an additional cost,
and accordingly reverse osmosis, like all the desalination
techniques, is used only in cases where no water of low salinity is
available.

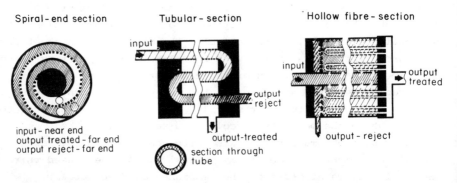

Spiral- end section   Tubular- section   Hollow fibre- section

input

input
output
treated

input- near end
output treated- far end
output reject- far end

output
reject

output-treated   section through
tube

output - reject

Fig 12. Reverse Osmosis Modules

Organic Matter & Reverse Osmosis   Although the most important long
term use of reverse osmosis will probably be the original one,
desalination, there is an extremely valuable subsidiary effect.

For the electronics industry, the purest obtainable water is
required.   Conductivities of $0.05 \mu S/cm$, virtually identical with
totally pure water, and representing 0.01 mg/1 ionised solids, are
regularly achieved.   Two problems arise however, First, there
can be found in treated water from certain raw water sources,
traces of ionised organic matter, which have passed through the
process of coagulation, filtration and ion exchange, and are not
removed completely by additional processes such as active carbon.
These materials are probably intermediate molecular weight
polycarboxylic acids which are too large to enter ion exchange
resins, and not large enough to give perfect coagulation.

A second type is neutral organic matter, probably fairly low in
molecular weight and possible of a sugar type structure resulting
from degradation of cellulose.   Concentrations of 0.1 mg/1 and
occasionally up to 1.0 mg/1 have been found in fully treated water.

Both these materials are rejected in a reverse osmosis system to
an extent which takes them down to the limit of detection.
Accordingly, for ultra pure water in the electronics industry,
reverse osmosis units are now included in the system, either
before the de-ionising stage or just before the mixed bed.

Evidence is also accumulating that the presence of this organic
matter is inadvisable in the highest pressure boilers.   The
extra cost of the reverse osmosis stage, on the vast scale of
boiler feedwater treatment, is less easily accepted than in the
electronics industry, and the situation is still under study.   It
is possible that the technique will come into general use in the
power industry by 1980.

The Sirotherm Process[12]   Weakly acidic and weakly basic ion
exchange resins have little action in neutral solution.   It was
shown however, in 1966 by Weiss et al CSIRO, Melbourne, Australia
that when the regenerated form of these resins are mixed, they do
have a level of de-ionising action, albeit to a much lower
capacity and treated water quality than the strongly acidic and
basic resin mixtures.   They further showed that if the mixed weak
resins are heated, their action is reversed, and they are auto-
regenerated without the use of chemical agents.

These two observations formed the basis of the Sirotherm Process.
Brackish water, alternately cold and hot, is pumped through a
bed of mixed resin. The cold water is partially de-ionised, the
salts appearing in the warm fraction. Severe chemical engineering
problems were initially encountered. It was necessary to minimise
not only liquid mixing, but also the extent to which the resin and
its container acted as a heat exchanger, transfering heat between
the fractions. Special resins were invented having greater
efficiency for this process than the commercial materials
originally used. Work has progressed to the point of first
industrial scale installations, and while much remains to be done,
the technique holds promise for the future.

9. Condensate Polishing[13]

The returned condensate from modern high pressure boiler-turbine
condensor systems can no longer be considered of suitable quality
for boiler feed water. It contains corrosion products, mainly
insoluble iron oxides with traces of soluble copper salts, and
also general salts arising from pinhole leaks in the condensor
system, permitting the ingress of cooling water.

The condensate is thus "polished", i.e., de-ionised and filtered.
Two opposing systems are in industrial use. In one, standard
mixed beds are used for de-ionising, while simultaneously acting
as filters. They become fouled with solids, but by vigorous
washing before regeneration, can be retained in active condition.
The reverse system, the "Powdex Process" is to use finely ground
mixed ion exchange resins as pre-coats on candle filters. They
give virtually perfect filtration and a high level of de-ionising
efficiency, but on exhaustion cannot be regenerated and must be
replaced. As this situation recurs at intervals of only a few
weeks the Powdex Process becomes expensive wherever condensate
leakage is significant. Nevertheless it is widely used.

10. Additives

Antiscalants. Until the advent of basic exchange softening and
de-ionising, antiscalant additives were universally used in
industry for low pressure boiler feedwater, cooling circuits, etc.
They still have a wide range of applications, particularly as some
types act both as antiscalants and anticorrosives.

Sodium Phosphate. Calcium and magnesium salts are converted to
insoluble phophates, which give non-adherent precipitates,
removed from the system by boiler blowdown, or cooling circuit
bleed. The quantities used are equivalent to the hardness salts
present and can therefore be high.

Alkaline Tannin Solutions. These act by distorting the crystal
structure of calcium carbonate, and producing loose sludge instead
of adherent scale. The quantities required are less than for
sodium phosphate.

Polymers. Sodium salts of polyacrylic, methacrylic, and
polymaleic acids act in a similar way to the tannin-based products.
The choice in any individual case is largely empirical.

Phosphonates. These are used mainly in highly concentrated
circuits, such as seawater cooled systems, or recirculatory
systems with high evaporative losses. Their efficiency falls
rapidly at temperatures over $90^{\circ}C$ and for seawater evaporators,
the new polymaleate materials are showing advantages.

Acid Dosing. Simple addition of H_2SO_4 equivalent to the
bicarbonate alkalinity of the water reduces scaling considerably,
and the system has been used in seawater evaporators. However,
it gives considerable corrosion problems, due to the free CO_2
formed, and its use is being superceded by the polymeric products.

Complexing Agents. Where lime softening is used, the reaction can
be incomplete, resulting in an unstable water which can be scale
forming under many conditions. The use of sodium ethylene diamine
tetra-acetate, sodium nitrilo tri-acetate and similar chelating
compounds complexes the residual calcium, and stabilises the water.
The quantity required is less than the stoichiometric amount, and
this system can, if required, be used directly on raw water.

<u>Anticorrosives</u>. Corrosion reduction is in general carried out by increase in pH, and reduction in dissolved oxygen. Some anti-scalants are themselves alkaline and serve two purposes.

<u>Sodium Sulphite</u>. Frequently used in conjunction with sodium phosphate as a combined feedwater/cooling water treatment.

<u>Alkaline Tannin</u>. This compound is an oxygen scavenger, and its use has been standard for low pressure boiler feedwater throughout this century. In spite of new developments, it remains very widely used.

<u>Volatile Amines</u>. For medium pressure boilers, in which dissolved solids are inacceptable, volatile alkaline compounds such as morpholine, and hydroxylamine are used.

<u>Vacuum de-aeration</u>. For high pressure power producing boilers, feedwater is totally de-oxygenated by means of a vacuum degasser.

<u>"Zero Treatment"</u>. One school of thought, particularly in Germany, considers that totally de-ionised and de-oxygenated water, being non-conducting, cannot support corrosion, and that the ideal internal treatment is therefore nothing. British practise assumes that trace organics may pass the treatment system, to become converted to CO_2 or other acids in the boiler. Accordingly, ammonia is used to maintain alkaline conditions.

<u>Filming Amines</u>. For low to medium pressure steam heating systems, with very long pipe runs, filming amines such as octadecylamine are added to the boiler feedwater, from which they condense over the entire system giving a protective alkaline film on all metal surfaces.

<u>Passivation</u>. Many anti-corrosive systems, particularly in closed circuit cooling and allied processes, are based on passivation of the metal surfaces. Sodium, and more particularly, zinc chromate are widely used for this purpose.

<u>Bactericides</u>. Finally, there is the problem of bacterial corrosion, caused by sulphate reducing bacteria in cold water systems, and giving characteristic iron sulphide tubercles, under which pipe perforation occurs. This problem is treated by chlorination, or the use of organic biocides such as the chlorinated phenols and quaternary ammonium compounds.

11. Water Recovery

Although industry uses only a small proportion of the total
precipitation , its supplies are often restricted in drought years,
as distribution rates are related to short-term reservoir levels,
rather than to total availability. Moreover, the combined cost of
water purchase and effluent discharge, which in 1970 totalled
under £0.04/m^3, is now approaching £0.20/m^3. The recovery of
waste water for re-use has accordingly become profitable in many
cases which were not previously economic.

The techniques used for water recovery are broadly similar to
those described for water treatment, but industrial effluents
differ so widely that it is difficult to use any logical method of
classification. Each one must be considered as a special case,
and any generalisations which can be made, must be broad and
subject to exceptions.

Basic Principles. Whereas raw water supplies are fairly uniform,
industrial effluents normally consist of a mixture of outgoings
from different processes in the same works, the streams varying
independently according to production requirements.

The first stage in analysing the problem is to decide whether it
is preferable to treat the existing heterogeneous mixture, or to
attempt to deal with its separate components at source. In most
cases, a compromise is desirable. Very frequently the works has
identifiable small outflows, or batch discharges, of highly
polluted effluent, together with very large volumes containing
only light contamination. The total mixture may be impracticably
expensive to treat, whereas the separation and separate disposal
of the small concentrated streams may permit the mixing of the
main flows for treatment by a single, relatively simple plant.

The chemical nature of the effluent is frequently not known with
any precision. A silver plating effluent can be analysed exactly,
but an abbatoir waste is an undefinable mixture. Accordingly,
methods for water recovery from effluents must be broad in their
action, and capable of giving constant results from variable
inputs. They are, therefore, often less precise, and less
chemically efficient, than the corresponding water treatment
process.

In considering water re-use systems for process waters, it is
important to realise that in a completely closed circuit system,

it is necessary to remove the dissolved matter from the water and
not merely obnoxious consistents. Anything added by the process
will otherwise build up to saturation point. Often it is
completely practical to de-ionise the effluent, but where this
process is uneconomic, a proportion of 25/30% can be bled from the
system to limit build up.

Cooling Systems. Large quantities of water are used for equipment
cooling only, and are discharged contaminated only with heat and
minor corrosion products. In cases such as coastal power stations,
there is no purpose in departing from the principle of "once
through" seawater cooling for the condensors, but inland power
stations use evaporatory cooling towers in closed circuit,
recovering over 90% of the circulating water. In smaller
industries, for the cooling of compressors, moulding machines,
reactor jackets etc., once-through cooling with mains water has
until recently been almost universal. Increasingly, this waste is
being eliminated by the use of closed circuit cooling, either with
packaged forced-draught evaporatory cooling units, which give
about 95% water recovery, or with totally enclosed systems using
air cooled heat exchangers. These units are more expensive to
operate, but give total water re-use and can also be sited to
recover the heat for space-heating purposes.

Where cooling is carried out by direct contact between water and a
process material, as in the edible oil industry, closed circuit
cooling systems can become blocked by fat carry-over, and are
therefore impracticable unless a reliable oil and fat separation
system can be included in the circuit.

Washing Processes - General. Industrial water which has been used
for physical dirt removal only (e.g. high pressure washing of car
bodies before painting), can be almost totally recovered by simple
filtration. Where the process uses soap or detergents, as in
laundering or allied activities, additional broad-action steps
such as carbon columns for removal of organic matter, and
re-chlorination for overall plant sterility, may be required.

Chemical Rinses. In the metal finishing industry, industrial
photographic processing etc., the treated items are rinsed to
remove reactant solutions. The rinse waters are contaminated
with toxic materials, and users of once-through systems are
faced with effluent treatment costs, in addition to water purchase
and disposal charges. De-ionising and recirculation

of process water gives a closed system at much lower cost, together with a pure water supply, instead of mains water, to the rinsing system, which often gives improved quality to the processed parts. The small volumes of regenerant effluents are treated intermittently by batch processes to give solid residues for disposal. In some cases, valuable metals are recoverable as by-products.

Organic Systems. Three main lines of attack are practised, separately or together.

a) Oil and Fat Separation. Major levels of unemulsified oils are relatively easy to separate in simple equipment. The problem becomes more difficult as the initial concentration decreases. Some of the main systems used are:-

Inclined Plate Separators, in which the oil-bearing water flows between a series of plates, on which oil globules coalesce, and climb to form a separated layer.

The Flocoil System, which pumps the oil bearing water through a vertical fibrous plastic mat, from which oil emerges as large globules which climb the downstream face of the mat. This system is particularly suitable for removal of last traces, and like the previous one, can obviously be used for liquid oils only.

Dissolved Air Flotation, suitable for both oils and solid fats, is extremely widely used for this purpose, in contrast with coagulation uses described above.

b) Heavy Coagulation and Filtration. Alum or ferric sulphate, used as in water treatment, but in much higher concentration up to 400 mg/l, forms massive, self-coagulating precipitates, which are directly filtered by sand or candle systems. The precipitates carry down both the insoluble oil, and a considerable proportion of emulsified and dissolved organic matter. The system, widely used for effluent treatment before discharge, is not universally applicable to water recovery, as it imparts an additional dissolved inorganic load which must be removed to permit repeated recycling.

c) <u>Biological Process</u>. For natural organic materials, derived
from food processing of all types, together with a wide range of
synthetic compounds, the standard biological treatment process as
used for sewage gives excellent results.

Very high concentrations are best handled by aerobic digestion,
in aerated columns, or by using the "deep shaft" principle, in
which a high hydrostatic head is used to increase oxygen
solubility and bacterial activity. Intermediate levels lend
themselves to treatment by classical trickle beds, or the modern
equivalent, plastic packed trickle towers, while lower concentrations are best treated by the activated sludge process, which
reduces organic content to 20 mg/l or less. In all cases, means
must be provided for disposing of the biological sludge, and the
treated water must be filtered and chlorinated for re-use.

d) <u>Domestic Sewage</u>. To a large extent, domestic sewage is
re-cycled, by passing from sewage works into rivers in which
further natural purification takes place before the water is
re-abstracted as the raw material for waterworks. Nevertheless,
a more direct re-use is possible. Treated sewage, after
filtration and de-ionising, is totally suitable for boiler
feedwater and allied purposes, and there are many cases where a
factory or power station could assure an increased water supply
by this means.

In general, increasing costs of water purchase and effluent
discharge are encouraging companies to install their own systems
for effluent treatment and re-use, and this trend will undoubtedly
increase during the next decade. It is to be hoped however, that
before the end of the century, it will be possible to replace a
proliferation of separate re-cycling plants with a national system
in which industrial effluent, after a minimum of local treatment,
can be totally purified at disposal works to be returned to
reservoirs and rivers for re-use. Taking into account the natural
purging of the total re-cycled system which must result from flows
of rivers to the sea, such a system is technically possible with
existing knowledge, and must in due course come into practise in
the interest of water and environmental conservation.

References.

1. W.S. Holden (Ed.) <u>Water Treatment and Examination</u>
    Churchill. London. 1970. Ch. 2,3.

2. W.O. Skeat (Ed. for I.W.E.S.) <u>Manual of British Water
                    Engineering Practice</u>
    Heffer. London. 1969. Vol. III. Ch. 5.

3. T.V. Arden. <u>Water Purification by Ion Exchange</u>
    Butterworths. London. 1968. Ch. 4.

4. U.S.P. 2,891,007. B.P. 806,107. 1955.

5. B.P. 1,158,730.

6. I.R. Higgins. ORNL 1907. U.S.A.E.C. 1955.

7. B.P. 987,021. 1960.

8. B.P. 1,070,251. 1963. B.P. 1,399,473. 1975.

9. Ref. 1. Ch. 5. P.266.

10. Ref. 3. Ch. 7. P.143.

11. W.J. Robertson. <u>Water Services</u> 1975. Jan. 16.

12. D.E. Weiss <u>et al</u>. <u>Aust. J. Chem</u>. 1966. <u>19</u> 561.
    D.E. Weiss <u>et al</u>. S.C.I. Conference. <u>The Theory and
                    Practice of Ion Exchange</u>
    1976. Paper 29.

13. P.E. Down and E. Salem. Ref.12. Paper 21.

14. J.J. Morse. <u>Water and Water Eng</u>. 1973. May. 161.

Paterson Candy International reverse osmosis plant
for cheese whey concentration — Holland.

Stella-Meta filters for precious metal
recovery — Africa.

Permutit-Boby continuous ion exchange plant —
Asahi-Dow Chemical Co., Japan.

Paterson Candy International water clarifiers —
Shek Pik, Hong Kong.

Permutit-Boby boilerfeed de-ionisers —
BP Chemicals, South Wales.

The Chlor-Alkali Industry

By R.W. PURCELL

ICI Ltd., Mond Division

The Chlor-Alkali industry refers to the manufacture of a related group of
heavy industrial chemicals, chlorine, caustic soda and sodium carbonate.
The first two products are linked by being produced simultaneously by the
electrolysis of sodium chloride: sodium carbonate is included in this
classification because it can be used interchangeably with caustic soda in
many end uses and because the one product can be made from the other. In the
past large quantities of caustic soda were made from sodium carbonate. Now
sodium carbonate is made from caustic soda on a limited scale. Another
obvious link is their dependence on a common raw material which frequently
means that a manufacturing Company with access to salt will make both
products, as for example, ICI Limited in Cheshire, Solvay SA at Dombasle in
France, and the Allied Chemical Company at Syracuse NY. These products are
classified as heavy chemicals because of the large scale on which they are
produced. The following table shows how they compare with some other
chemicals made in large quantities. The statistics relate to the USA.

Table 1

Production of some large volume chemicals in the USA in 1974 Quantities in millions of tonnes			
Sulphuric Acid	30	Sodium Tripolyphosphate	0.8
Ammonia	14	Titanium Dioxide	0.7
Ethylene	11	Phosphorus	0.5
Caustic Soda	10	Potassium Hydroxide	0.2
Chlorine	10	Sodium Metal	0.2
Sodium Carbonate	7	Sodium Chlorate	0.2
Nitric Acid	7	Hydrogen Peroxide	0.1
Sodium Sulphate	2		
Ethanol	2		
Vinyl Chloride	2		

Total current annual World production of chlorine, caustic soda and soda ash
are 29 million, 31 million and 26 million tonnes respectively, that is,
every day the World produces and consumes about 80 000 tonnes of each of
these products. Of this enormous quantity about half the sodium carbonate
is used for making glass, a quarter of the chlorine for PVC and over half
the caustic soda is used in making other chemicals. The following table
itemises some of the uses in more detail: the data are for the USA.

Table 2, Percentage of Production for Stated End Use

Chlorine		Caustic Soda		Sodium Carbonate	
VCM	18	Inorganic Chemicals	21	Glass containers	34
Solvents	22	Organic Chemicals	17	Flat glass and	11
Propylene Oxide	5	Soap	4	glass fibre	
Chloromethanes	10	Pulp and Paper	14	Sodium Phosphate	12
Inorganics	8	Alumina	7	Silicates	5
Pulp and Paper	11	Rayon	4	Pulp and Paper	4
Other	26	Neutralisation	12	Alkaline Cleaners	5
		Unspecified	21	Unspecified	29

The atomic weights of chlorine and caustic soda are 35.5 and 40 so that the
electrolysis of sodium chloride produces 1.13 tonnes of caustic soda for
every tonne of chlorine. Since most of the chlorine is made this way,
production of caustic soda is not necessarily equal to demand for caustic
soda. Up to ten years ago, production was less than demand and the balance
was made by the Lime-Caustic Process in which lime was reacted with a
solution of sodium carbonate to provide a dilute solution of caustic soda from
which the precipitated calcium carbonate was removed by settling and
filtration.

$$Ca(OH)_2 + Na_2CO_3 \rightleftharpoons CaCO_3 + 2NaOH$$

However, during the 1960's, demand for chlorine was growing rapidly and
overtook the demand for caustic soda, first in the USA, and soon after in
Europe. The need for a process to produce additional caustic soda ceased
and the main Lime-Caustic plants in the World had closed down by 1970.
Chlorine production continued to grow more rapidly than the demand for
caustic soda so that caustic soda in excess of demand is now being converted
into sodium carbonate in Europe and Japan, and is also replacing sodium
carbonate in the manufacture of sodium salts and in neutralisations.

When a substantial amount of caustic soda was being made from sodium
carbonate the price of caustic soda was naturally higher than that of the
equivalent amount of soda ash, but now in Europe the prices are about the
same. One tonne of caustic soda (which would normally be purchased as two
tonnes of solution of concentration 50%) costs £69 ex Works; it is
chemically equivalent to 1.32 tonnes of sodium carbonate at £55 per tonne,
which costs £73. When the extra transport costs to the consumer of the
caustic soda solution are allowed for there is little difference in the
delivered price.

Ten years ago, the demand for chlorine in Europe was growing at 10% a year,

and of caustic soda at about 7% a year, and it was apparent that if these
rates continued a large excess of caustic soda would arise to displace the
manufacture of an equivalent amount of sodium carbonate. The demand for
sodium carbonate was growing relatively slowly even without displacement and
it seemed possible then that the independent manufacture of sodium carbonate
might cease in the foreseeable future. However, the demand for chlorine is
not now growing so fast; its growth was sustained by the penetration of PVC
into many new end uses, a process which has inevitably slowed down. Other
developments which have reduced the demand for chlorine have been the
development of a new direct oxidation process to replace in part the
chlorhydrin route to propylene oxide and the curtailment of some chlorine
derivatives because of their ecological effects. It is possible that the
supply and demand of caustic soda are now in balance though more likely that
chlorine is still growing somewhat faster. It is clear however that
independent sodium carbonate manufacture will be required for many years more.

The Manufacture of Chlorine

More than 90% of the chlorine used in the World is made by the electrolysis
of an aqueous solution of sodium chloride (brine). The rest is made by the
electrolysis of molten sodium chloride (co-producing sodium metal), aqueous
solutions of potassium chloride (co-producing potassium hydroxide), and molten
magnesium chloride (co-producing magnesium metal). Hydrogen chloride is a
by-product of many hydrocarbon chlorinations and this may be reconverted into
chlorine by electrolysis of an aqueous solution or by oxidation in the
Kelclor process.

C. W. Scheele is credited with the discovery of chlorine in 1774 which he
obtained by the reaction of hydrochloric acid and manganese dioxide. It is
probable that the gas had been observed by the alchemists who knew how to
make the mineral acids and describe experiments with these in which chlorine
is likely to have been one of the products. Scheele described the bleaching
effect of chlorine and it was this that eventually led to the demand for
chlorine on a scale sufficient to justify its industrial manufacture. The
same textile industry that generated a demand for the Leblanc process alkali
also provided a use for the hydrochloric acid that was made in the first
stage of that process and had earlier been wasted.

$$2NaCl \ + \ H_2SO_4 \ \xrightarrow{\text{heat}} \ Na_2SO_4 \ + \ 2HCl$$

The hydrochloric acid was oxidized with pyrolusite (natural manganese

dioxide) to chlorine and the resultant manganous chloride was wasted until
Weldon (1866) showed how it could be recycled. The Weldon process competed
with the Deacon process (1868) in which the acid was oxidised by air on a
catalyst of cuprous chloride.

Electrolysis of brine to produce chlorine was described as early as 1800
by Cruickshank,and in 1834 the fundamental laws of electrolysis were
formulated by Faraday. The application of electrolysis to the industrial
production of chlorine and caustic soda were the subject of many patents
from the mid 19th century onwards. The development of the dynamo by Gramme
(1872) allowed some of these inventions to be tested on a suitable scale
and this revealed the problems of anode wear, low current efficiency,
corrosion of cell materials, the poor mechanical strength and variable
porosity of diaphragms, all of which had to be overcome to produce a cell
suitable for industrial use. The large number of patents and developments
in the next two decades make it difficult to decide who made the significant
discoveries leading to the first commercial production. It is probable that
the first in production was a discontinuous diaphragm cell based on the patents
of Mathes and Weber, which was operated at Frankfurt in 1891. Hargreaves
and Bird were in operation at Widnes with a continuous diaphragm cell in
1897, using as do all present day cells, asbestos as the diaphragm material.
The principle of the mercury cathode had been proposed by Nolf in 1883 and
its development proceeded simultaneously with that of diaphragm cells;
Castner, an American chemist working at Birmingham, and Kellner, an Austrian
in Vienna, independently developed very similar versions of the mercury cell.
Their combined patents were used by the Castner Kellner Alkali Company in a
factory at Runcorn which began production in 1897.

In the USA, the first diaphragm cell plant, also for continuous operation,
designed by Le Sueur was operated commercially at Romford Maine in 1893 and a
plant using Castner cells was started at Saltville Va in 1896. Clearly there
was at this period an international recognition of the possibilities
opened up by developments in electricity generation and in electrochemistry.
The potential of carrying out in one step, the transformation of salt to
chlorine and caustic soda promised considerable profits to those who were
successful, and helps to explain the speed and resourcefulness which these
early technical entrepreneurs brought to their research, development and
production. At the time, the following operations were required to make
chlorine and caustic soda –

React sulphur or pyrities, air and water to make sulphuric acid
(chamber process)
React sulphuric acid with salt to make hydrochloric acid and sodium
sulphate
React the sodium sulphate with coke and limestone to make sodium
carbonate
Convert limestone to lime by heating
React lime with sodium carbonate to make dilute caustic soda
Oxidise hydrochloric acid by Weldon or Deacon process

These 19th century electrolytic plants had about 300 kW of DC power available
and made about 2 tonnes per day of chlorine. The biggest modern plants make
about 1000 tonnes a day.

Cell Types

In a description of the industry in 1939 it was found necessary to distinguish
ten different types of diaphragm cell and four different types of mercury
cell, then in operation. The scene is now much simpler, with only one
type of mercury cell and two types of diaphragm cell being installed. The
first problem which designers of brine cells faced was to keep the products
from reacting. If two electrodes are placed in a concentrated sodium
chloride solution and a current passed between them, the sodium hydroxide
formed at the cathode reacts with the chlorine liberated at the anode to
produce sodium hypochlorite. The two distinct solutions to this problem
which appeared at the very beginning of cell development were the separation
of the cell into two compartments by a porous diaphragm and the alternative
of using as the cathode, mercury with which the liberated sodium forms an
amalgam.

In the diaphragm cell a porous partition, the diaphragm, is placed between
the electrodes and a continuous brine flow is maintained from the anode side
to the cathode side to inhibit the diffusion of hydroxyl ions from cathode to
anode. The principle is illustrated in Figure 1.

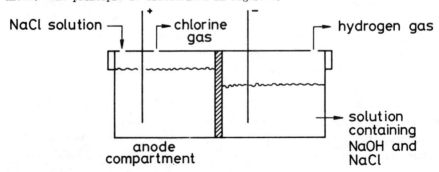

Figure 1 Diaphragm cell

Note that hydrogen ions are discharged at the cathode rather than sodium
and that the solution which leaves the cathode compartment contains
undecomposed salt. In the practical application of diaphragm cells less
than half the salt is converted to caustic soda. Steam-heated evaporators
are provided to concentrate the solution, which is commonly about 11% NaOH
and 16% sodium chloride, to 50% NaOH. As the concentration rises in the
evaporator, crystalline salt separates. The composition of the final
solution is 50% caustic soda and 1% sodium chloride.

The mercury cell uses a stream of mercury as the cathode. Sodium ions are
discharged on the cathode surface as sodium metal which amalgamates with the
mercury and is prevented from reacting with water by the high hydrogen over-
voltage. The amalgam flows from the cell into a separate vessel, the
denuder, in which the amalgam reacts with water. The denuder is packed with
pieces of graphite so that the reacting system is in effect a separate cell:

Graphite H^+ OH^- NaHg   (hydrogen discharged at the anode)

It is a cell which is short circuited because the two electrodes, amalgam and
graphite, are in contact. The net effect is that sodium hydroxide is formed
and hydrogen is drawn off from the top of the vessel while mercury, free from
sodium, is pumped back to the cell.

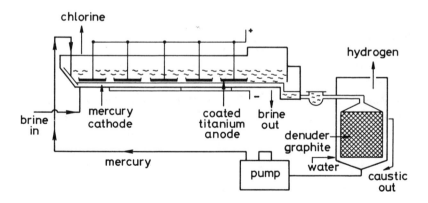

Figure 2   Mercury cell

Note that sodium not hydrogen is discharged at the cathode and that a caustic
soda solution at 50% concentration, free from salt, is provided directly from
the denuder.

The way in which Figures 1 and 2 relate to industrial designs will be

discussed, but first some of the quantitative aspects of brine electrolysis
must be considered.

Cell Current

Faraday found that 96 494 ampere seconds (coulombs) liberate 1 gramme
equivalent of a substance. In an industrial chlorine plant a careful account
is kept of the weight of product produced (usually the caustic soda because
it is easier to measure accurately) and the current consumed. Theoretically,
to produce one metric tonne of chlorine per day in one cell should require:

$$\frac{10^6}{35.5} \quad x \quad \frac{96\ 494}{60 \ x \ 60 \ x \ 24} \quad = \quad 31,458 \text{ amps} \quad \text{Say } 31.5 \text{ kA}$$

In practice at least 32.4 kA are required and this would be expressed by
saying that the current efficiency is not greater than 97%.

The current efficiency is less than 100% because reactions other than the
desired one go on. For example, in a mercury cell, a small part of the
chlorine liberated at the anode finds its way to the cathode surface and
reforms sodium chloride. A small amount of hydrogen is liberated at the
cathode, and oxygen at the anode, a reaction which not only wastes current
but contaminates the chlorine produced.

$$2H_2O \rightleftharpoons 2H^+ + 2OH^-$$

$$4H^+ + 4e \longrightarrow 2H_2$$

$$4OH^- \longrightarrow 2H_2O + O_2 + 4e$$

In diaphragm cells the electrolyte in the cathode compartment (the catholyte)
is a mixture of sodium chloride and sodium hydroxide. Hydroxyl ions tend to
move back through the diaphragm against the flow of brine, helped by their
high mobility. These add to the hydroxyl ions already present in the water
and as their concentration rises some may discharge at the anode as oxygen
and some form hypochlorite and chlorate, which re-enter the catholyte
compartment and pass out of the cell in solution with the caustic soda :

$$2OH^- \longrightarrow H_2O + O_2 + 2e$$

$$2OH^- + Cl_2 \longrightarrow 2OCl^- + H_2$$

$$3ClO^- \longrightarrow ClO_3^- + 2Cl^-$$

The higher the concentration of hydroxyl ion is allowed to rise in the
cathode compartment the more will this process take place with a consequent

reduction in the current efficiency. Minor amounts of current are also
wasted in discharging small amounts of impurity present in the brine supply.

Cell Voltage

The standard electrode potentials of the elements met with in brine
electrolysis are

Standard Electrode Potentials	
	Volts
Sodium	+2.714
Hydrogen	0
Oxygen	−0.401
Chlorine	−1.359

From a consideration of these potentials, it might be thought that if the
voltage across a diaphragm cell was increased steadily from zero, the
electrolysis would begin at 1.359 volts and similarly for a mercury cell at
4.073 volts.   In practice the decomposition or reversible voltage is found
to be 2.15 volts for a diaphragm cell and 3.05 volts for a mercury cell.
These differences are to be explained by the very different conditions in
which the products are discharged in commercial cells and those under which
the standard is determined.

In both types of cell the reversible anode potential is not very different
from the standard; it tends to be slightly less because the concentration
of brine is about 5.3 molar (compare 1 molar for standard) and the
temperature about 85°C (compare 25°C standard).

In the diaphragm cell the reversible potential at the cathode is considerably
greater than standard because hydrogen is being discharged from a strongly
alkaline solution (about 3 molar) compared with 1 molar acid. The potential
increases by about 60 millivolts for every pH unit increase. In the mercury
cell the reversible potential at the cathode is 0.91 volts less than the
standard (about −1.80 volts compared with −2.71 volts) because the formation
of sodium amalgam is exothermic, that is, the energy required to discharge
sodium ions as sodium metal is greater than the energy required to discharge
them as a component of an amalgam. The reduction of the voltage due to this
effect may be derived from the heat of reaction (more strictly the free
energy change, but ΔF differs from ΔH by only a small amount).

$$\Delta H = 82.9 \text{ kJ mol}^{-1}$$

$$\Delta EMF = \frac{82.9 \times 10^3}{96494} = 0.86 \text{ volts}$$

The difference between 0.86 volts and 0.91 volts is accounted for by the higher concentration of brine compared with standard.

When the voltage applied to a cell is increased above the decomposition voltage current flows and electrolysis takes place. A substantial further increase in voltage is required to sustain the formation of the products at a steady rate. This extra voltage is required to drive the current through the resistances of the contacts to the cell electrodes and within the electrodes themselves, through the electrolyte (which is broken up by gas bubbles), through the diaphragm, and to overcome the "over-voltage" at the electrodes.

The voltage across a working cell with current I is thus

$$V = V_d + IR + P$$

where R is the sum of the electrical resistances and P the over-voltage to the electrodes. Most of the resistance is within the cell and is inversely proportional to the area of the electrode, that is

$$R = r/a \quad \text{where } r \text{ is the resistance per unit area, and}$$

$$IR = rI/a = rD$$

I/a is the current density D. Over the range at which commercial cells work the over voltage may also be assumed to be constant so that

$$V = V_o + KD \qquad\qquad \text{where } V_o = P + V_d \qquad\qquad (1)$$

The over-voltage is affected by the material and nature of the electrode surface as well as by temperature and current density, for example, the chlorine over-voltage on platinum is lower than on graphite. There is no well developed explanation of over-voltage but it seems to involve concentration effects on or near the electrode surface governing the diffusion of ions towards the electrodes and kinetic effects which require an activation energy barrier to be surmounted.

The very fact that mercury cells are useful, that is, that they discharge sodium rather than hydrogen, is due mainly to the very low sodium over—voltage (~ 0.01 volts) and the very high hydrogen over-voltage (1.4 volts) together with a high pH in the boundary layer near the cathode.

In table 3, typical voltage and power data, representative of large modern cells are given.

	Mercury	Diaphragm
Current/kA	400	150
Anode Area/m^2	35	55
Current density/kA m^{-2}	12	2.7
Decomposition Voltage/V	3.05	2.15
Over voltage at the anode/V	0.06	0.03
Over voltage at the cathode/V	0.01	0.30
Voltage to overcome cell resistance/V	0.80	0.50
Voltage to overcome resistance of connections and electrodes/V	0.40	0.20
Voltage to overcome resistance of diaphragm/V	-	0.30
Working Voltage/V	4.32	3.48
No of cells if 500 volts is permissible across the series	115	143
Output per cell/tonnes per day	12.2	4.6
Output from series/tonnes per day	1400	650
Current efficiency (%)	96	96
DC Power for Electrolysis/kWh per tonne	3400	2740

Factors Determining Electrolytic Plant Design

Interelectrode Gap

The cost of power is the biggest single factor in the cost of chlorine and caustic soda so that the designer will first of all seek to reduce all the contributions to cell resistance. In particular he will arrange for the lowest possible interelectrode gap. In mercury cells the gap is typically 3 mm. This is about the minimum which can be achieved without risking occasional short circuiting due to ripples forming on the mercury or disturbances to the flow caused by 'thick mercury', a phenomenon in which impurities introduced with the brine cause a change in the viscosity of the mercury which can accentuate the risk of a short circuit. The gap in a diaphragm cell is wider, having to accommodate the diaphragm; the structure of the cell also makes it difficult to assemble the gap to close tolerances.

Electrode Material

Until recently, graphite was the preferred anode material being both cheap and resistant to corrosion by chlorine. However, its electrical resistance is higher than a metal and in use it slowly wears by oxidation to CO_2 due to

the small amount of the oxygen evolved at the anode. The search for an
alternative anode has been in progress for a long time. It had been known
that titanium had adequate corrosion resistance but a high over-potential.
Recently, it has been found that the over-voltage can be reduced below that of
graphite by coating the titanium with an oxide of the platinum group of
metals and this discovery has led to the replacement of graphite in most
existing cells and its universal use in new cells. The conductivity of
titanium is about 30 times that of graphite. An anode which does not wear
is particularly valuable for the diaphragm cell in which the design is such
that the anode cannot be adjusted while the cell is in service.

The economic choice for the cathode in a diaphragm cell is woven steel wire.

Cell Voltage

When the electrode gap has been decided and the thickness of the electrodes
and the method of connection to the intercell bus bars has been designed the
ohmic resistance of the cell can be calculated. The nature of the electrode
material and its surface finish determine the over-voltage which can be
estimated at a range of current densities from laboratory data or past
experience. The constant K in equation (1) is thus determined.

Current Density

The voltage of a cell and therefore the power consumption to make a tonne of
chlorine increases as the current density is increased but the capital cost
of the cell is reduced. The optimum current density which gives the minimum
total cost is 10 - 15 kAm^{-2} for a mercury cell and 2 - 3 kAm^{-2} for a diaphragm
cell, the figure being highest when the unit cost of power is lowest. It
may be found that the current density optimised in this way leads to the
production of more resistance heating than the cell can lose without boiling.
In this case the operating current density would have to be lowered below
the optimum.

Size Of Cell

The capital cost of building plant to make a given amount of chlorine is
reduced as the size of the individual cells is increased. The maximum size
of a cell is usually fixed by engineering considerations, for example, the
ability to produce a flat base plate for a mercury cell and support it so
that it stays flat in service begins to be a formidable problem when the
weight of the base exceeds 10 tonnes.

Number of Cells

When the size (area) of the cell is fixed and its current density has been
decided, the output per cell follows. The ohmic resistance and the over-
voltage at the operating current density determines the voltage of each cell.
A factory will have many cells connected in series with what is considered a
safe maximum total voltage (now about 500 volts) across them. Usually the
rectifiers are placed at one end of the cell room and the series in two lines
with the current going down one line and returning along the other. The
total voltage divided by the cell voltage gives the number which can be
accommodated in a series. Table 3 shows that the peak output from typical
cell rooms is 1400 and 650 tonnes per day for mercury and diaphragm cells.

Description of Typical Cell Installations

Mercury Cell Process

Typically, the electrolyser is a long narrow trough (1-2.5m wide and up to
15m long) made of steel. The sides are lined with hard rubber to prevent
corrosion, the bottom is a thick flat slab of steel on which the mercury
flows. The flow is induced by giving the cell a slight slope of about 1 in
100. The gas-tight cover is usually also made of rubber-covered steel. The
anode stems pass through the cover where they are sealed by flexible rubber
skirts which permit enough movement to allow the anodes to be adjusted to the
specified distance away from the mercury cathode. Electrical connection is
made to the cell base externally by thick copper bars and to the anodes by
flexible copper ribbons. In modern cell rooms it is common to have an
automatic system for adjusting the anodes to give the lowest possible power
consumption.

The amount of mercury in a cell and its flow rate relative to the chlorine
production determine the sodium concentration in the amalgam, which it is
desirable to keep low; typically amalgam leaves the cell containing 0.4%
sodium. If the proportion of sodium were allowed to rise, hydrogen
evolution at the cathode would increase and reduce the current efficiency,
and would also lead to the possibility of an explosive reaction between
hydrogen and chlorine. A dangerous situation can arise if the mercury pump
stops so that the base plate drains and is left exposed. The current must be
shut off immediately to avoid the base plate acting as a cathode for the
evolution of hydrogen.

Purified saturated brine enters the cell at a concentration of 310 g/l and is
allowed to fall to 250 g/l before being taken out. It is dechlorinated by

blowing air through it, salt is added to resaturate it and its pH adjusted
to 5 by the addition of hydrochloric acid, after which it is recycled to the
cell.

At each end of the electrolyser a barrier maintains the separation of mercury
and brine layers so that the mercury amalgam is taken out under the barrier
and flows free of brine to the denuder. Formerly, the denuder was a
horizontal steel trough under or along side the electrolyser; in new
installations it is a vertical tower made of mild steel. The tower
contains pieces of graphite supported on a screen over which the amalgam
and water flow countercurrently. Hydrogen is collected from a pipe in the
top cover and the caustic soda solution is taken off by another pipe above
the graphite. Mercury leaving the bottom of the denuder, still containing
about 0.003% of sodium, is returned to the cell by a pump.

Diaphragm Cell Process
One of the aims of a cell designer is to provide the maximum electrode area
inside a given cell volume while ensuring that the electrode gap is as small
as possible. Early diaphragm cells used either a simple horizontal sheet
mesh or a simple vertical cylindrical mesh on which they deposited the
asbestos diaphragm. The cathode area in these was small and the output per
cell low. In an effort to increase the cathode area a corrugated mesh was
substituted for the flat mesh and this required the anodes to be shaped to
follow the contours of the corrugations. In the next evolutionary development
corrugations became large pleats and the anodes were sheets of graphite
sitting between but not touching the pleats. This pleated form evolved into
a cell in which the pleat was sealed at its open end and the anode inserted
along the length of the pleat rather than through the top. Each sealed
pleat formed a separate catholyte compartment from which the caustic and
hydrogen were collected through connections to the frame of the cell.

The latest diaphragm cells are made of three parts :

   a metal base to which a row of vertical metal sheet anodes are rigidly
   attached.

   a cathode box which is a hollow metal frame across which pairs of steel
   gauze sheets are fixed. The space between each pair of gauzes forms the
   catholyte compartment when the gauze is covered with a diaphragm of
   asbestos fibres. Each pair of gauzes communicates with the hollow frame
   so that caustic soda solution and hydrogen can flow out.

   a cell cover which forms a gas-tight lid and carries the offtake pipe for
   the chlorine and the brine inlet.

When the cell is assembled each catholyte compartment fits between a pair of anodes. The complete cell is roughly a 2.5m cube. Figure 3 shows this principle in a vertical section.

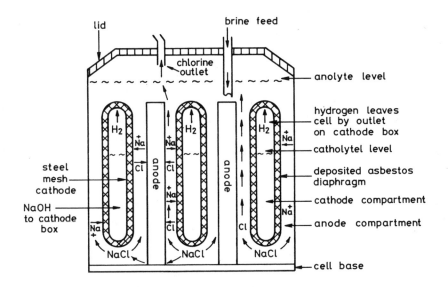

Figure 3 Cell for chlorine and caustic

The level of brine stands higher in the anode space to provide the pressure to drive the brine through the pores in the asbestos.

During construction of the cell the asbestos is deposited in a carefully controlled way by immersing the cathode box in a slurry of asbestos fibres and sucking the slurry through the mesh for a specified time. This procedure lays down a mat of fibres ~ 2 mm thick on the outside of the mesh.

Bipolar Diaphragm Cell

The diaphragm cell already described is called monopolar because the cathodes and anodes are separate assemblies within each cell. An alternative design, called bipolar is arranged so that one side of each electrode acts as a cathode and the other side as an anode. A number of these electrodes is clamped together in a frame leaving a space between the cathode side of one element and the anode side of the next. This space forms the cell or more precisely, the anolyte chamber. The catholyte chamber is incorporated in each electrode element and is constructed of steel mesh and asbestos in a similar way to that described for monopolar cells.

The Glanor electrolyser is a typical bipolar cell. It consists of ten
electrode elements clamped together by tie rods between two end electrode
elements, forming a sealed electrolyser module of eleven cells. The
assembly is 5.3 metres long, 3.5 metres wide and 2.7 metres high and can
produce up to 27 tonnes of chlorine per day. The anode side of each element
is made of coated titanium and the cathode is steel wire mesh.

Other Electrolytic Products Associated with Chlor-Alkali

Potassium hydroxide (caustic potash) and chlorine are made by the electrolysis
of potassium chloride in the same diaphragm and mercury cells as are used for
sodium chloride. The mercury process is preferred because the smaller demand
for caustic potash means that the size of the evaporator plant to accompany
a diaphragm process is relatively expensive.

Sodium

Sodium is made by the electrolysis of a molten mixture of 40% sodium chloride
and 60% calcium chloride, this mixture being chosen to have a melting point
of $600^{o}C$. At this temperature, it is possible to avoid two consequences of
the use of molten sodium chloride at its melting point of $800^{o}C$; the sodium
vapour pressure is reduced to a safe level and the liberated sodium does not
dissolve in the melt. The first is important because sodium vapour ignites
spontaneously in air and the second because a solution of sodium metal
conducts electronically rather than ionically and so electrolysis stops.

The cell in which this reaction takes place, called the Downs cell after its
American inventor, consists of a circular steel casing about 1.5m in diameter
and 2.5m high lined with firebrick. The central anode is a thick graphite
rod projecting upwards from the bottom of the cell and is surrounded by a
cast steel cathode. A gauze screen is inserted between the electrodes.
Sodium formed at the cathode melts and floats to the surface, being prevented
by the gauze from recombining with the chlorine being evolved at the anode.
The sodium is collected in an inverted circular trough to which a riser pipe
is attached. The molten sodium is less dense than the electrolyte and is
pushed upwards through this pipe into a collecting tank. In operation, the
heat produced by the resistance of the electrolyte is sufficient to keep the
contents molten.

The molten sodium is packed by casting it into steel drums. It may also be
transported in 40 tonne rail tankers in which the metal is allowed to
solidify; at its destination it is remelted by passing hot oil through pipes
within the tanker.

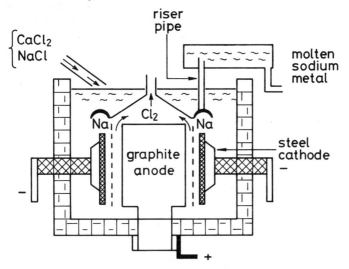

Developments

The membrane cell is a new type now in an advanced state of development. It seeks to improve the quality of the caustic soda solution produced by the diaphragm cell by excluding chloride ions from the catholyte compartment. Thus, it is hoped that it may be able to produce a marketable caustic soda solution directly from the cell and avoid the need for evaporation. In principle, the membrane cell is like a diaphragm cell in which a cation exchange membrane has been substituted for the asbestos diaphragm. Sodium ions can pass through but chloride cannot.

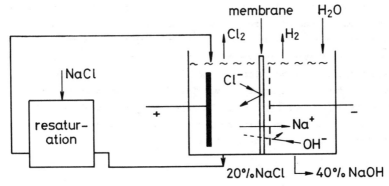

Note that since there is no percolation of solution through the membrane, water has to be added to the cathode compartment.

Membranes have to be able to survive attack by chlorine and the large scale tests at present being carried out in Japan and the USA use polymerised perfluorinated organic sulphonic or carboxylic acids. The present membranes do not give such a high current efficiency as a diaphragm cell

because imperfections in the membrane allow some migration of hydroxyl
ions backwards through the membrane, which carry significant
amount of the current as the concentration of the catholyte
reaches commercially useful levels of around 40%. Membranes are prepared in
large sheets which lend themselves to incorporation in bipolar rather than
monopolar designs. The first users of membrane cells seem likely to be the
wood pulp industry which needs both chlorine and caustic soda and can use
more dilute solutions than the 50% of commerce.

A less fundamental development in diaphragm cell technology is the use of
fibres of synthetic polymers rather than asbestos. Polytetrafluoroethylene
is now coming into use, either in the form of a porous sheet or as chopped
fibre which can be deposited on the cathode mesh in the same way as asbestos.
When graphite was used as the anode material the life of the anode and the
life of an asbestos diaphragm before its porosity fell away, were similar
(about a year). Coated titanium anodes last several years and this is the
incentive to find a better diaphragm material.

Mercury vapour is toxic and strict precautions have always been taken in
cell rooms to avoid concentrations in the air which would harm the plant
operators. As a result, the health record of the industry is excellent.
In the last five years, however, it has become known that small quantities
of inorganic mercury discharged from industrial processes can be biomethylated
and concentrated in the marine food chain. Methyl mercury is considerably
more toxic than the metal and consequently the regulatory authorities are
setting limits on the amount of mercury which can be discharged by industrial
plants. The extra cost of these measures and the unwillingness of
manufacturers to be further involved with a metal which has earned so much
public disapproval is causing chlorine producers who formerly preferred
mercury cells to turn to diaphragm cells. The change over to diaphragm
cells has been made a legal requirement in Japan. In other countries it is
also likely that all new large chlorine plants will use diaphragm or
possibly further in the future, membrane cells.

Ancillary Plant

The cells used in chlorine manufacture have been described in some detail
but the cell room represents only about 25% of the total cost of establishing
a diaphragm cell plant.

Evaporators, which may be triple or quadruple effect, depending on the
price of fuel, concentrate the caustic soda from 11% to 50%.   The suspension
of salt in caustic solution which leaves the evaporators is centrifuged,
and the salt is recycled to make more brine.

The chlorine produced from the cells is saturated with water vapour and
contains a trace of hydrogen, oxygen and usually also air which leaks into
the cells.   The gas is very corrosive when wet so the first operation is
to cool it to condense out the water, then remove by filtration any mist
carried from the condenser.   The remaining moisture is taken out by countercurrent scrubbing with concentrated sulphuric acid.   The dry chlorine may
then be used as such, in the manufacture of eg Ethylene dichloride or it
may be liquified both to enable it to be pransported or to purify it.   At
30°C the vapour pressure of liquor chlorine is 8.8 stm absolute so that it
can be readily liquified by moderate compression and refrigertion since
this limits the pressure rise required and therefore the heat of compression.
When hot, dry chlroine becomes corrosive, which makes sealing and lubrication
difficult.

The gas in equilibrium with the liquified chlorine contains all the hydrogen
and air present in the cell gas, but now at a greater concentration.   Care
has to be taken that the liquifaction is not carried so far that this "tail
gas" contains sufficient hydrogen to reach the lower explosion limit for
the reaction of hydrogen and chlorine.   To be on the safe side the hydrogen
concentration is always kept below 4%.   The "tail gas" is also used for
chlorination, where the concentration of the chlorine is not important, for
example in the manufacture of sodium hypochlorite.

The Ammonia Soda Process for the Manufacture of Sodium Carbonate

This process is also referred to as the Solvay process after Ernest Solvay who
first succeeded in operating it successfully at Couillet in Belgium in 1869.
The chemistry of the process had earlier been investigated by Fresnel (of
optical fame) in 1811 and by Vogel in 1822; Thom in 1836 actually produced
on a near commercial scale in Glasgow for a short time and in 1838 H G Dyar
and J Hemming were granted a British patent for the process and erected a
factory at Whitechapel which failed to make money and was abandoned after
two years. Factories were also built and quickly abandoned by Muspratt at
Newton-le-Willows in 1840 and by Schlosing and Rolland at Pukeaux near Paris
in 1857. The cause of their failure, usually cited, was the inability to
recover and recycle all the ammonia, then a relatively expensive intermediate
but the early French plant also suffered from the imposition of the salt tax,
especially onerous because unlike the Leblanc process the Solvay process
converts only 75% of the salt input. However, if these early entrepreneurs
had had more capital available they might have persevered long enough to
solve sufficient of their problems to begin making some cash to keep their
works going. Solvay brought to the process adequate capital, a systematic
way of tackling production problems, a capacity for inventing items of plant
to work continuously rather than batch wise and a determination to succeed
in his aim to produce sodium carbonate more cheaply than the users of the
Leblanc process.

The problems of the pioneers in this field can hardly be appreciated by
todays chemical engineers, who have a library of catalogues from which they
can select well designed standard valves, pumps, motors, gas compressors and
heat exchangers in a range of sizes and materials. Ludwig Mond who obtained
the license to use Solvay's patents in England and built the first English
plant found that "everything that could break down did break down and
everything that could burst did burst." The reaction vessels in this, as
indeed in many subsequent plants, were made of cast iron which he found were
protected from corrosion by the sulphides introduced accidentally with the
ammonia which at that time was derived from making coke. Corrosion in a
chemical plant is important not only because it obviously causes the
equipment to wear out but also spoils the colour and purity of the product.

The overall reaction exploited by the ammonia soda process is :

$$2\ NaCl\ +\ CaCO_3\ \longrightarrow\ Na_2CO_3\ +\ CaCl_2$$

which will of course, go in the reverse direction in solution as the
permanence of the White Cliffs of Dover testifies.   It is carried out in the
following steps which are illustrated in Figure 6.

Brine Preparation and Purification

A saturated solution of salt is prepared by injecting water into an underground
salt deposit and recovering the brine (eg in Cheshire) or by mining rock salt
and dissolving it in water (eg in Pakistan), or from Solar salt made by
pumping sea water into a series of shallow ponds where water is progressively
removed by the sun's heat until salt crystallises out, is recovered, drained,
washed and dissolved in fresh water (eg in South Australia).

In all cases the solution will contain some calcium and magnesium ions which
in the conditions of the Ammonia Soda process would precipitate to form
scales of calcium and magnesium carbonates which would quickly block up
pipes and columns. These unwanted ions are therefore removed from the brine
before it is fed to the process. The usual method is to add sodium carbonate
and sodium hydroxide to the solution in slight excess and in such a way as to
form a flocculated precipitate of calcium carbonate and magnesium hydroxide
which can be separated from the brine by settling.

The brine also contains sulphate ions which at one point in the process form
a calcium sulphate scale but the cost of removing the sulphate ion from the
brine is greater than the reduction in the cost of de-scaling.

Absorption of Ammonia

The first process step is to dissolve ammonia in the brine but before the
brine enters the absorber in which this takes place, it is used to scrub small
quantities of ammonia from a number of gas streams before they are rejected to
atmosphere. This scrubbing is commonly carried out in packed towers worked
countercurrently and serves both to prevent atmospheric pollution and
conserve expensive ammonia. The absorber is commonly a bubble cap column
down which brine flows countercurrent to the recycled ammonia gas. The
solution process is exothermic and since to obtain the maximum yield of sodium
bicarbonate later in the process an ammonia concentration of about 6N is
required, it is necessary to cool the absorber to achieve this concentration at
ambient (atmospheric) pressure. Fortunately the cooling requirement is
somewhat less stringent than might be expected because the recycled gas
contains some carbon dioxide as well as ammonia which tends to reduce the
total vapour pressure of the solution. Water vapour also present in the

The Interox hydrogen peroxide plant at Jemeppe, Belgium

recycled gas dilutes the ammoniacal brine; some dilution is necessary to prevent salt being precipitated (salt is less soluble in a solution of ammonia than in pure water), but care is taken to control the composition of the gas because too extreme dilution would reduce the eventual yield of sodium bicarbonate.

Carbonation

The carbonation of the ammoniacal brine is carried out in tall towers fitted with "pasette" plates to break up the gas flow into bubbles, yet with smooth contours which discourage the bicarbonate from settling out. Water-cooled tubes traverse the tower to remove the heat of the exothermic reaction. The crystallization of the bicarbonate is unfortunately not confined to the suspended phase but also forms on the metal surfaces and over a period of about four days the tower is gradually reduced in cross section to the point where it must be taken off "making" duty and put on "cleaning". The "cleaning" tower receives the ammoniacal brine and a small amount of carbon dioxide is blown in to provide agitation and some heating by reaction; the cooling water is turned off. The bicarbonate scale dissolves in the ammoniacal brine which after leaving the "cleaning" tower is distributed to a group of "making" towers. Dilute carbon dioxide is injected at the bottom (and in some Works at higher levels) of the making tower and is absorbed as it passes upwards forming ammonium bicarbonate which reacts with the sodium chloride present to form sodium bicarbonate. It is desirable that the crystals of bicarbonate should be easy to wash free of mother liquor and to filter to a dry cake which is to say that they should be large and smooth. However, even with careful control the crystals are small and show considerable twinning. The control system aims to prevent the supersaturation rising to the point where excessive nucleation takes place by regulating the cooling water flow to different levels, so that the reaction is made to progress smoothly as the solution descends the tower.

The tower is about 26m high and is operated full of solution, hence the highest partial pressure of carbon dioxide is available at the bottom to help achieve the highest possible conversion of sodium chloride to sodium bicarbonate. In England it is practical to cool the solution to about $25^{\circ}C$; at this temperature, the solubility relationships in the five component system NH_4^+, Na^+, Cl^-, HCO_3^-, H_2O, are such that at most 75% of the salt is converted to sodium bicarbonate. If carbonation is pressed beyond this point other phases (chiefly ammonium bicarbonate) would precipitate. The suspension of sodium bicarbonate flows from the tower to

a rotary drum vacuum filter or centrifuge on which the cake is washed with
water to displace the mother liquor and to avoid the presence of chlorides
in the final product. The filter cake, "crude sodium bicarbonate", is fed
to the calciners; the filtrate is pumped to the ammonia recovery plant.

Calcination
In this operation the crude bicarbonate is progressively heated to drive off
the adherent ammonia and water and then to decompose the bicarbonate to
carbonate with the evolution of carbon dioxide :

$$2 \; NaHCO_3 \quad \xrightarrow{150^{\circ}C} \quad Na_2CO_3 \; + \; CO_2 \; + \; H_2O$$

The equipment in which this reaction is carried out has evolved in stages
from a batch-fed closed pan heated by a coal furnace and manually stirred with
rakes inserted through the pan cover. The latest equipment is a
rotating drum heated by internal steam pipes which enable the heat input to be
regulated with considerable precision. saving fuel and avoiding fire damage
to the shell. A sealing arrangement is provided at both the feed and
discharge end of the calciner drum to avoid air leaking in and diluting the
carbon dioxide which is re-cycled through condensers to remove water and
thence to the gas compressor which feeds the carbonating towers.

The product from the calciner is a fluffy solid with a packing density of
about 0.5 g/ml and is known in commerce as Light Soda Ash. Its crystalline
form is that of the bicarbonate from which it is derived and its lightness
due to its shape and the void left by the evolved carbon dioxide and water.
For many purposes a denser product is desirable and this is made by
recrystallising the Light Soda Ash as sodium carbonate monohydrate and then
dehydrating the mono.

Ammonia Recovery
The filtrate after the removal of crude bicarbonate contains ammonium
chloride, ammonium bicarbonate, sodium bicarbonate and sodium chloride. It
is heated to decompose the bicarbonates :

$$NH_4HCO_3 \quad \xrightarrow{heat} \quad NH_3 \; + \; H_2O \; + \; CO_2$$

$$NaHCO_3 \quad \xrightarrow{heat} \quad Na_2CO_3 \; + \; H_2O \; + \; CO_2$$

$$Na_2CO_3 \; + \; 2NH_4Cl \quad \longrightarrow \quad 2NaCl \; + \; 2NH_3 \; + \; CO_2$$

and the solution is stripped of all the CO_2 before lime is added. If this
were not done some of the lime would be used up in forming calcium

carbonate. The third reaction plays only a minor part in
the release of ammonia. The next stage is carried out in the 'distiller',
a bubble cap column which has to be designed so that it does not block up
when handling the grit (which milk of lime always contains) and which is also
easy to descale from the calcium sulphate found over a period of years. The
reaction of lime with ammonium chloride releases the "fixed" ammonia and forms
calcium chloride :

$$Ca(OH)_2 \ + \ 2NH_4Cl \ \longrightarrow \ CaCl_2 \ + \ 2NH_3 \ + \ 2H_2O$$

The recovery of the ammonia from the solution is increased by the injection
of steam into the bottom of the distiller, both to raise the temperature and
to assist stripping by providing another component in the gas phase. This
series of reactions is carried out in a number of vessels through which
ammonia, carbon dioxide and steam travel countercurrent to the filtrate, so
that cool gas leaves at one end and a hot solution containing calcium
chloride and sodium chloride leave at the other.

The vessels are placed one on top of the other making a tower over 45 m high.
The top-most vessel acts as a heat exchanger heating up the incoming filtrate
and cooling the outgoing gas so that its water content is reduced to a
tolerable level before the gas is recycled to the absorber.

The demand for calcium chloride as a product is much smaller than the demand
for sodium carbonate so that only a minor amount of the calcium chloride is
recovered for sale in the UK; the rest forms an effluent from the process.
Before the effluent is rejected its solid content (derived from the
impurities in the lime) is removed by settling and disposed of into the
underground cavities left in the salt measures by the brine making operation
described above.

Lime Burning
The lime needed for the recovery of the ammonia and the carbon dioxide
required as a raw material for the sodium carbonate are both provided by
"burning" (ie heating to about 1100°C) limestone in a lime kiln.

$$CaCO_3 \ \xrightarrow{\text{heat}} \ CaO \ + \ CO_2$$
$$C + O_2 + (4N_2) \ \longrightarrow \ CO_2 \ + \ (4N_2)$$

Limestones differ in their calcium carbonate content; they frequently contain
substantial amounts of magnesium carbonate, quartz and clay. These impurities
are highly undesirable for a number of reasons. The magnesium carbonate

absorbs expensive heat in changing to magnesia and carbon dioxide but the
magnesia reacts so slowly with ammonium chloride that it is all discharged
unchanged with the effluent from the distiller. At the high temperature in
the kiln, some of the quartz reacts with the lime as it is formed, converting
it to an unreactive calcium silicate containing three moles of lime to one
of silica wasting what would otherwise be useful lime. Clay is undesirable
because at the temperature of the kiln it may combine with lime and silica to
produce compounds which melt at the high temperature and stick the pieces of
lime together preventing the smooth flow in the kiln. The limestone used in
the UK comes from Buxton and is among the best in the world for this process,
both because it is pure and because it occurs in massive deposits near the
surface, so that it can be extracted by quarrying on a scale which allows the
process to be highly mechanised. The quarried limestone is crushed and
washed free from clay before being loaded into 40 ton steel hopper wagons for
its rail journey to Northwich, 35 miles away.

The lime kilns are cylindrical empty towers 3 to 5 metres in diameter and 20
to 25 metres high. They are charged through the top with a mixture of coke
and limestone. Air for combustion enters at the bottom and is drawn through
the kiln by fans at the top. The cold air entering cools the lime being
drawn out and is itself warmed up to combustion temperature. At the top of
the kiln the charge moves countercurrent to the hot mixture of gases flowing
upwards and out of the kiln. The proportion of coke added to the charge is
slightly less than is required to decompose all the calcium carbonate. As a
result, at the bottom of the kiln, some of the pieces of lime still have a
core of unchanged limestone. However, this procedure avoids the sudden rise
in temperature of each lump which occurs when the endothermic reaction is
completed. The rise in temperature is accompanied by a phase change which
makes the lime less reactive; it is said to be "over burnt".

The lime drawn from the kilns is treated with hot water in rotating drum
slakers to produce the boiling suspension of calcium hydroxide (milk of lime)
used in ammonia recovery. The unslaked material is sieved from the milk of
lime and recycled to the kilns with the fresh limestone feed.

The mixture of gases is scrubbed free from lime dust by passage through a
packed tower countercurrent to a stream of water, and then is compressed and
mixed with the gas from the calciners before being fed to the carbonating
towers.

In some parts of the world coke is not available and kilns have been designed to use oil or natural gas as fuel. These are not so satisfactory for the ammonia soda process because less carbon dioxide is produced, part of the heat from the combustion of the hydrocarbon being associated with the production of water rather than carbon dioxide.

$$CH_4 + 2O_2 + (8N_2) \longrightarrow CO_2 + 2H_2O + (8N_2)$$

Thus, the weight of carbon dioxide produced from the fuel in association with a given weight of lime is reduced. This might not seem to matter since the carbon dioxide produced from the decomposition of the limestone is chemically equivalent to the carbon dioxide needed to make the sodium carbonate product. However, it is not feasible to avoid the loss of a certain amount of carbon dioxide along with the nitrogen which leaves the top of the carbonating towers and the loss is made up by the carbon dioxide derived from the fuel in the kilns. As well as the quantity of carbon dioxide being less from a hydrocarbon fuelled kiln, the concentration is also lower (even after the water is condensed and removed), because of the additional nitrogen in the air used to burn the hydrogen.

Steam and Power

The ammonia soda process needs a considerable heat input at relatively low temperatures. Steam is supplied to the distiller at about 2 atm absolute, and to the calciners at 12 to 20 atm absolute. The process also needs energy for compressing carbon dioxide and pumping large volumes of cooling water as well as for mechanical drives, pumping brine and conveying solids. This energy is most economically provided by raising the steam needed in high pressure (60 – 80 atm) boilers and allowing it to expand through turbines to drive compressors and alternators, before being used at the lower pressures required in the process.

Developments

The equipment used in the process has grown in size in response to the substantial scale of manufacture noted in Table 1. Advances in plant fabricating technique and the availability of resistant metals like titanium and of plastics have also led to design changes aimed at reducing capital and maintenance costs.

Control of a process having so many stages has always been a problem. Not only has each unit to be operated at the optimum flow, pressure, temperature and composition but the rates of the units have to be matched and adjustment

made for the gradual changes brought about by scaling. It is not surprising
therefore that these factories pioneered control systems. The first
application of three-term hydraulic control was in a UK soda ash plant in the
thirties as was the first application of a digital computer to chemical plant
control in 1962.

Trona

Sodium carbonate has been obtained, from pre-historic times, from naturally
occurring sodium sesquicarbonate found in dried up lake beds particularly in
Egypt (Wadi Natron). A natural source which is still actively producing
trona by precipitation is currently being worked at Lake Magadi in Kenya.
However, the most substantial recent developments are of the underground
deposits in Wyoming. Discovered in 1938, during exploration for oil, the
value of the ore took some time to be recognised but commercial production
began in 1952 and is now on a very substantial scale. The cost of production
is considerably less than that of the ammonia soda process and as a result
six out of the nine ammonia soda plants in America have now been closed, and
it has been predicted that in five years time, none may be left. The trona
ore which contains over 80% of sodium sesquicarbonate is mined by conventional
methods at a depth of 460 metres. The ore is crushed and heated in large
horizontal calciners to decompose the sesquicarbonate

$$2Na_2CO_3 . NaHCO_3 . 2H_2O \xrightarrow{heat} 3Na_2CO_3 + CO_2 + 3H_2O$$

The crude sodium carbonate is dissolved in recycled hot water, the insoluble
material originally present in the ore is settled out and the supernatant
solution filtered. In addition to these insoluble components the trona ore
contains a very small amount of organic material which is removed by passing
the filtered solution through active carbon. If allowed to remain, it would
contaminate the product and also modify the crystal habit at the next stage,
where water is removed in a multiple effect evaporator, in which a suspension
of sodium monohydrate crystals is maintained. As these crystals grow they
are removed and centrifuged from their mother liquor. The crystals are then
heated to dehydrate and dry them in steam-heated rotary dryers which are 13
feet in diameter and 90 feet long.

Of the 7M tonnes per year of sodium carbonate made in the USA, 5M is now made
from trona. Reserves of trona in Wyoming alone have been estimated to be
60000M tonnes.

Sodium carbonate is also made at Searles Lake in California from naturally occurring alkaline brine. The Lake is a mass of minerals, dry on the surface but having water present at lower levels. A saturated solution containing about 7% alkali as sodium carbonate and bicarbonate, derived from the solution of trona, together with other sodium potassium and lithium salts, is pumped from the Lake. In the two older plants this solution is worked for a range of chemicals including sodium carbonate, sodium sulphate, potassium chloride and lithium chloride. In a large new plant being constructed the main product will be sodium carbonate. In the production of sodium carbonate, the solution from the Lake is reacted with carbon dioxide to precipitate sodium bicarbonate. The sodium bicarbonate slurry obtained is then treated in the same way as in the ammonia soda process to produce sodium carbonate. The process is much simpler than the ammonia soda process because of the absence of ammonia absorption and recovery, though a supply of supplementary carbon dioxide is not then automatically available. The two solutions adopted are, on the small scale, to burn limestone and sell the lime as a product and on the large scale, where the local demand for lime would not be sufficient to use dilute (about 10%) carbon dioxide from boiler flue gases, and accept the penalty of compressing very large volumes.

Chlorine and Chlorination

By A. CAMPBELL

ICI Ltd. Mond Division

Chlorine is by far the most abundant of the four halogens. It amounts to
about 0.15% of the earth's crust where it occurs almost entirely as chloride
ion. It is the major inorganic constituent of sea water (1.9% w/v chloride)
where its presence has had a profound effect on the evolution of all living
creatures, and the dried up residues of former seas are found today as
massive deposits of crude salt which form the principal source of commercial
chlorine as well as providing a plentiful source of chlorides. The biggest
deposits of salt in the UK are in northern Cheshire where they provide the
raw material for the manufacture of chlorine by ICI Mond Division which is
the biggest UK manufacturer with about 1 million t/year capacity, although
BP Chemicals, AOC and Staveley Chemicals have substantial outputs also. Not
all the salt which is used is converted to chlorine, the largest quantities
of salt seen by the average person are probably the roadside tips of crushed
rock salt which in a severe winter may be used at the rate of about a million
t/year for clearing roads of snow.

The atmosphere also contains traces of chlorinated products apart from
particles of dried wind blown chlorides. The principal component is methyl
chloride derived from natural sources and in the stratosphere there are very
low concentrations of hydrogen chloride, chlorine monoxide and chlorine atoms.

Elementary chlorine exists as a toxic yellow gas, bp -34.6°C at normal
pressures and it contains two isotopes with atomic weights 35 and 37 in the
proportion of 75.4% to 24.6%, giving chlorine an average atomic weight of
35.457. There is also an unstable isotope with atomic weight 36 which is
widely used as a radioactive tracer. Chlorine was discovered by Scheele in
1774 and first shown to be an element by Davy in 1810, and with the development of the alkali industry it became a major commercial product, initially
because of the need to find outlets for the co-product of caustic soda
manufacture but later because its high reactivity and unique properties led
to the development of a wide range of new products and processes throughout
the whole of industry. The presence of chlorine in these products had a great
effect on their properties and it was the growth of understanding of the way
in which manufacture of these products could be carried out and the reasons
why their properties were so greatly affected by the chlorine which they
contained which led to the present position of chlorine as a major industrial
raw material. This review aims to explain why some of these effects are
obtained, and to describe some of the chemistry of chlorine on which these
processes are based and summarise the more important ways in which chlorine
is used today.

Commercial Position

Figure 1 shows the interaction of chlorine and alkali manufacture which are
discussed in the last paper and it gives the major outlets for chlorine and
HC1. It also shows how closely the various markets involving chlorine are
interlocked so that changes in one area can affect the situation in quite a
different business area.

FIGURE 1:   Interaction of chlorine and alkali processes

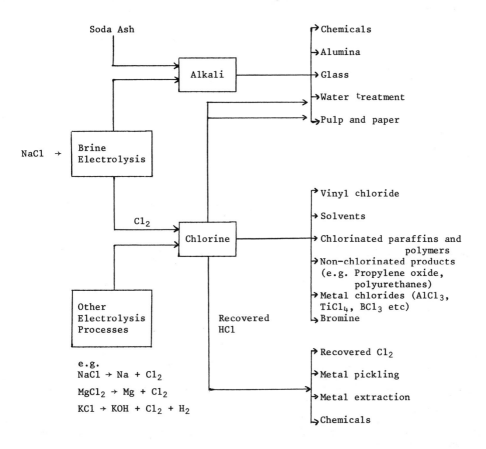

At the beginning of 1976 the overall world capacity for chlorine was about
32 million tons per year of which about 13.6 million were in Europe. About
two thirds of European chlorine capacity is in the EEC,with Germany being
the major supplier and the UK share being about the same as that of France
and Italy, Figure 2.[1]

FIGURE 2:   Chlorine Capacity in Europe 1976

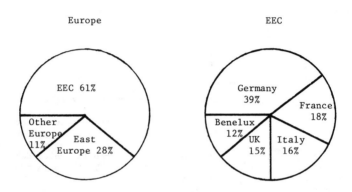

EEC capacity is expected to grow at about 4%/year over the next ten years,
which is equivalent to the construction of more than one 1000 t/day plant
every year, and as the cost of a single plant of this type may approach £100M,
this represents a large investment of new capital although some will undoubtedly
come from expansion of existing plant.

The major outlets for chlorine are shown in Table 1 and these are discussed
separately, but there are also many smaller processes such as the manufacture
of speciality dyestuffs which are very important to their users but which
cannot be covered in a review of this length.

Table 1:   Uses of Chlorine in the EEC

	% chlorine consumed
Vinyl chloride monomer	31
Chlorinated solvents	14
Organic intermediates	16
Propylene oxide	8
Other	31

The table shows that most chlorine is used in organic chemicals but it would
be an artificial separation to exclude these products from a review of the
inorganic chemicals industry because of the many inorganic aspects of their
process chemistry. An interesting economic aspect of the manufacture of
chlorinated organic chemicals is the way in which sales of organic products
with a high chlorine content have been favoured in recent years because of
the steep rise in cost of oil-based intermediates. For example crude oil
has risen by about 850% from about £6/t to £52/t between 1973 and 1977 whereas

in the same time chlorine has roughly doubled from about £33/t to £70/t and as
a result the costs of products with a high carbon content have been affected
more than the costs of products with a high chlorine content. The change in
the price of oil has been exaggerated by the fall in the value of the £ which
has had less effect on the cost of UK-based chlorine manufacture.

The following sections of the review give some of the more important chemistry
and technology of chlorine itself and of the processes and products in which
it is used.

Chemical Properties of Chlorine

Chlorine exists in its molecular form Cl_2, as chloride ion Cl^- and transiently
as chlorine atom $Cl\cdot$ and the positive ion Cl^+. The chlorine atom has seven
electrons in its outer shell so it has to acquire only a single electron to
make up its octet and its affinity for electrons is very high, for example
the electron affinity of a chlorine atom is 3.61 eV, whereas oxygen,which
resembles chlorine in many respects has an affinity of 1.47 eV. In mechanistic
terms Cl_2, $Cl\cdot$ and Cl^+ are all strongly electrophilic and can abstract electrons
from a wide range of sources. Cl^- on the other hand is a weak nucleophile.

Once chlorine has acquired its additional electron the electronegativity of
the atom is not abnormally high, for example the values for fluorine, chlorine
and oxygen are 3.9, 3.0 and 3.5, so chlorine is able to share its electrons
fairly readily to form one or more bonds in addition to its familiar single
bond, examples being the chlorine-oxygen bonds in chlorates and the chlorine
bridges formed in various metal chlorides. Essentially the atom has unoccupied
$\underline{d}$-orbitals and by promoting electrons into these orbitals up to four sigma
bonds may be formed.

Partly because of its lower electronegativity and partly because of the
distribution of its orbitals, products containing chlorine show less tendency
to form intermolecular complexes such as hydrogen bonds than for example
compounds containing oxygen,and this leads to the properties of low boiling
point and low viscosity coupled with high molecular weight which, as will be
seen later, are such useful features of the chlorinated solvents. However
some association does exist, for example linear complexes involving hydrogen
bonds form in liquid HCl, and chlorine will form charge-transfer complexes
with both aromatic and olefinic hydrocarbons which play an important part in
the early stages of the reaction of chlorine with these materials. Some
well known products such as 'chlorine hydrate' a crystalline product stable
below 9°C which contains about 7.5 molecules of water per chlorine molecule,
which might be thought to be a complex really shows little interaction between
the chlorine and water, being a gas clathrate in which chlorine is trapped in
the cages formed by water molecules.

Chlorine Handling Technology[2]

In temperate climates the vapour pressure of chlorine varies from about
2.75 bars in winter to 7 bars in summer, so it can be conveniently handled
either as gas or liquid. At ordinary temperatures metals such as aluminium,
tin and titanium are attacked but a good range of metals are available which
show little reaction with dry chlorine, such as iron, copper, steel, lead,
nickel and silver, and as a rule mild steel is used for handling dry gas or
liquid. However moist chlorine is corrosive to all common metals principally
because of hydrolysis to hydrochloric acid, and attack on metals is made
easier by the relatively small size of the chloride ion which can penetrate
the protective film which coats many metals, much more rapidly than large
ions such as sulphate.

At higher temperatures dry chlorine will react readily with many metals, for example carbon steel has been reported to ignite at $250^\circ C$ and accidents have occurred from local overheating which has led to rapid attack on reactor walls. Nickel and its alloys are most resistant to chlorine and as a rule these are employed for temperatures up to about 500° beyond which brick or ceramic linings are usually employed in reactor construction.

Reactions of chlorine with hydrocarbons are strongly exothermic and the rate of reaction may accelerate to a level at which explosion occurs, so careful definition of the explosion limits are required in the development of any chlorination process. However many millions of tons of chlorine are used every year without incident in processes where these factors have been taken into account.

Reaction of Chlorine with Carbon

A very large number of commercially useful products ranging from insecticides to solvents and from polymers to anaesthetics can be made from carbon, hydrogen, and chlorine; however, the largest single product is vinyl chloride $CH_2=CH_2Cl$ which is used almost exclusively in the manufacture of PVC. ICI Mond Division is the largest UK manufacturer with about 300,000 t/year capacity at Runcorn and Fleetwood and BP Chemicals also have substantial capacity at Baglan Bay.

Vinyl Chloride

Vinyl chloride was originally made from acetylene which in turn was made from coal-based calcium carbide, the acetylene being treated with HCl over a mercury catalyst.

$$CH \equiv CH + HCl \quad \rightarrow \quad CH_2 = CHCl$$

This process was operated on a large scale in the 1950s, the HCl being obtained by burning stoichiometric equivalents of hydrogen and chlorine together in a brick-lined reactor, and a row of HCl burners was a familiar sight in Works which manufactured chlorinated hydrocarbons. However, acetylene was expensive and its manufacture consumed large amounts of energy. The availability of cheap naphtha in the 1950s allowed the development of cheap ethylene which proved to be highly competitive with acetylene, and which has replaced it almost completely in the last two decades for chlorinated hydrocarbon manufacture. Ethylene is likely to remain the favoured feedstock for vinyl chloride for the remainder of the century although the steady escalation of oil prices will narrow the economic gap and may eventually favour a return to coal-based processes.

A disadvantage of ethylene, C_2H_4, is that it contains one more hydrogen than vinyl chloride and by conventional chlorination processes this hydrogen appears as by-product HCl. It would be completely uneconomic to discard this because a 300,000 t/year vinyl chloride plant could product about 175,000 t/year HCl and the principal approach used by industry has been to develop technology by which chlorine can be recovered from the HCl or to use it directly in the chlorination process.

Three of the major chlorine recovery processes are

1   Electrolysis of aqueous HCl to give chlorine and hydrogen. Novel cells for this purpose have been developed by Uhde in Germany [3] and De Nora.[4]

2   The 'Kel-Chlor' process developed by Kellogg[5] in which HCl is oxidised by a mixture of oxides of nitrogen and oxygen to chlorine and water, the nitrogen oxides being recycled

$$2 \text{ NO} + O_2 \rightarrow 2 \text{ NO}_2$$

$$\text{NO}_2 + 2 \text{ HCl} \rightarrow \text{NO} + Cl_2 + H_2O$$

To force the equilibrium to the desired side and prevent loss of nitrosyl chloride by hydrolysis the whole reaction is carried out in 80% sulphuric acid at temperatures up to 250-325°C.

3   The Deacon process. A long established process in which HCl and oxygen are used to oxidise a bed of cuprous chloride to water and cupric chloride which is then thermally dissociated to chlorine and cuprous chloride. The process is discussed in a later section.

However, the principal outlet for by-product HCl in present vinyl chloride technology is its internal consumption by the vinyl chloride manufacturing process, using technology usually called 'oxychlorination'. A modern vinyl chloride process therefore usually contains three stages:

1   Chlorination of ethylene with chlorine to give 1,2-dichloroethane
2   Thermal cracking of 1,2-dichloroethane to vinyl chloride and HCl
3   Reaction of the HCl from stage 2 with oxygen and further ethylene to give 1,2-dichloroethane by oxychlorination.

Each stage may involve reactors in which single units have outputs of well over 100,000 t/year and between them these stages cover a wide space of commercially important chlorination chemistry which is summarised below.

Ethylene Chlorination

Ethylene chlorination can be carried out by either radical or ionic mechanisms.[6] In the radical route chlorine forms a bridge complex with ethylene which results in homolytic splitting of the chlorine molecule with formation of a radical and a chlorine atom. The radical then attacks a chlorine molecule to initiate a conventional chain reaction which gives 1,2-dichloroethane and also unwanted by-products, as shown below:

$$\text{Cl-Cl} + \begin{array}{c} CH_2 \\ \| \\ CH_2 \end{array} \rightarrow \text{Cl} \begin{array}{c} CH_2 \\ \| \\ CH_2 \end{array} \rightarrow \dot{C}H_2CH_2Cl + \cdot Cl$$

$$\dot{C}H_2CH_2Cl + Cl_2 \rightarrow CH_2ClCH_2Cl + \cdot Cl$$

or $\cdot Cl + CH_2ClCH_2Cl \rightarrow CH_2Cl\dot{C}HCl + HCl$) by-product formation by
) overchlorination
$CH_2Cl\dot{C}HCl + Cl_2 \rightarrow CH_2ClCHCl_3 + \cdot Cl$)

Alternatively an ionic mechanism may be followed in which the chlorine is split heterolytically to give Cl^+ and Cl^-. The former adds to the ethylene to give the ion $\dot{C}H_2CH_2Cl$ which reacts with chloride to give 1,2-dichloroethane. This reaction does not continue beyond the dichloro product because the ionic addition only occurs across the double bond.

Ferric chloride is an excellent catalyst for ethylene chlorination because it catalyses the heterolytic splitting of chlorine through its activity as a Lewis acid, promoting the ionic mechanism shown below:

$$\text{FeCl}_3 + Cl_2 \rightarrow \left[\text{FeCl}_4\right]^- Cl^+ \rightarrow \left[\text{FeCl}_4\right]^- \overset{+}{C}H_2CH_2Cl \rightarrow CH_2ClCH_2Cl + \text{FeCl}_3$$

and it simultaneously destroys chlorine atoms by reaction to give chlorine and ferrous chloride, so inhibiting the radical chains which lead to cover-chlorination. In the presence of only 100 ppm ferric chloride almost the sole product of ethylene chlorination is 1,2-dichloroethane.

Dehydrochlorination of 1,2-Dichloroethane

The thermal dehydrochlorination of 1,2-dichloroethane is a conventional radical chain process which is carried out at about 450-500°C and which can be catalysed by chlorine or other sources of chlorine atom.[7]

$$CH_2ClCH_2Cl + \cdot Cl \rightarrow CH_2Cl\overset{\cdot}{C}HCl + HCl$$

$$CH_2Cl\overset{\cdot}{C}HCl \rightarrow CH_2{=}CHCl + \cdot Cl$$

It can be complicated by the various polymerisation reactions which are commonly encountered in processes of this type.

Oxychlorination of Ethylene

Reaction of the HCl produced by the above thermal dehydrochlorination is usually carried out by passing ethylene, HCl, and oxygen under pressure into a reactor operating at about 250°C, which contains a fluidised bed of copper chloride catalyst on an inert support, usually alumina. The ethylene is chlorinated to ethylene dichloride by the cupric chloride which is reduced to the cuprous state then the cupric salt is regenerated by reaction of the cuprous chloride with the HCl and oxygen giving a continuous process with a net make of dichloroethane and water.

Many attempts have been made to develop other metal oxychlorination catalysts but copper chlorides have retained their place as the most widely used primary catalyst although various advantages have been claimed for the addition of other metal chlorides to the copper. Kominami[8] has offered the simplest explanation of the high activity of copper which he relates to metal-chlorine bond energies and so to the rate at which various chlorides can carry out the two stages of the oxychlorination reaction.

1 $\quad CH_2{=}CH_2 + 2\ MCl_x \rightarrow CH_2ClCH_2Cl + 2\ MCl_{x-1}$

2 $\quad 2\ MCl_{x-1} + 2\ HCl + O \rightarrow 2\ MCl_x + H_2O.$

He showed that with increasing bond energy the rate of chlorination decreased whereas the rate of oxidation increased. The two rates intersect as shown below, Figure 3, with the continuous line being rate determining. Copper gives the same rate for each reaction and therefore has optimum performance.

FIGURE 3: Rate of chlorination and catalyst reoxidation

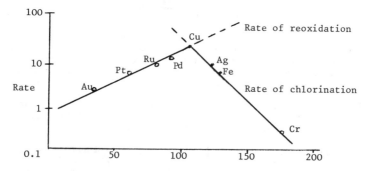

Bond energy as heat of formation of metal chloride (kJ/gm atom Cl)

A modern vinyl chloride plant

A modern vinyl chloride plant

On this perhaps over-simplified explanation there is no possibility of obtaining another single metal catalyst as effective as copper, but as Kominami
points out a mixture of metals of higher and lower bond energy could have a
better overall performance, and many mixtures of this kind have been patented.

Chlorinated Solvents

Other major products derived from carbon hydrogen and chlorine are the
chlorinated solvents, in particular trichloroethylene $CHCl=CCl_2$, perchloroethylene $CCl_2=CCl_2$, 1,1,1-trichloroethane CH_3CCl_3, and methylene chloride
CH_2Cl_2. The EEC capacity has been estimated as:

	tons per year
Trichloroethylene	} combined, > 300,000
Perchloroethylene	}
1,1,1-Trichloroethane	200,000
Methylene Chloride	250,000

In the UK ICI Mond Division is the sole manufacturer.

The major advantages of the chlorinated solvents are derived from their
chlorine content. They have very low flammability because on overheating they
decompose to give chlorine atoms and these disrupt the radical chain reactions
of combustion processes. Their solvent power particularly for oils and greases
comes from their low intermolecular attraction, which results in low surface
tension and good spreading, and gives them solubility parameters[9] which
are close to those of the greases which they dissolve so readily. They have
high molecular weights so they have high vapour densities with resultant
low losses by evaporation during use, but at the same time their low intermolecular attraction allows them to be evaporated at relatively low temperatures
($\sim 100°C$) so operations such as dry cleaning and recovery of spent solvent
for re-use are readily carried out.

Trichloroethylene and perchloroethylene can be made from ethylene by successive
chlorination and dehydrochlorination processes usually carried out in a single
reactor at about 350-500°C often over a catalyst[10] or an inert fluidised
support[11] to act as a heat transfer medium. If this type of process is used
the by-product HCl is recovered for processing elsewhere. On the other hand
oxychlorination processes have been developed in which ethylene[12] or
ethylene dichloride[13] can be fed to fluidised bed reactors usually containing copper catalysts in which direct reaction to mixtures of trichloroethylene and perchloroethylene occurs.

$$4 \ CH_2ClCH_2Cl + 3 \ Cl_2 + 1\tfrac{1}{2} \ O_2 \rightarrow 2 \ CHCl=CCl_2 + 2 \ CCl_2=CCl_2 + 3 \ H_2O$$

As a rule the reactor temperatures are higher than those used for 1,2-dichloro-
ethane manufacture in order to minimise the formation of saturated chloro-
alkane by-products. As with vinyl chloride the processes have the advantage
of being self-contained.

Routes to 1,1,1-trichloroethane are more complex because of the difficulty of
locating three chlorine atoms at one end of the molecule. Ethylene or ethane
may be used as feedstock and the following figure shows the choice of routes
that can be used by building up permutations of the three steps, chlorination,
hydrochlorination and dehydrochlorination.

FIGURE 4:   Routes to 111-Trichloroethane

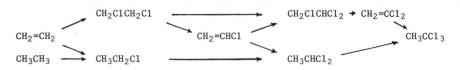

Hydrochlorination is usually effected by catalysis with Lewis acids such as ferric chloride and dehydrochlorination by thermal cracking or by the action of base.   The chlorination stage can either take place in the liquid phase where solvation[14] and cage effects can affect the product distribution or in the gas phase[15].   In either phase reaction can be initiated by radical sources.

Industrially a careful balance has to be struck between the economics of the various choices.

Manufacture of methylene chloride (dichloromethane) and the other chlorinated methanes presents fewer problems of chemistry since no isomers exist, although reactor design can give complex engineering problems.   Methylene chloride can be made either by the direct reaction of methane or methyl chloride with chlorine[16] or by the oxychlorination of methane[17] where loss of methane by oxidation can complicate the process.

Propylene is the other low molecular weight hydrocarbon extensively used in chlorination processes, especially for propylene oxide.   In a typical process[18] chlorine and 90% propylene are reacted in dilute aqueous solution to give the chlorhydrin together with some propylene dichloride and by-products.   The reaction mixture is then contacted with limeslurry to give propylene oxide which is separated by fractional distillation from the by-products.

Reaction of Chlorine with Oxygen

The outer octet of chlorine is available for bonding and the best known compounds in which more than one bond is formed are the chlorine oxygen derivatives.   Up to four oxygens can be reacted with the chlorine atom and as some of the bonds are necessarily weak the best known characteristic of the group of compounds is ready release of oxygen, which in commercial terms leads to bleaching and the possibility of destroying a wide variety of unwanted impurities and contaminants.

All compounds containing only chlorine and oxygen atoms are relatively unstable and some can be explosive.   They form two groups in which the central atom may be either oxygen or chlorine and the simplest of the first group is chlorine monoxide Cl_2O, a symmetrical molecule $Cl-O-Cl$, formed by reacting mercuric oxide with chlorine.   Chlorine heptoxide $O_3Cl-O-ClO_3$ which is probably the most stable of all the chlorine oxides, is the other well known member of the group and is formed by reacting perchloric acid with phosphorus penta-chloride.   The other group, which is more important commercially has a central chlorine surrounded by oxygen and the simplest member is chlorine dioxide $O \leftarrow Cl \rightarrow O$, which is a highly reactive yellow gas.   It is an unusual molecule in

that it has a surplus electron, in addition to those shown in the above
structure, which might be expected to be located on the chlorine to give a
radical which would readily combine with another to give the dimeric form

$$2 \cdot ClO_2 \rightarrow O_2Cl-ClO_2$$

However its tendency to dimerise is small, unlike chlorine hexoxide which is
almost wholly in the dimeric form $O_3Cl-ClO_3$. Probably the loss of resonance
arising from the reaction of three electron pairs on each chlorine locates
the electron on the Cl and leads to chlorine-chlorine bonding. The hexoxide
can be made by ozone treatment of chlorine dioxide.

Chlorine dioxide is widely used for bleaching and water treatment so several
methods for its manufacture have been developed, which all involve the acidic
reduction of chlorate[19]

$$HClO_3 + HCl \rightarrow HClO_2 + HOCl$$

$$HClO_3 + HClO_2 \rightarrow 2 ClO_2 + H_2O$$

complications follow because the HClO may then react with HCl to produce
chlorine

$$HOCl + HCl \rightarrow Cl_2 + H_2O$$

giving a mixture of ClO_2 and Cl_2. A reducing agent is therefore added to
convert HOCl to HCl for further reaction with the chlorate, although
reduction of $HClO_3$ to $HClO_2$ can also occur and a careful balance is required.

The oxo acids referred to in the last paragraph are an important group consisting of

HOCl	hypochlorous acid
$HClO_2$	chlorous acid
$HClO_3$	chloric acid
$HClO_4$	perchloric acid

Hypochlorous acid is readily formed when chlorine is passed into water

$$Cl_2 + H_2O \rightleftharpoons HOCl + H^- + Cl^-$$

and in dilute solution and above pH4 conversion to HOCl is almost complete.
Dissociation of the HOCl is limited below pH7, and extensive above pH8, and
stable alkali metal salts are formed between the metal ion and OCl^-. Calcium
hypochlorite was formerly made on a very large scale as 'bleaching powder'
by the direct chlorination of lime but is now little used, having been
extensively replaced by the 'oxygen' bleaches based on peroxides, but sodium
hypochlorite is a popular cheap bleach for domestic use and for outlets such
as swimming pools sterilisation where chlorine handling would be impracticable
for the average user. It is made on a large scale by direct reaction of
chlorine with caustic soda, although when small quantities are needed it may
be made more conveniently by direct electrolysis of sodium chloride solution
using cells with both electrodes in the same compartment.

Chlorous acid is little known although chlorites are made by the reaction of
chlorine dioxide with bases, and the

$$2 ClO_2 + 2 OH^- \rightarrow ClO_2^- + ClO_3^- + H_2O$$

salts have some use as bleaches. Chloric acid is also unstable as the free
acid although the salts are well known, and they are all powerful oxidants.
Domestically this property has led to the widespread use of sodium chlorate as

a weedkiller, but chlorates have outlets in a variety of specialist uses such
as match heads and fireworks. They are usually made by electrolysis of a
mixture of the appropriate chloride and chlorate in a diaphragmless cell
at 35-45°C where a number of oxidation reactions occur simultaneously but the
chlorate ion is thermodynamically the most stable so disproportionation of
the partially oxidised intermediates occurs, giving overall

$$NaCl + 3 H_2O \rightarrow NaClO_3 + 3 H_2$$

Perchlorates are also made by electrochemical oxidation of chlorates. In a
typical process[20] 60% w/v sodium chlorate solution is fed to a cell fitted
with platinum anode and mild steel cathode at 40-45°C and an anode current
density of ∿ 31 amps/sq dm. Other salts can then be made by reaction of the
appropriate chloride with the sodium chlorate. The perchlorates are
reasonably stable and they are quite widely used as explosives and solid fuel
propellants.

The largest outlet of the chlorine oxides is undoubtedly in bleaching and
water treatment where they are used in conjunction with chlorine itself. In
the UK about 2-2.5% of chlorine manufacture is used in water treatment,
primarily for potable water and industrial cooling water together with a little
for swimming baths. Potable water sources are becoming poorer as the water
demand rises and the normal practice is to add about 10 ppm chlorine to
domestic drinking water then to dechlorinate with sulphur dioxide or sulphite
leaving about 0.25-0.5 ppm in drinking water to keep it sterile. The chlorine
forms hypochlorous acid which will oxidise sulphur compounds to sulphate and
ammonia to chloramines, which are themselves quite bactericidal. Bacteria and
other organisms are rapidly killed in chlorinated water, its toxicity being
proportional to the concentration of undissociated HOCl which is present, for
it is able to penetrate the bacterial cell wall. Nitrites and miscellaneous
organic impurities can also be destroyed by oxidation and chlorine, although
if phenolic compounds are present chlorine dioxide is more effective than
chlorine in avoiding the residual taste of chlorophenols.

In industrial cooling water one of the main applications is for biological control, for example coastal power stations use sea water and condenser blockage
by mussels growing in the circulating water can be severe. About 0.25 ppm
chlorine added to the water can overcome the problem completely. Many
applications of this type exist, for example the leakage of juice from sugar
beet when these are floated in water for quick transport, can promote rapid
growth of microorganisms and these are similarly controlled by traces of
chlorine. Flocculation of suspended colloids by destruction of their stabilising charges is another important result of chlorine treatment.

The whole subject of water treatment is of great social importance but its
consumption of chlorine is relatively small and to avoid repetition of the
earlier full discussion, it will not be considered further in this review.

There is another important industrial product which is in a different category
from the products containing chlorine and oxygen discussed above, but which
for simplicity can be included in this part of the review. It is phosgene,
$COCl_2$ which is usually made by the direct reaction of carbon monoxide and
chlorine over charcoal but which is also a common by-product of many chlorinated
hydrocarbon oxidation processes. Phosgene is a valuable intermediate which
can be converted to a range of materials such as chloroformates and ureas
which are widely used in pest control and weedkiller applications, but its
largest use is undoubtedly for the manufacture of polyurethanes which are now
widely used for foam plastics and paints, and its manufacture represents an
appreciable part of the unspecified outlets for chlorine listed under the
heading 'Other' in Table 1.

Apart from oxygen, the principal industrially important elements in which a
chlorination stage is used as part of their process technology are phosphorus,
sulphur, and nitrogen.

The principal phosphorus chlorides are phosphorus trichloride, PCl_3, phosphorus
oxychloride $POCl_3$, and phosphorus pentachloride PCl_5, although phosphorus
chlorides containing sulphur instead of oxygen are also known. Phosphorus
trichloride is usually made by direct reaction of chlorine and phosphorus in
a continuous reactor charged with an initial batch of trichloride, and the
pentachloride by further reaction of trichloride dissolved in an inert solvent
such as carbon tetrachloride. Lead-lined steel equipment is frequently used
for reactor construction. The products are principally converted to esters and
used for low flammability plasticisers, oil additives and stabilisers, and in
recent years salts of alkyl phosphates have become of increasing importance as
detergents and wetting agents.

Chlorine will also react readily with sulphur to give various chlorides in which
the sulphur tends to remain associated, but there are no chlorine sulphur
equivalents of the $Cl \rightarrow O$ bond. The most common reaction product of sulphur and
chlorine is sulphur dichloride $Cl-S-S-Cl$ formed by chlorination of molten
sulphur and excess chlorine at room temperature but at $-80°$ SCl_4 can be obtained
as yellow crystals. By treating either of these with excess sulphur long
polysulphide chains $Cl-S-(S)\underline{n}-S-Cl$ can be obtained which contain up to 100
sulphur atoms. More important industrially are thionyl chloride $SOCl_2$ and
sulphuryl chloride SO_2Cl_2 which are respectively used in the fine chemical
industry for converting $-OH$ groups to $-Cl$ and for more selective chlorination
than can be obtained with chlorine alone.[21]

Nitrogen-chlorine compounds are usually called 'chloramines' industrially. The
term includes all compounds which contain one or more N-Cl bonds, especially
chloramines $RNHCl$, chloroamides $RCONHCl$, and N-chlorosulphonamides RSO_2NHCl.
They have applications as bleaches principally because the N-Cl bond is hydrolysed in water to give $HOCl$, which is responsible for their bleaching and
disinfecting properties. They are generally regarded as safe bleaches
usually of rather low activity, of which the chlorinated cyanuric acids are
probably the most widely used. Another industrially important N-Cl compound
is nitrogen trichloride NCl_3, not because of its commercial applications for
it is very unstable and explodes readily, but because of its effects on the
reactivity of chlorine. It is found in commercial chlorine because salt beds
normally contain traces of nitrogenous compounds formed from residues of
organic matter and these are leached into the brine used for the preparation
of chlorine by electrolysis, and in the anode compartment of the cell they
are broken down to nitrogen trichloride which is sufficiently volatile to
leave the cell with the chlorine. If the chlorine is used for radically
induced chlorination at temperatures below $100°C$ the trichloride acts as an
effective trap for chlorine radicals,

$$2\ NCl_3 + 2\ \cdot Cl\ \rightarrow\ N_2 + 4\ Cl_2$$

reducing the reaction rate by breaking the chlorination chain and requiring
the use of a higher level of initiating catalyst. On the other hand at higher
temperatures, $\sim 200°C$ upwards, nitrogen trichloride is itself unstable and
decomposes to give additional chlorine radicals and acceleration of the rate
of chlorination occurs. As the level of the trichloride in chlorine may vary,
to obtain optimum control of the rate of chlorination it is often advisable to
remove it by one of the established catalytic methods such as passage of the
chlorine over charcoal.

Another important inorganic with which chlorine usually reacts readily,
which is in a different category from the above group is bromide. Bromides

frequently occur with chlorides in salt deposits and bromine is preferentially released during brine electrolysis and becomes concentrated in the chlorine. Essentially the same reaction occurs in the preparation of bromine by the chlorination of sea water, a process used by AOC at Amlwch in North Wales, where chlorine is passed into sea water then the bromine is recovered by air blowing. The process is possible because the reduction potential of bromine is lower than that of chlorine.

	Reduction potential/ volts
$Cl_2 + 2 e \rightarrow 2 Cl^-$	+ 1.358
$Br_2 + 2 e \rightarrow 2 Br^-$	+ 1.065

so that the equilibrium

$$Cl_2 + 2 Br^- \rightarrow Br_2 + 2 Cl^-$$

lies heavily to the right hand side. Although bromine becomes concentrated in chlorine during electrolysis the amount of bromine normally present in commercial chlorine is very small and it exists almost entirely as Br-Cl, reacting with hydrocarbons as if it were chlorine but contributing to some instability in chlorinated organics if it is not removed, because of the lower strength of the carbon-bromine bond.

	kJ per gm bond
C-Cl	323
C-Br	269

Reaction of Chlorine with Metals and Metal Salts

Since chlorine is strongly electrophilic it will readily oxidise metals and ions to their highest valency state, and this gives rise to the chlorine handling problems reviewed in an earlier section. There is also a highly developed technology associated with metal chlorides and depending on their varying properties. Most metal chlorides are regarded as mainly ionic but in fact they show a gradation from purely ionic structures such as sodium chloride which has discrete ions in the lattice, through partially covalent systems to lattices consisting of almost completely covalent molecules. Covalency increases with increase in the charge/radius ratio of the ion so in metals with a variable oxidation state the higher state is more covalent. The extent of covalency has an important effect on the reactions which can be carried out, thus the removal of magnesium from alloys during the recovery of aluminium from scrap can be achieved by oxidation of the magnesium to the chloride with chlorine, while the aluminium is unaffected. Reaction occurs because of the difference between the oxidation potentials of the metals, the aluminium chloride being more covalent.

	Oxidation potential/volts
$Mg - 2e \rightarrow Mg^{2+}$	2.375
$Al - 3e \rightarrow Al^{3+}$	1.706

In covalent chlorides the lattice consists of separate molecules held together by weak van der Waals forces and the lattice energy is low, so covalent chlorides are generally volatile compared with ionic chlorides and they frequently form chloride bridges between adjacent metal ions,

```
    Cl                          Cl
   /                           /  \
  M                      M         M
           M      →       \      /
        /                   Cl
      Cl
```

the metal-chlorine bonds being identical in the above structure.

Industrially important metals falling into this category are broadly those of
the B sub group of Groups IV, V and VI of the Periodic Table especially
titanium, zirconium, vanadium, chromium, molybdenum, tungsten, aluminium,
silicon, and antimony, with aluminium chloride probably still the most important halide although others, especially titanium tetrachloride, have become
important intermediates and catalysts. Usually the metal chlorides are made
by chlorinating the metal or its oxide possibly after addition of carbon to
react with the oxygen, and some years ago the technology was reviewed by
Piester and Muren[22] and Skrzec and Lowry[23]. Specialised equipment during
the manufacture of these chlorides is usually required because of the severely
corrosive nature of the chlorides and their tendency to produce blockages
by sublimation.

Some higher valency metal chlorides decompose fairly readily on heating to give
the lower valency state of the metal and free chlorine, and this is the basis
of the well known but little used Deacon process for the recovery of chlorine
from by-product HCl. The HCl is mixed with oxygen and passed over a copper
chloride catalyst usually at a temperature of 400°C or higher. Partial
dissociation of copper(II) chloride occurs with release of chlorine and
formation of copper(I),then reoxidation of the copper(I) takes place probably
via the copper(II) oxychloride CuO, $CuCl_2$ which then reacts with the HCl
regenerating copper(II) and forming water. Volatilisation of the copper
chloride is reduced by addition of potassium chloride which form complexes
such as $KCuCl_3$ and K_2CuCl_4. The chemistry of the process has been studied in
depth[24] and many variations have been developed, but since the introduction of
oxychlorination technology it has been given relatively little attention.

Future Prospects

Chlorinated products have now reached a highly developed state both in their
process technology and their market position and it appears unlikely that their
market share will be seriously challenged in the near future. The efforts that
have been put into resolving the environmental problems of vinyl chloride have
been largely successful and it has been predicted[1] that the annual growth rate
of this, the largest outlet for chlorine, will remain at 6% + 3%. Sales of
chlorinated solvents however are expected to grow less rapidly because of
present trends towards lowering the levels of discharge to the atmosphere and
a growth of 4% + 2% has been suggested with 1,1,1-trichloroethane taking a larger
share of the solvents market[1]. The diversified uses of chlorine in the
inorganic market such as inorganic products, pulp and paper, water treatment,
hydrochloric acid, and agricultural chemicals are however likely to show very
limited growth, and the overall EEC growth rate of $\sim$ 4%/year for chlorine
mentioned in the earlier section therefore appears to be a reasonable prospect.

Most of the advantages of scale that were sought in the last two decades by
increasing the size of production units have now been realised and the major
efforts in chlorination technology during the next decade are likely to be in
seeking improvements in chemical efficiency and above all in seeking to reduce
the ever-increasing cost of the energy requirements of modern chlorination
processes.

References

1   J.G. Helfenstein, 'Chloro-alkali Products in Europe. Current Situation and a View of the next Decade'. Talk to EMCRA Conference Madrid 30 Oct 1976.

2   'Properties and Handling of Chlorine', ICI, Mond Division.

3   B. Strasser Chem Eng World 1970, 5, 21.

4   F.M. Berkey 'Chlorine. Its Manufacture, Properties and Uses', (Reinhold, New York 1962), Chapter 7, p 204.

5   C.P. van Dijk and W.C. Schreiber Chem Eng Progress 1973, 69 (4), 57.

6   M.L. Poutsma, Science, 1967, 157, 997.

7   D.H.R. Barton and K.E. Howlett. J Chem Soc 1949, 148, 155.

8   N. Kominami. J of Japan Petroleum Inst 9, 100

9   J.H. Hildebrand and R.L. Scott 'Solubility of Nonelectrolytes' (Reinhold, New York 1950) Chapter 23, p 424.

10   BP 673,565 'Chlorinated Ethylene Derivatives', Diamond Alkali Co 11 June 1952.

11   BP 1,275,700 'Manufacture of Trichloroethylene and Perchloroethylene Diamond Shamrock.

12   BP 1,370,344 'Production of Chlorinated Hydrocarbons' J.S. Sproul and R.J. Gleason to FMC Corporation 24 Sept 1971.

13   BP 1,144,842 'Production of Trichloroethylene and Perchloroethylene' PPG Industries 22 December 1966.

14   E.S. Huyser 'The Chemistry of the Carbon-Halogen Bond' (Wiley and Sons, London 1973) Chapter 8 p 569.

15   BP 1,281,541 and 1,286,807 'Manufacture of 1,1,1-Trichloroethane A. Campbell and R.A. Carruthers to ICI Limited, 10 Oct 1968 and 23 Jan 1969.

16   R. Landau and S.N. Fox. Ref 4 Chapter 12, p 334.

17   BP 894,137 'Process for the Production of Chlorinated Hydrocarbons' E. Makris and J.E. Millam to Columbia Southern 5 June 1959.

18   Anon. Petrochemical Industry 22 Sept 1959.

19   W.H. Sheltmire, Ref 4 Chapter 17, p 537

20   M.P. Grotheer and E.H. Cook, Electrochem Technol 1968, 6(5-6), 221.

21   G.A. Russell, J Amer Chem Soc, 1958, 60, 5002

22   W. Piester and A.P. Muren Ref 4, Chapter 25, p 761.

23   A.E. Skrzec and L.B. Lowry Ref 4, Chapter 27, p 805.

24   D.M. Ruthven and C.N. Kenney J Inorg Nucl Chem 1968, 30, 931.

Hydrofluoric Acid, Inorganic Fluorides and Fluorine

By H.C. FIELDING and B.E. LEE

ICI Ltd., Mond Division

Introduction

What links the steel industry with atomic energy, the aluminium industry with aerosols, refrigeration with anaesthetics, non-stick frying pans with fire-fighting agents, the petroleum industry with glass-etching? The answer is they all depend on the inorganic fluorine industry for important raw materials.

In this paper, we shall review those branches of the inorganic chemical industry which supply these raw materials, namely calcium fluoride, hydrofluoric acid, inorganic fluorides and finally fluorine itself, a rather specialised branch of the industry.

General History

As early as the 16th century, it was known that fluorspar would lower the melting point of minerals and as long as 100 years ago this property was being used in the open-hearth process for the manufacture of steel. Crude hydrofluoric acid was probably first prepared by Scheele in 1771, although the property of etching glass with the acid gases from fluorspar was recognised long before. Isolation of substantially anhydrous hydrofluoric by Frémy in 1856 was followed by the preparation of fluorine by Henry Moissan in 1886.

For many years, only fluorspar found wide use, mainly as a fluxing agent in metallurgical processes. It was only in the 1930s, with the production of anhydrous hydrofluoric acid in commercial quantities that the demand for fluorspar expanded rapidly, until today around half the fluorspar mined is used in the production of anhydrous acid.

Occurrence of Fluorine

Although it has been established that fluorine is 19th in the list of elements arranged in order of relative abundance, only three minerals are of any commercial importance (Table 1).

Table 1. Principal fluorine-containing minerals

Mineral	Composition	% fluorine
Cryolite	$3NaF.AlF_3$	54
Fluorspar	CaF_2	49
Fluorapatite	$CaF_2.3Ca_3(PO_4)_2$	4

Cryolite. The only commercial deposit of cryolite is in Greenland, but very little is mined. However, large tonnages of synthetic cryolite are manufactured to supply the aluminium industry and will be considered later.

Fluorapatite. Although this mineral only contains 3-4% of fluorine, the vast tonnages processed as 'phosphate rock' by the fertiliser industry suggest an enormous potential source of fluorine. Interest in utilising this fluorine is growing and will be discussed when we consider the manufacture of hydrofluoric acid.

Fluorspar. Fluorspar is a common mineral widely distributed in nature, with workable deposits found all over the world. As can be seen from Table 2, Mexico and Western Europe are the largest producers. The United States is the largest consumer, using around 1.1 million tonnes in 1975, 80% of which was imported from Mexico.

Table 2. Fluorspar : World Production by Countries, 1975[1]

Country	millions tonnes
Mexico	1.10
Western Europe	1.00
Eastern Bloc	0.90
Far East	0.65
Africa	0.32
South America	0.13
United States	0.12
Canada	0.10
	4.32

The United Kingdom ranks about eighth amongst fluorspar-producing countries, production in 1975 being around 200,000 tonnes[2]. The deposits are located in Derbyshire and there are three major producing companies, Laporte Industries, Weardale Lead (a subsidiary of Imperial Chemical Industries) and the British Steel Corporation. All three companies consume most of their own production in fluorine chemical manufacture or in steel making.

Fluorspar is a translucent or transparent glossy mineral, varying in colour from white, amber, green and blue to purple. The fluorspar is invariably associated with other minerals such as quartzite, calcite, dolomite, barite and galena, but mostly with quartzite. For most purposes mining of the mineral is followed by beneficiation, the degree of concentration depending on the nature of the product to be marketed. The major uses for fluorspar are

(a) as a fluxing agent in the metallurgical and related industries,

(b) in the glass and ceramics industry, and

(c) in the chemical industry for manufacture of hydrofluoric acid.

Three grades of fluorspar are marketed according to these end uses. The cheapest grade, selling at around £20-£30/tonne, is the metallurgical grade containing from 60 to 80% calcium fluoride and concentrated by such rudimentary techniques as gravity separation. Next, at around £30-£40/tonne, is the ceramic grade containing 85 to 95% calcium fluoride and concentrated by flotation. Finally, there is acid grade, selling at around £40-£50/tonne and containing a minimum of 97% calcium fluoride[3]; this grade is also concentrated by a flotation technique.

The increasing importance of fluorspar as a basic commodity for modern industry can be appreciated from an analysis of the world production over the last 20 years (Table 3).

Table 3. Fluorspar : Growth of World Production[1,4]

Year	1954-58 (average)	1960	1964	1968	1972	1974	1975
Production (tonnes x 10^6)	1.63	2.0	2.1	3.14	4.56	4.54	4.28

As can be seen from the table, production has levelled out in recent years. However, demand for fluorspar is expected to recover gradually as the effects of the widespread economic recession recede, although growth will probably be slower than was predicted at the beginning of the decade.

Reserves. Exploration efforts over the last few years have resulted in a welcome expansion of world fluorspar reserves to about 300 million tonnes, more than double the known reserves of 1970 (Figure 1).

Assuming an average 30% CaF_2 content, these reserves should last for at least 20 years at present consumption although environmental considerations, together with continuing exploration for new deposits, almost certainly make this a rather conservative estimate.

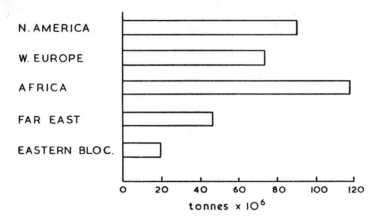

Figure 1   World Proven Fluorspar Reserves

A further factor in the estimated life of these reserves is the
possible exploitation of the fluorine content of fluorapatite, the
phosphate rock which supplies the fertiliser industry. However, there
are still economic problems associated with utilising this potentially
massive source of fluorine, which we will discuss when we consider the
manufacture of hydrofluoric acid.

Uses for Fluorspar

Fluorine is the element of value in fluorspar, whether it is
used as calcium fluoride or is extracted as hydrofluoric acid. World-
wide, the production of hydrofluoric acid has recently become the
largest consumer of fluorspar, although use patterns vary somewhat
from country to country.

Just under half of the world's fluorspar output is consumed as a
metallurgical fluxing agent, primarily in the manufacture of steel.
Metallurgical grade spar is used to remove impurities during melting
and also to improve separation of metal and slag in the furnace by
increasing the fluidity of the slag, resulting in better steel
recovery and lower fuel costs. Consumption of fluorspar in the steel
industry has increased substantially in recent years because of
increased steel output and changing technology. Steel makers have
shifted increasingly from the basic open-hearth process to the basic
oxygen process, and as can be seen from Table 4, the oxygen process
uses three times as much fluorspar for each tonne of steel produced
compared with the open-hearth process, with consumption in the electric-
arc furnace lying in between.

Table 4. Fluorspar : Requirements for Steel Manufacture[5]

Steel Process	kg/t
Basic open-hearth	1.80
Electric arc	4.05
Basic oxygen	5.40

Environmental considerations will be important in assessing the future consumption of fluorspar in the steel industry, with pressures increasing to limit the amount of fluoride emitted to atmosphere.

Ceramic grade fluorspar is used as an opacifier in enamels and opal glass and also as a flux in clear glass manufacture. However, these outlets together with several other minor uses, for example in cement manufacture, account in total for less than 5% of world fluorspar production.

Hydrofluoric Acid

Just over half of the world fluorspar requirement is for the acid grade and is used in the manufacture of hydrofluoric acid, the most important manufactured compound of fluorine.

Anhydrous hydrofluoric acid is a colourless, fuming liquid with a boiling point of 19.4°C. That the boiling point is anomalous and unexpectedly high can be seen from the following table:-

Compound	HF	HCl	HBr	HI
B.p. °C	19.4	-83.7	-67	-36.4

The abnormally high boiling point is a result of hydrogen bonding between the molecules and both the liquid and the gas show large departures from ideal behaviour.

Years of experience in the manufacture and use of hydrofluoric acid have shown that the compound can be handled safely, provided its hazards are recognised and the necessary precautions are taken. The anhydrous acid, together with the higher concentrations of aqueous acid, are extremely corrosive to skin, eyes, mucous membranes and lungs. Sometimes, burns of the skin may not be immediately evident and may be manifested later by deep-seated ulceration, very slow to heal.

The classical, quite limited uses of the acid, such as glass etching and foundry scale removal, predominated until the 1930s. The

enormous increase in demand during the 1930s came with requirements
to manufacture fluorides for the rapidly expanding aluminium industry,
followed soon after by the development of chlorofluorocarbons as
refrigerants, then as aerosol propellants. The growth of all these
industries, together with other uses to be discussed later, brought the
volume of hydrofluoric acid production up to the current high level.

Manufacture of Hydrofluoric Acid

From Fluorspar. Hydrofluoric acid is produced by heating together a
mixture of acid grade fluorspar (above 98% CaF_2) and concentrated
sulphuric acid (above 98%). As mentioned previously, the fluorspar is
benificiated by flotation to achieve the high calcium fluoride content
required (in general, two to three tonnes of ore must be mined to
produce one tonne of acid grade spar).

$$CaF_2 + H_2SO_4 \rightarrow 2HF + CaSO_4$$

As the reaction is endothermic heat must be supplied to complete
the reaction in a reasonable time. Inevitable impurities in the spar are
calcium carbonate, silica and sulphides and these undergo reactions
along with the calcium fluoride, as follows:-

$$CaCO_3 + H_2SO_4 \rightarrow CaSO_4 + H_2O + CO_2$$

$$SiO_2 + 4HF \rightarrow SiF_4 + 2H_2O$$

At a later stage in the process, at a lower temperature, the
silicon tetrafluoride reacts with aqueous hydrofluoric acid to form
fluorosilicic acid.

$$SiF_4 + 2HF \rightarrow H_2SiF_6$$

The silica in particular is highly undesirable as each mole
consumes 6 mol of hydrofluoric acid, giving production losses
equivalent to 2 tonnes HF per tonne silica in the fluorspar. The calcium
carbonate in the fluorspar increases sulphuric acid usage and also
produces carbon dioxide, a non-condensable gas which has to be removed
from the product. Sulphur-containing impurities in the fluorspar can
generate hydrogen sulphide and sulphur dioxide, which in turn can react
together to form sulphur, leading to deposits in the gas-handling
equipment. For these reasons, it is desirable to use as high a purity of
acid-grade fluorspar as possible.

Plant and Process. In modern practice, anhydrous hydrofluoric acid is
generated continuously in either a horizontal, externally fired rotary

kiln or a horizontal, externally heated conveyor–mixer. The reaction time is normally 30-60 minutes at 200-250°C, with the hydrofluoric acid leaving the generator at 100-150°C. Obviously, the lowest operating temperature at which good yields are obtained is desirable, to minimise corrosion problems.

The most widely used technology is that developed by Buss AG of Basle, Switzerland[6], a leading builder of HF units. For a single unit producing up to 20,000 tonnes per annum, the steel kiln could be around 30 metres long, the whole rotating at around one revolution per minute.

A typical raw material requirement per tonne of hydrofluoric acid is fluorspar 2.2 tonnes, sulphuric acid 2.6 tonnes. The spar and acid are fed continuously into a pre-reactor, from which the reactants pass as a paste to the heated rotary kiln main reactor. Gases evolved in the kiln leave through the feed end into the disengaging section, while calcium sulphate solids are discharged at the downstream end.

The remainder of the plant consists of facilities for drying and pulverising the feed fluorspar, neutralising and handling of the calcium sulphate, scrubbing, cooling, condensing and purification of the HF product and elimination of fluorine compounds from the vent gases. The whole process is shown schematically in Figure 2.

The hot reactor gases enter a column (the precondenser) where they are cooled and scrubbed free of entrained solids and sulphuric acid mist using incoming sulphuric acid. This acid, after the addition of oleum to combine with any water, then becomes the feed to the pre-reactor. The HF vapour leaving the pre-condenser is condensed from the gas stream using refrigeration, to give a 99% product which is distilled to give the final product of 99.9% purity. The gaseous effluent is now largely SiF_4, which is absorbed into water and recovered as 30% fluorosilicic acid. The calcium sulphate residue from the kiln is usually discarded after neutralisation with lime, although some is sold for manufacture of anhydrite cement.

From Fluorapatite. During the processing of phosphate rock for the production of fertilisers, the fluorine content of the fluorapatite is liberated as SiF_4. Briefly the wet acid process entails the decomposition of ore by sulphuric acid.

$$CaF_2, 3Ca_3(PO_4) + 10H_2SO_4 \rightarrow 10CaSO_4 + 6H_3PO_4 + 2HF$$

Silica impurities in the rock react with the HF to form volatile silicon tetrafluoride:

$$SiO_2 + 4HF \rightarrow SiF_4 + 2H_2O$$

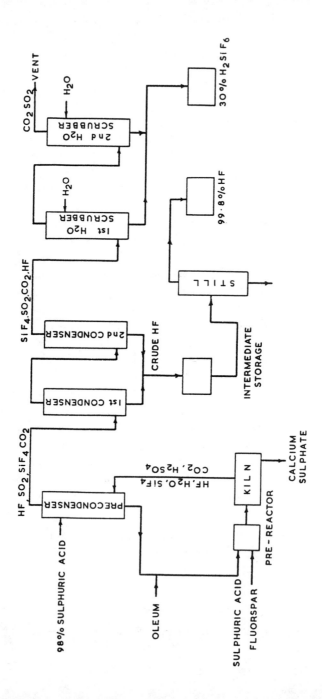

Figure 2   Flow Diagram for Manufacture of Hydrofluoric Acid

The silicon tetrafluoride has to be removed from the waste gases to avoid pollution and is recovered as aqueous fluorosilicic acid by reaction with aqueous HF.

$$SiF_4 + 2HF \rightarrow H_2SiF_6$$

It has been estimated that around half a million tonnes of fluorosilicic acid is discarded by the US phosphoric acid industry each year, equivalent to about 1 million tonnes of fluorspar, enough to supply the nation's entire hydrofluoric acid requirements.

A number of processes have been developed in recent years in an attempt to use this potential source as a raw material for production of hydrofluoric acid or calcium fluoride. For example, the US Bureau of Mines[7] has a process to CaF_2 based on the following sequence

$$H_2SiF_6 + 6NH_3 \rightarrow 6NH_4F + SiO_2$$

$$6NH_4F + 3Ca(OH)_2 + 3H_2O \rightarrow 3CaF_2 + 9H_2O + 6NH_3$$

The H_2SiF_6 feed solution is neutralised with NH_3 to form NH_4F solution and SiO_2 is precipitated. The reactor slurry is filtered and the SiO_2 filter cake discarded. The filtrate is treated with hot $Ca(OH)_2$ to form a slurry of CaF_2, the NH_3 liberated being recycled. The CaF_2 is filtered and dried as 'acid-grade' fluorspar.

Other processes, for example those developed by Davy Powergas[8], RHT[9] and Buss[10], all depend on the decomposition of H_2SiF_6 to give HF and SiO_2.

$$3H_2SiF_6 \rightarrow 6HF + 3SiF_4$$

The HF is either condensed out of the gas stream or is selectively absorbed into a solvent, the SiF_4 being recycled to produce more H_2SiF_6 feed by reaction with water. The silica produced at this stage is filtered out of the solution and is discarded.

$$3SiF_4 + 2H_2O \rightarrow 2H_2SiF_6 + SiO_2$$

As fluorspar prices increase and environmental pressures demand the recovery of H_2SiF_6 from the fertiliser plants, processes of this type will become increasingly important, although very little HF or CaF_2 is recovered by these processes at present.

In addition to its potential use as a source of HF, there are several other outlets for H_2SiF_6. One such outlet is its use in water fluoridation, another is the use of its salts in the ceramics industry. A third outlet of growing importance is in the aluminium industry for the manufacture of aluminium fluoride and synthetic cryolite.

Structure of the Industry

While aqueous HF has been used commercially on a small scale for many years, the anhydrous material has been available in the United States only since 1931 and in the UK from about 1942. Virtually all modern developments in fluorine chemistry are based on anhydrous HF. Its first commercial production was for use in making fluorocarbon refrigerants, and since then, expansion in demand for fluorocarbons and the requirements of the aluminium industry have been largely responsible for the rapid increase in production of the anhydrous acid.

From statistics on fluorspar consumption[3], world production of HF in 1975 is estimated to have been around 1 million tonnes, with the United States accounting for about 30% of production at around 300,000 tonnes.

Hydrofluoric acid is produced in most advanced countries of the world and Table 5 shows some of the important chemical companies involved, although the list is far from complete[11]. The reported capacities are probably already outdated, but they do give an indication as to the size of the industry. The table does not include the aluminium companies, a number of which produce HF for their own internal use.

Table 5. Major Chemical Companies Producing Hydrofluoric Acid

Company	Reported Capacity (tons per year)	Company	Reported Capacity (tons per year)
United States		West Germany	
Allied Chemicals	100,000	Bayer	30,000
Du Pont	90,000	France	
Penwalt	23,000	Ugine Kuhlmann	30,000
Stauffer	16,000	Italy	
United Kingdom		Montedison	50,000
ICI	40,000	Japan	
Imperial Smelting	20,000	Asahi Glass	10,000
Laporte Industries	10,000	Hashimato Chemicals	25,000
		Daikin Kogyo	6,000

At the present time (1976), list prices for anhydrous HF are around £400-£500/tonne for tank-car quantities, although a high proportion of HF produced (around 80%) is used captively.

Rotary kilns on an anhydrous hydrofluoric acid plant

Crystalline fluorspar

Large scale industrial fluorine cells in the uranium hexafluoride plant at the Springfield Works of British Nuclear Fuels Limited (Photo courtesy of BNF Ltd)

Uses for Hydrofluoric Acid

Because of the high proportion of HF used captively, use patterns are hard to fix precisely, and they also vary from country to country. However, from Table 6, which shows an estimated end-use pattern, it can be seen that about 80% of total production is split roughly between fluorinated organics and aluminium manufacture with the remaining 20% being taken by a wide variety of uses.

Table 6. Estimated Utilisation of Hydrofluoric Acid %

Fluorocarbon manufacture	40%
Aluminium manufacture	40%
Petroleum alkylation	4%
Fluoride salts	4%
Stainless steel pickling	3%
Uranium processing	2%
Miscellaneous	7%
	100%

Fluorocarbon Manufacture. World production capacity for fluorocarbons is presently estimated at about 1 million tonnes/year, with more than 90% of this located in the Western world and about 60% of the total production in the United States.

Fluorocarbons are produced principally by the catalytic reaction of HF with chlorinated hydrocarbons, usually in the liquid phase using antimony catalysts, or in the gas phase over chromia catalysts. For example, of the three most important fluorocarbons, two are made from carbon tetrachloride and one from chloroform.

$$CCl_4 + HF \rightarrow CCl_3F + HCl \text{ (trichlorofluoromethane, fluorocarbon 11)}$$

$$CCl_4 + 2HF \rightarrow CCl_2F_2 + 2HCl \text{ (dichlorodifluoromethane, fluorocarbon 12)}$$

$$CHCl_3 + 2HF \rightarrow CHClF_2 + 2HCl \text{ (chlorodifluoromethane, fluorocarbon 22)}$$

These three fluorocarbon gases dominated the industry, probably accounting for 90% of the fluorocarbons made. The remaining 10% of the industry is made up of a number of products.

Fluorocarbon liquids such as trichlorotrifluoroethane ($CF_2Cl.CFCl_2$, fluorocarbon 113) find use as inert solvents, for example in the cleaning of electronic equipment and in dry-cleaning, but they account for only a small proportion of total fluorocarbon production.

Fluorinated resins are another important product of the industry, although their production accounts for very little of the HF produced. For example, tetrafluoroethylene, the monomer for PTFE, is manufactured by pyrolysis of chlorodifluoromethane, fluorocarbon 22.

$$2CHClF_2 \rightarrow C_2F_4 + 2HCl$$

Although small in the amount of HF they consume, several other fluorinated organics are important as speciality products. For example, as vaporising liquids for fire-fighting (e.g. bromochlorodifluoromethane, $CBrClF_2$), as fluorochemical surface active agents in fire-fighting foams, as fluorinated resins to impart oil and water repellency to textiles and as fluorinated anaesthetics (e.g. 'Fluothane' $CF_3CHBrCl$). The present end-use pattern of these products is shown in Table 7. However, with increasing environmental pressures on the use of fluorocarbons, particularly as aerosol propellants, this pattern of use will almost certainly change in the future.

Table 7.   Consumption Patterns of Fluorocarbons

Aerosols	50%
Refrigerants	28%
Plastics	10%
Solvents and degreasing	5%
Foam blowing agents	5%
Others	2%
	100%

The Aluminium Industry. At the present time, about 40% of all HF produced is used in the aluminium industry to produce aluminium fluoride and synthetic cryolite, used as the molten electrolytes in the electrolytic reduction of alumina to aluminium. The aluminium fluoride is produced by reaction of alumina with HF, the cryolite by reaction of alumina with HF and sodium hydroxide.

$$Al_2O_3, 3H_2O + 12HF + 6NaOH \rightarrow 2Na_3AlF_6 + 12H_2O$$

The aluminium pot-line is normally started with cryolite as the electrolyte, with fluorine losses being made up by addition of aluminium fluoride. About 25 kg of each electrolyte is used per tonne of aluminium produced, equivalent to around 60 kg of acid grade fluorspar or 30 kg of HF.

As this fluorine is largely lost to the environment in various effluents, increased recovery and recycling of fluorides by aluminium smelters should lead to a reduction in the amount of HF used per tonne of aluminium in the future.

A further factor in the consumption of hydrofluoric acid by the aluminium industry has already been mentioned, namely the growing use of fluorosilicic acid as a source of fluorine to produce aluminium fluoride and synthetic cryolite[12].

$$H_2SiF_6 + 2Al(OH)_3 + 2H_2O \rightarrow 2AlF_3.3H_2O + SiO_2$$

Several plants are reported to be in operation for this purpose, in the USA and Canada for example, but not as yet in the United Kingdom.

Other Uses. As we saw from Table 6, other uses for HF are restricted to a few percent of any one outlet. In the petroleum industry, hydrogen fluoride is used as an alkylation catalyst, to produce high octane blends for use in petrol. Small amounts are also used for stainless steel pickling, for the manufacture of fluoride salts and in uranium processing. Miscellaneous uses include manufacture of speciality metals, etching and frosting of glass products such as light bulbs and television tubes and some small uses in the electronics industry. Of these outlets, only the manufacture of inorganic fluorides, the processing of uranium and the production of fluorine will be considered further.

Inorganic Fluorides

Although a large number of inorganic fluorides are available at the industrial level, and even more at the laboratory level, the total volume of production is small with the exception of those already discussed, namely fluorspar, aluminium fluoride, cryolite and fluorosilicic acid.

From Table 6, it can be seen that only about 4% of HF produced is used to manufacture inorganic salts, equivalent on a weight basis to around 40,000 tonnes HF. Some examples of these salts and their uses are as follows:

Sodium fluoride, NaF. Although its main use is still in water fluoridation, it is being replaced in this application by fluorosilicic acid and its salts; this trend will probably continue. Sodium fluoride also finds use as a fluxing agent in the steel industry, as a wood preservative and in the formulation of insecticides and fungicides. It is also used to remove HF from gaseous fluorine in fluorine manufacture.

Sodium bifluoride, $NaHF_2$. Finds use in two main areas, the
electro-tinning of steel and as a laundry sour, of which the first is
the more important.

Boron trifluoride. The major uses of BF_3 are as a catalyst in organic
synthesis and as an accelerator in epoxy resins.

Fluorosilicic acid and its salts. **Fluoro**silicic acid has been mentioned
previously, particularly as to the large amounts available as a waste
product of the fertiliser industry. Apart from its growing use in the
manufacture of aluminium fluoride and cryolite, it is the main agent for
fluoridation of water. It is also used in laundry sours, glass and
ceramics manufacture and in metal ore treatment.

Stannous Fluoride. Probably one of the best known uses of inorganic
fluorides as far as the public is concerned is the use of stannous
fluoride in toothpaste to prevent dental decay.

No other fluorides, apart from those prepared from elemental fluorine,
are of any real commercial importance.

Fluorine

Fluorine is a pale yellow gas, b.p. $-188^{\circ}C$. It was first isolated
in 1886 by the French chemist Henri Moissan, after twentyfive years of
continuous research by several other chemists, including such well known
names as Davy and Ampère. Fluorine is the most reactive member of the
periodic table and reacts with practically all organic and inorganic
substances. Despite this reactivity, engineers have learned how to harness
fluorine and cope with the problems it presents, although it wasn't until
the 1940s that production on an industrial scale began, with the need for
elemental fluorine in the manufacture of uranium hexafluoride.

Manufacture. Fluorine is produced by the electrolysis of a solution of
anhydrous potassium bifluoride electrolyte containing various
concentrations of free HF. The fluoride ion is anodically oxidised to
elemental fluorine whilst at the same time, hydrogen is liberated at the
cathode. Anhydrous HF itself cannot be used directly as the electrolyte,
because of its low conductivity. However, an electrolyte approximating
to $KF.2HF$ has high conductivity, a melting point of around $70^{\circ}C$ and has
a relatively low HF vapour pressure, the latter ensuring that a minimum
amount of HF leaves the cell with the electrolysis products.

<u>Cell Design</u>. Although anode materials such as nickel and graphite have been evaluated in the past, all manufacture of fluorine today uses carbon anodes and steel cathodes, in steel cells operating at around 90°C. Effective cooling is necessary to keep the cell at working temperature and this is achieved by water cooling pipes immersed in the electrolyte or by a cooling jacket around the cell. The cooling water is at 75°C, to avoid any risk of solidifying the electrolyte. Figure 3 is a simplified line diagram showing the arrangements of the components of a fluorine cell. The skirt separating the anode and cathode compartments is essential to keep the hydrogen and fluorine apart. For this reason, it is important that the electrolyte is maintained at a definite level, in order to maintain a seal between the fluorine and hydrogen compartments of the cell.

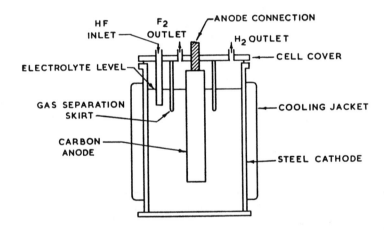

Figure 3   Arrangement of Components in a Fluorine Cell

The hydrofluoric acid feed to the cell is introduced through a dip pipe leading into the electrolyte; the actual HF content of the electrolyte is maintained between 40 and 45% by intermittent addition of liquid HF.

The cell cover has holes for the release of hydrogen and fluorine; the gases from each anode compartment are fed into the fluorine manifold, the gases from the cathode compartments into the hydrogen manifold.

In a commercial cell, a number of carbon anodes are suspended along the length of the cell. For example, in a 3 x 0.8 x 0.6 metres

cell holding around 1,000 kg of electrolyte, there might be twelve
anode assemblies, each holding two anode blocks.

A cell of this size would be rated at 4,000 to 6,000 A and would
operate at 8–12 V, generating up to 3 to 4 kg of fluorine per hour.
In a fluorine plant, cells of this size would be duplicated many times,
depending on the capacity of the plant. For example, a plant
generating 10 tonnes/day fluorine would need over 100 such cells.

Both the fluorine and the hydrogen leaving the cells are
contaminated with about 4% HF (by volume). Caustic soda liquor is
normally used to remove acid from the hydrogen stream. Sodium fluoride
is used to remove HF from the fluorine stream, by formation of the
bifluoride $NaHF_2$. Refrigeration can also be used to remove the bulk of
the acid, with sodium fluoride lowering the remaining HF to the
required 0.2%.

The gaseous fluorine can be used directly from the plant or it
can be liquefied for storage. Alternatively, the fluorine can be
converted to chlorine trifluoride, which is easier to handle and
transport than fluorine. Chlorine trifluoride, b.p. 11.7°C, is
prepared by direct reaction of chlorine and fluorine at around 300°C.

$$Cl_2 + 3F_2 \rightarrow 2ClF_3$$

Handling and Safety Factors. Most fluorine generated is used
captively, although it can be transported, either as a gas in steel
cylinders under a pressure of 400 psi or in bulk as liquid fluorine.
Bulk transport is now confined to the US, although in the past it has
been transported this way in the UK or as chlorine trifluoride. The
liquid fluorine is contained in an inner vessel surrounded by liquid
nitrogen, which boils 8°C lower than fluorine, thereby ensuring that
fluorine losses are minimised.

Fluorine is undoubtedly a dangerous material, but it can be
handled with proper precautions. It is highly reactive and extremely
corrosive and irritant to skin tissue. It has a sharp penetrating
odour detectable at around 0.02 ppm, well below the recommended
threshold limit value of 0.1 ppm set by the ACGIH (Americal
Conference of Government Industrial Hygienists).

Uses for Fluorine

The separation of uranium 235 from uranium 238 by gaseous
diffusion of uranium hexafluoride created the first major use of
elemental fluorine. Uranium tetrafluoride is first prepared from
uranium dioxide by reaction with hydrofluoric acid

$$UO_2 + 4HF \rightarrow UF_4 + 2H_2O$$

The UF_4 is then fluorinated either directly with fluorine or using chlorine trifluoride

$$UF_4 + F_2 \rightarrow UF_6$$

$$3UF_4 + 2ClF_3 \rightarrow 3UF_6 + Cl_2$$

In this last reaction, the chlorine is recycled to generate more chlorine trifluoride by reaction with fluorine.

The uranium hexafluoride produced is fed to a diffusion plant or to a centrifuge, in which the 235 UF_6 is separated from the 238 UF_6. The product, enriched in the 235 isotope, is then used to prepare the fuel for use in the nuclear electric-power industry.

In 1973, the US Atomic Energy Commission used an estimated 6,000 tonnes of hydrofluoric acid for production of fluorine although at the height of atomic weapon manufacture, the consumption was probably very much higher[13]. During the present decade, hydrofluoric acid consumption for this application has been increasing, reflecting the growing importance of nuclear power generation.

In the UK, British Nuclear Fuels at Preston operate fluorine cells for the preparation of uranium hexafluoride, using anhydrous hydrofluoric acid supplied by Imperial Chemical Industries and Imperial Smelting Corporation.

Other Uses. Sulphur hexafluoride, SF_6, prepared by the direct fluorination of sulphur, is used as a gaseous insulator in high voltage equipment such as power transformers and X-ray equipment. Its high dielectric strength, inertness, thermal stability and ease of handling make it ideal for this type of application. However, present tonnages are only small, the main outlets being in the United States.

The reaction of sulphur with fluorine is highly exothermic and the reaction is self-sustaining without application of heat. The crude gas from the reactor is pyrolysed to disproportionate any disulphur decafluoride present, then washed with caustic alkali to remove sulphur tetrafluoride

$$S_2F_{10} \xrightarrow{500^\circ C} SF_6 + SF_4$$

Fluorine is also used to prepare perfluoroalkanes for use as stable, inert fluids. At present, the small tonnages manufactured are used in the electrical industry, although exciting possibilities are emerging for their biological use in outlets such as artificial blood,

blood oxygenators and organ perfusion. In the UK, Imperial
Smelting Corporation have a facility for manufacturing such compounds,
of which perfluorodecalin $(C_{10}F_{18})$ is an example.

Another recently developed use for fluorine is in the
preparation of carbon monofluoride, by direct fluorination of graphite.
The product, a pale coloured solid of formula $(CF)n$, is finding use
as a dry lubricant and as a mould release agent. In Japan, the
Matsushita Electrical Industrial Company is developing high
performance batteries using carbon monofluoride as the anode and
lithium as the cathode[14].

Conclusions

We have briefly reviewed an industry which is younger than many
of the people engaged in its pursuit. During its forty or so years,
many problems have been tackled and overcome, including the
manufacture and handling of the most reactive element, fluorine. The
Industry has grown at a tremendous rate and its products are used in
a wide range of other industries and applications. Now we are
undoubtedly seeing a levelling off in the growth rate and
environmental considerations are perhaps the current area for attention.

However, overall demand for the products of the Industry will
probably continue to grow, although not as spectacularly as in the past.
Almost certainly, end-use patterns will also change, both from country
to country and worldwide, with some present uses declining and new ones
being introduced.

Acknowledgement

The authors thank their colleagues in Mond Division for their
help in the preparation of this paper.

References

1   B L Hodge, Mining Annual Review (Mining Journal, London 1976),
     'Fluorspar', p.111

2   Business Monitor, Minerals 1974 and 1975 (HMSO, London, PA 1007, p.14)

3   Industrial Minerals, 1977, 112, 51

4   'Fluorspar', Minerals Yearbook (US Bureau of Mines, Washington,
     published annually)

5   H H Gossling and H W A McCulloch, Mineral Sci. Engng., 1974, 6, 206

6   W R Rogers, Chem. Eng. Progr., 1963, 59, 85

7   H E Blake, 'Utilisation of Waste Fluorosilicic acid'
     (US Bureau of Mines, 1971), Rep. Inves. No RI 7502

8   W R Parish, 164th National Meeting of the Am. Chem. Soc.,(New York,1972)

9   R S Reed, 75th National Meeting of the Am. Inst. Chem. Eng.,
     (Detroit, 1973)

10   British Patent No 1,262,571 to Buss (Feb, 1972)

11   From Chemical Week, Chemical Marketing Reporter,
     European Chemical News and Industrial Minerals

12   E Steininger, 164th National Meeting of the Am. Chem. Soc.
     (New York, 1972)

13   'Hydrofluoric acid', Chemical Economics Handbook, (Stanford
     Research Institute, California, 1975)

14   J H Holloway, Chem. Brit. 1977, 13, 50

The Bromine and Bromine-Chemicals Industry

R. B. McDonald
W. R. Merriman

Great Lakes Chemical Corporation

Objectives and Summary

Bromine is an important raw material used widely in a variety of applications
with annual consumption worldwide of 250,000 - 300,000 tons. During the last
several years, the increasing amount of regulation designed to protect the
consumer, especially in the United States, has resulted in counter-balancing
influences on consumption of bromine. The concern for improved air quality
has required reduced automobile exhaust emmissions; in the U.S. this
reduction has been accomplished largely by using catalytic converters which
cannot tolerate lead thus reducing the demand for ethylene dibromide. On the
other hand, increasing concern for consumer protection against flammable apparel
and home appliances has dramatically increased the need for fire retardants,
many of which are based on bromine. Finally, the growing pressure for
increased food production to meet the needs of a burgeoning world population
have increased the useage of certain bromine derivatives and intermediates as
methods of counter-acting certain soil borne pests. As a result, the industry
has experienced modest overall growth in the last ten years. During the same
interval, new and much more economical sources of bromine in Arkansas and the
Dead Sea have begun to displace the more dilute Michigan brines and sea water.
Taken together, the bromine industry has undergone dynamic change which is
expected to continue during the next several years. The purpose of this
paper is to provide the research chemist a brief review of the chemistry of
bromine, along with a summary view of the industry with emphasis on
production and end uses.

Introduction

Bromine was first isolated and recognized as an element by the French experimenter, Balard, in 1825. He was carrying out an examination of the residues of salt manufactured from seawater, using a chlorine treatment, and noticed an unexpected liberation of a brown material which could be distilled and could be dried over calcium chloride to give a reddish brown liquid. The Greek name "bromos" was chosen to reflect the "smell" of the new substance.

The ability of the material to react with many elements and compounds was soon recognized, as was the tendency of some of its compounds to decompose under certain conditions leading to its early use in photography and medicine. Bromine is the second most common halogen and is found as dissolved salts in seas, oceans and subterranean brines, and in small quantities in natural occurring rocks. It is a reactive, dark red, heavy, corrosive, toxic liquid (one of only two elements that are liquid under ambient conditions) that requires specialized technology to handle, store, transport, and use. The use of lead alkyls as octane improvers for motor fuel and the subsequent discovery that ethylene dibromide was the scavenger of choice to prevent the deposit of lead during the combustion, led to large scale production of bromine, especially from sea water and from Michigan brines during the 1920's and 1930's. In the U.S. during the past 25 years, this end use has declined from 90% of the total consumption of bromine to less than 60% presently. In recent years, bromine derivatives used for flame retardants, oil-well drilling completion fluids, chemical intermediates and for soil treatment have become increasingly important. The early uses of bromine in dye-stuffs, pharmaceuticals and photographic chemicals still remain as important, but not high volume consumers of bromine.

Bromine exists naturally in nearly equal proportions of the two stable
isotopes of weight 79 and 81 with an average atomic weight of 79.916. It is in
Group V11$_x$B$_x$. three of the periodic table and has physical properties
intermediate between chlorine and iodine. It is highly electronegative and
shows valences of -1, +1, and +5. It is a moderately strong oxidant and will
readily oxidize numerous metallic and non-metallic elements to form salts or
covalently bonded compounds. In its reactions with organic compounds it is
less reactive than chlorine, but more than iodine, and forms most compounds
either by addition reactions to unsaturated linkages or by replacement
reaction, usually of hydrogen. In the latter case, hydrogen bromide is a
by-product and must normally be utilized, most often by recycle to bromine,
for purposes of economy and to avoid problems of waste disposal. Because of
its greater size and mass, bromine reacts more slowly and with less vigor
than does chlorine; steric hindrances are frequently encountered, so that
the bromine analogs of several chlorine compounds are synthesized with
difficulty or not at all.

On the other hand, the bromine-carbon bond in its organic derivatives is
weaker than its counterpart chlorine-carbon bond, a factor which causes
bromine compounds to be more effective flame retardants, less persistent in
the environment, more reactive in chemical displacement reactions and more
biologically active.

The reactivity of bromine with organic matter is especially important because
it represents serious hazards to human beings. Exposure of skin to elemental
bromine can cause serious chemical burns which are very slow to heal. Bromine
vapor is also hazardous and it has been reported that exposure to
concentrations of 500 - 1000 ppm (by volume), even for a short time is fatal.
It has also been reported that exposure to 40 - 60 ppm for 30 minutes may be
dangerous to life. Some health authorities place atmospheric limits at
around 1 ppm for continous exposure, but even at this low level, the odour is
strong enough to produce an uncomfortable environment and effective warning.

A lime slurry is found to be an effective skin antidote, and treatment with ammonia vapors can be helpful in preventing serious lung damage from inhalation.

Production and Producers

Bromine is widely dispersed in nature, though in very small proportions. The search for the best possible sources for bromine has narrowed the field to two or three which currently are being exploited.

The major natural source is the ocean, and when very large quantities of bromine were first required for satisfying the demand of EDB for motor fuel, it was here that the industry built its major increase in production capacity. Ordinary seawater contains about 65 ppm bromine in soluble salts, but the concentration is higher on the bottom arising from production of salt from the ocean. Currently the most economical sources of bromine are the underground brine deposits in Arkansas, U.S.A. and the waters of the Dead Sea where the brines are preconcentrated in Potash production. Bromine is also recovered as a by-product from potash mines in Germany and elsewhere, but the quantity available is directly dependent upon the production and market for the potash.

The concentrations of bromide ion in the commercially important sources are:

Seawater	65 - 70 ppm
Michigan Brines	1,300 - 2,100 ppm
Arkansas Brines	3,800 - 5,000 ppm
Dead Sea	4,500 - 5,000 ppm

In the United States, Arkansas now accounts for more than 75% of the total U.S. production and Michigan all of the remainder.

All commercial processes for the manufacture of bromine begin with a solution of bromide. Four basic steps are involved : 1) oxidation of bromide to bromine, 2) separation of bromine vapor from the brine, 3) condensation of the vapor (or absorption), and 4) purification.

The recovery process generally used for brines containing more than about 1,000 ppm bromide is the "steaming out process". A typical steaming out plant for bromine is depicted in figure 1. Fresh brine, which may be preheated through one or more heat exchangers is introduced into the top of the tower; chlorine and steam are injected into the bottom of the tower. The spent brine from the bottom of the tower is neutralized and returned to disposal wells or to the brine source an appropriate distance from the intake, usually through the heat exchangers to conserve the heat content. The mixture of bromine, chlorine, and water vapors from the top of the tower is condensed and the condensate separated into crude bromine and an aqueous phase. The crude bromine is distilled and separated from high boiling impurities, mostly halogenated hydrocarbons. Vapors from this column are condensed and the bromine is fed to a stripping column where remaining traces of chlorine are stripped away for return to the tower, and refined bromine is removed from the bottom of the column; subsequently, the last traces of moisture are removed and the end product is packaged. In theory, about 0.44 lbs. of chlorine is required to release 1.00 lbs. of bromine; in practice, substantially more chlorine is required with the excess used in the oxidation of other components of the brine, particularly organic materials and hydrogen sulfide, and in some instances to acidify the brine. Recoveries of 95% or more of the contained bromine in the brine are achieved routinely. When brines of low bromine concentration (e.g. sea water) are used, the steaming out process is uneconomic.

It has been mentioned previously that many organic bromination processes produce by-product hydrogen bromide. By-product bromides also occur when a bromine compound is used to attach one moiety to another one. For economic and environmental reasons, it is desirable to recovery this bromine, and the process used for this operation is very similar to that used to recover virgin bromine from concentrated brines.

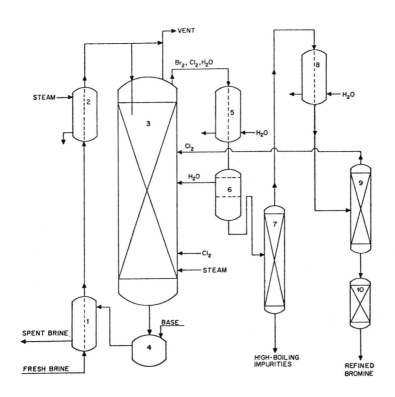

Figure 1

Typical Bromine Recovery Plant

Key No.	Description
1, 2	Preheaters
3	Bromine extraction column
4	Spent brine neutralization
5	Condenser
6	Decanter
7	Distillation column
8	Condenser
9	Chlorine stripping column
10	Drying column

The seawater plants differ in a number of respects. They must process very
large volumes of water and are therefore larger in physical size. The
bromine is swept out of the brine after liberation by chlorine, using large
quantities of air. The bromine laden air is treated in a variety of ways, all
of which involve the absorption and reaction of the bromine, with an agent
from which it is subsequently released. It has been estimated that 200,000
cubic feet of air may be required for a single ton of bromine, though for
bromide-weak brines the figure could be much higher with resulting rise in
cost of energy per ton of bromine. The bromine in the air stream cannot be
collected by condensation, but must be absorbed chemically and subsequently
released. Usually, then, two chlorine treatments are required.

It will be noted that the recovery of bromine is an energy intensive process,
both for the heating of the brine and for the production of the chlorine used
to oxidize the bromine.

According to the U.S. Bureau of Mines, the world production of bromine (in
metric tons) in 1975 by country was as follows :

France	15,000
Germany (West)	4,300
India	270
Israel	18,000
Italy	3,500
Japan	11,200
Spain	450
U.K.	28,200
U.S.	185,100

Worldtrade in Bromine

It is only in the last few years that a significant world-wide trade in elemental bromine has developed. Historically, only small quantities have been shipped by water, and those were packed in individual 3 kg. net bottles, This is a safe method since in the event of breakage, only a small quantity would be released and the bottles are always packed with sufficient absorbent to take up most of the bromine.

During the 1960's, however, a small lead-lined demountable tank was developed for use within the European market. More recently, returnable ISO lead-lined tanks of 14 metric ton capacity were developed by Great Lakes Chemical and with U.S. Coast Guard approval, are being shipped routinely across the Atlantic.

The diversity of government legislation on the transport of bromine is illustrated by the discouragement of forms of transport other than rail in certain European countries. In the U.K., unlike most E.E.C. countries, bromine is widely moved by road. Ships sailing under various flags will also have different safety regulations. Certain countries limit the 14 ton ISO tank to deck cargo, two only per ship and with quick repair kits permanently at hand. These regulations may change slightly as more shippers develop experience in the handling of the 14 ton ISO tank.

In addition to the large ISO container, bromine can also be shipped in non-returnable steel cylinders having a capacity of 2,800 lbs. net weight of bromine; however, there is a small amount of iron dissolved during the shipment, with concentrations reaching up to 500 ppm.

In some organic brominations, the iron acts as a catalyst and causes no problems. In other cases, it is necessary to vaporize the bromine to leave the iron salts behind.

End Use Pattern

According to a composite of various estimates, the end use pattern for

bromine in 1975 was as follows (in thousand metric tons) :-

Motor fuel additive	150
Agriculture	50
Flame retardants	25
Cleaning Agents	10
Dyestuffs	10
Drilling Fluids	5
Photography	5
Water Sanitation	10
Pharmaceutical	5
Misc. synthesis and other	10
Total	280

Further comments on each end use area will be found in the following

paragraphs.

Ethylene dibromide (EDB) is one of two key ingredients used in motor fuel

additives to combine with the lead so as to make it volatilize rather than

depositing in the engine. (The other material is ethylene dechloride).

The history of EDB closely parallels the history of the bulk bromine industry

itself. Bromine was a small volume item of commerce until it was established

in 1922 that lead tetraethyl could be used to prevent preignition of gasoline

in the internal combustion engine, but that it needed a scavenging additive

to remove its deposits from the engine. The remarkable expansion of this

application led up to a peak consumption in 1973 of about 190,000 tons which,

at that time, accounted for 75 percent of the world production of bromine.

By 1975 this application accounted for 55% and by 1980 may consume only 40%

of the world's bromine production.

The use of lead compounds represents the method of choice for improving

octane rating of motor fuel, since this permits the most efficient

utilization of crude oil, allows for lower cost techniques for refining, and

minimizes capital investment. On the other hand, the volatile lead compounds

do constitute a part of exhaust emissions and all auto exhaust emissions are

being reduced in the U.S. to conform to increasingly strict regulations.

This has led to a conflict between the need to conserve resources and the

desire for cleaner air. Projections being made by knowledgeable sources

suggest that the use of lead in motor fuel will decline in the U.S. over the

next several years. The repidity and end point of this decline are not yet

clear.

<u>Agricultural Chemicals</u>. Bromine compounds are widely used in agriculture

directly; examples are methyl bromide, ethylene dibromide,

dibromochloropropane, and certain proprietary pesticides. Bromine is also

used to produce a number of pesticides in which bromine does not appear in

the final product. Methyl bromide is one of the three largest volume bromine

compounds used in agriculture and one of the most effective nematocides known.

In addition, it is a good all-around pesticide (i.e. herbicide, fungicide and

insecticide). It is used widely for the fumigation of soil, which is to be

planted with high value crops, such as tobacco, strawberries, tomatoes, peanuts,

vegetables and some flowers. It may be applied to soil in greenhouses as well

as to open fields. In addition, gelled methyl bromide is now available in the

United States. This new form permits the application of methyl bromide with

plastic film at 50 - 70% of the former rate. Methyl bromide is a colourless,

highly toxic gas. Normally 2% chloropicrin is incorporated with the methyl

bromide to serve as a warning agent. Even so, it should only be used by

qualified professionals. In some countries the use of this material is

restricted to those trained in government approved schools and regularly

requalified as methyl bromide operators.

A second bromine compound used in large volume in soil fumigation is EDB. This compound is easier and cheaper to apply than methyl bromide, but is limited to nematode control capabilities. It is finding increasing use in fumigation of soil which is to be planted in lower value crops, such as soy beans, cotton or potatoes, and other grain crops. U.S. industry will soon be required to provide further proof of EDB's safety for use as a soil fumigant. Particular emphasis will likely be placed on the rapidity of degradation and the character and destination of the fragments. Great Lakes Chemical does have data that shows that at least a dozen crops grown in EDB fumigated soil do not contain EDB residues by methods sensitive to 5 PBB. The third bromine compound used widely in soil fumigation is dibromo-chloropropane (DBCP) which is unique in that it has low phytotoxicity, meaning that it can be applied to living plants in sufficient concentration to control the nematodes without permanently damaging the plant. However, questions have also been raised about the safety of this compound for use in soil fumigation and there is some evidence, although sketchy at this time, that the product is assimilated in the plant and significant DBCP has been found in the fruit.

Flame Retardants. Significant losses of life and property from fire occur every year. Plastic materials and textiles have been judged by governmental authorities to constitute fire risks in the home, office and public buildings. Legislation varies, but increasingly requires that the consumer be given more protection. Compounds containing the halogens, phosphorus and antimony have flame retardant properties with bromine being about 2.5 to 4 times as effective as chlorine on a weight basis. Bromine compounds are higher priced on a weight basis, but due to their greater efficiency, are the flame retardants of choice in many applications because they can be used at a lower loading level, thus having less effect on the desirable properties of the substrate.

With plastics and with synthetic fibres, it is possible to incorporate the halogen either in the basic polymer chain or, alternatively as an additive blended in with the polymer. The terms "reactive" and "additive" fire retardants are widely used to describe these two approaches.

In general, the use of an additive fire retardant in a plastic may result in a product where mechanical or other desirable properties are less affected than if a reactive material is used. Also, it will often prove more expensive to use the reactive approach as some design changes in the equipment to make and fabricate the polymer may be needed. Bromine and bromine compounds can be corrosive so the choice of the appropriate molecule to meet all the processing conditions is an important one.

For textile application, two distinctly different approaches have been used - "fibre inclusion" and "after treatment". As the names imply, the fire retardant may be incorporated into the substrate during the manufacture of the fibre, either as a co-monomer during the polymerization or as an additive during the spinning. The "after treatment" depends on the application of the fire retardant to the textile with techniques similar to dyeing. Until recently, the most widely used textile fire retardant was tris (dibromopropyl) phosphate, a product that could be included during the fibre spinning operation for acrylic and acetate fibres or that could be applied as an after treatment to polyester or acrylic textiles. It has recently come under a cloud since it was shown to be a mutagen on the AMES test. Further work by the National Cancer Institute in the United States has shown that it may cause tumors in test animals at high dosages, thus indicating the possibility that it is carcinogenic. During 1976 and early 1977, this has resulted in a major decrease in the use of this chemical in textile applications. Efforts continue, especially in the United States, to find other flame retardants which can be used to meet the various regulations associated with the Flammable Fabrics Act.

Great Lakes Chemical and others have identified promising alternatives to "Tris" but significant additional work will be required before the compounds are confirmed safe for use.

Other compounds which are used as flame retardants for textiles include the co-mononers, vinyl bromide, and the bis (2-hydroxyethyl ether) of tetrabromobisphenol A. Both compounds have been used on a commercial basis, the former for incorporation into acrylic fibres and the latter for incorporation into polyester fibres.

For plastics, both additive and reactive flame retardants are used; additives are widely used in thermoplastics and reactives in thermosets.
The only major thermoplastic that can use a reactive flame retardant are the polycarbonates. The major flame retardant for this application is tetrabromobisphenol A.

For the other thermoplastic polymers, reactive flame retardants are not widely used, due principally to the significant changes in properties which result from making co-polymers instead of homopolymers and the potential liberation of corrosive hydrogen bromide during the polymerization. Polystyrene and ABS are the most widely used thermoplastics incorporating brominated fire retardants via the additive approach and are used principally for electrical/ electronic applications. Since these plastics are processed at high temperatures, good thermal stability is required; only brominated aromatic compounds are suitable. The commercially important brominated flame retardants for these applications include tetrabromobisphenol A, decabromodiphenyl oxide, octabromodiphenyl oxide, pentrabromoethylbenzene and decabromo-biphenyl. In earlier years, hexabromobiphenyl was used in ABS, but a high order of persistence, coupled with relatively high chronic toxicity, has all but eliminated its use and it has virtually disappeared from the market.

Thermoset resins usually incorporate reactive flame retardants, the most important ones being tetrabromobisphenol A in epoxy resins, and tetrabromophthalic anhydride in unsaturated polyester. Additive type products are used to a limited extent in thermosets, and are mainly fully brominated aromatic compounds, such as pentabromotoluene, pentabromoethylbenzene, and decabromodiphenyl oxide.

The market for brominated flame retardants will respond to changes in regulations. As the regulations require increasingly severe performance tests and/or the number of consumer products covered expand, so also will the demand for brominated flame retardants. This class of compounds represents the most cost effective means of achieving this level of consumer protection.

<u>Oil Well Completion Fluid.</u> Another major newly developed application for a bromine compound is the use of calcium bromide (and perhaps other inorganic bromides) as a high density fluid for completion of new oil and gas wells and for work-over of existing wells. After a production oil well has been drilled, it has to be "completed". That is, the section of the hole that runs through the oil-bearing strata must be modified to allow oil to flow in optimum quantities.

Clear heavy fluids are often preferred over normal drilling muds since these muds are suspensions and may result in plugging of the formation. The inorganic bromides can be used to produce liquids having a density of greater than 18 lbs. per U.S. gallon, and by combining calcium chloride, calcium bromide and/or zinc bromide, fluids having a density between 11.5 and 18 lbs. per gallon can be obtained. The density required is directly dependant upon the down-hole pressure.

The useage in a single well can be up to 1,000 barrels of this high density fluid, so work is well under way to find methods to recycle and reuse or to minimize the quantity required by confining the heavy fluid to the zone of interest.

The practice of "working over" a well after a few years of production can
also require heavy brines, although if reservoir pressure has fallen
drastically, heavy brines are not needed. It is too soon to forecast the
changes that may occur in down-hole pressure in the North Sea Oil fields,
so future useage in this area is subject to errors in projections. However,
the wells in the Gulf of Mexico are projected to maintain high bottom hole
pressures, meaning that high density fluids will continue to be required for
workover. It is anticipated that the search for oil will continue to lead to
increasingly deeper wells with prospects for higher downhole pressures and a
continued growth in the use of calcium bromide and other inorganic bromides.

Future Developments

The dynamic change that has been occurring in the bromine industry over the
last several years will continue. Brines containing low concentrations of
bromide will be under increasing economic pressure from the more economic
sources, and only national interests or valuable co-products can justify the
continued commercial importance of these low concentration raw material sources.

The use of EDB in motor fuel will continue to decline, although the rapidity
and ultimate use level is difficult to predict. Increasing quantities of EDB
will be used for soil fumigation, although this demand will be inadequate in
the near term to compensate for the decline in the U.S. motor fuel market.
Other brominated compounds will also find increasing usage in soil fumigation,
and the potential in this application is substantial since only a very small
proportion of total cropland, even in the U.S., is being treated now.
Bromine derivatives will be used in increasing quantities as flame retardants,
but at a growth rate much lower than that of the past ten years. Use of
bromine to produce reactive intermediates is a significant growth area.
Water treatment represents a huge potential market, with one source
identifying a possible market of more than 250,000 tons annually in the U.S.
alone. Bromine has the advantage of forming less persistent and less toxic
organic compounds when compared to chlorine.

The Modern Sulphuric Acid Process
By A. PHILLIPS
Sim-Chem Ltd.

Introduction

As a brief introduction to the subject of sulphuric acid plant design let us first consider a few facts related to the uses of sulphuric acid, the size range of plants and the type of plant which is to be featured in this paper.

The approximate breakdown of users of sulphuric acid and oleum in the U.K. is shown in Table I.

*APPROXIMATE BREAKDOWN OF USERS OF SULPHURIC ACID AND OLEUM IN THE U.K.

FERTILISERS	32%
CHEMICALS	16%
PAINTS/PIGMENTS	15%
DETERGENTS	11%
FIBRES/CELLULOSE FILM	9%
OTHERS	17%

* Based on 2nd and 3rd quarter figures for 1976 given by The National Sulphuric Acid Association Ltd.

Table 1

The production of sulphuric acid/oleum in the U.K. is between 12,000 and 13,000 tonnes per day, of which more than 96%[1] is manufactured from sulphur and almost all of it produced in plants built to the 'Contact' process. Most of the plants are of a small capacity (less than 300 tonnes per day H_2SO_4) although some larger plants built in recent years have capacities in the range of 500-750 tonnes per day H_2SO_4. In contrast the large plants built in the United States in recent years have typically had capacities ranging from 1500 to 2100 tonnes per day H_2SO_4.

All new plants built in the U.K. must now conform to the latest air pollution controls defined in the 'Best Practicable Means Policy' of the Alkali Inspectorate[2]. Essentially, this means that the sulphur emitted with the depleted gases from the plant stack must be less than 0.5% of the sulphur burned, i.e., more than 99.5% conversion of sulphur dioxide to sulphur trioxide, and that the stack gas be 'substantially free from persistent acid mist'.

In order to comply with the stack gas limits, most modern plants have been built according to the Interpass Absorption (IPA) process also known as the double absorption or double catalysis process.   Approximately one third of U.K. acid is produced on plants designed to the interpass absorption process, the rest being produced in single absorption or conventional plants.

It will suffice at this stage to define in simple terms that a conventional plant is one in which the conversion of sulphur to acid is generally 98% to 98.5%, whereas, on an interpass absorption plant the conversion efficiency is in excess of 99.5%.   It should also be mentioned here that there are alternatives to interpass absorption including tail gas scrubbing but these will not be discussed in this paper as they have been dealt with in some detail in other published papers.

Sim-Chem have built more than 300 sulphuric acid plants throughout the world under licence to the Monsanto Enviro Chem sulphuric acid process and almost all of these utilise Monsanto Vanadium Catalyst.   Enviro Chem have designed more than 600 sulphuric acid plants all but of which 67 are IPA plants (as of December, 1976).

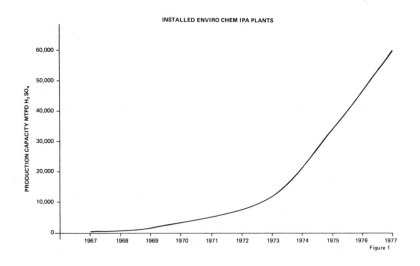

Figure 1 illustrates the marked trend over the past 5 years from conventional to IPA plants, which reflects the growing anti-pollution feeling and subsequent legislation in the more developed countries of the world.

Description of the Process

Gas System.   A typical simplified flow diagram of a double absorption plant
is shown in Figure 2.

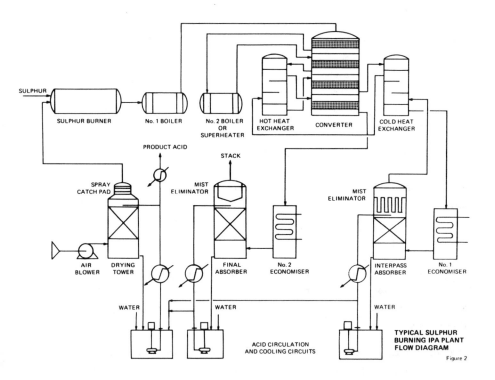

TYPICAL SULPHUR BURNING IPA PLANT FLOW DIAGRAM

Figure 2

Molten sulphur is fed to a horizontal brick-lined combustion chamber where it
is burned in dry air to produce a gas containing between 10 and 10.5% sulphur
dioxide by volume at a temperature of 950°C to 1000°C.   The gas is cooled in
a firetube boiler to around 420°C before entering the converter.

The converter is a vertical cylindrical vessel containing four layers of
vanadium pentoxide catalyst.   The gas passes through the first layer or pass,
where 60% to 65% of the SO_2 is converted to SO_3.   The reaction is exothermic
and the gas leaving at 600°C must be cooled indirectly to the optimum
temperature for further conversion to take place in the second catalyst pass.
This is normally achieved by passing the gas through a second boiler or steam
super-heater.   More of the SO_2 is converted in the second pass accompanied
again by a temperature rise.   The gas leaving is cooled in the 'Hot' gas/gas

heat exchanger before re-entry to the converter. Further conversion takes place in the third pass by which time 95% to 96% of the initial SO_2 has been converted to SO_3.

It is at this point that the process differs considerably from a conventional plant. In the latter case the gas would be passed directly to a fourth catalyst pass, where a cumulative conversion of about 98% would be achieved, after which the gas would be cooled and passed through a sulphuric acid absorption tower where the SO_3 produced would be absorbed in circulating concentrated sulphuric acid. The balance of the gas – oxygen, nitrogen and of course the residual sulphur dioxide would be passed directly to the atmosphere through a stack.

In the IPA process, the principle is to remove the SO_3 already produced before passing the gas to the fourth and final converter pass, thereby shifting the reaction ($SO_2 + \frac{1}{2}O_2 = SO_3$) more to the right allowing a new much higher equilibrium conversion to be approached. This is done by first cooling the gas in the 'Cold' gas/gas heat exchanger and an economiser before absorbing the SO_3 in a primary or interpass absorption tower over which sulphuric acid is circulated. The gas at a relatively cold temperature ($80^{o}C$) leaving the tower then passes through a mist eliminator, where any entrained acid droplets or mist are removed, and is re-heated up to reaction temperature ($420^{o}C$ to $430^{o}C$) by passing it through the shell side of the 'Cold' and 'Hot' heat exchangers.

The hot gas enters the fourth and final catalyst pass where most of the remaining SO_2 is converted to SO_3 to give an overall cumulative conversion of more than 99.5% of the original sulphur dioxide. The gas is then cooled in a second economiser to about $200^{o}C$ after which it passes through the second and final absorption tower where the SO_3 produced in the fourth pass is absorbed in circulating sulphuric acid. The depleted gas, this time containing less than 0.5% of the original sulphur dioxide is discharged into the atmosphere through a mist eliminator and stack.

Acid System. The acid circulated over the two absorbing towers is in the range 98% to 99% H_2SO_4, this being the optimum strength for effective absorption of the SO_3 without excessive formation of mist. The acid circulated over the drying tower, however, can be as low as 94% if this is required as product, although for the same temperature the higher the strength, the better will be the drying efficiency. In order to prevent the strength of the absorbing acid from increasing as SO_3 is absorbed into it, water is

added into both absorbing acid pump tanks at a controlled rate. As the level
of acid increases in the tanks it is bled off to the drying acid section from
where it is pumped to storage. The heat extracted from the gas and the heat
of reaction in the drying and absorbing systems is removed from the acid in
heat exchangers cooled either by water or air. The acid is pumped from the
pumping tanks through the coolers by means of vertical submerged pumps. The
cooled acid is fed into the acid distribution system over the top of the
packing in each tower. The acid returns from the towers by gravity back to
the respective pump tanks.

Steam System. Deaerated boiler feed water is normally fed through the No.1
and No. 2 economisers in series and a steam/water mixture enters the steam
drum of the boiler. More steam is then produced in the boiler by exchanging
heat with process gas. The steam is often superheated to temperatures over
$400^{o}C$ in a superheater usually located between first and second converter
passes. A small quantity of steam is normally consumed by the plant for
deaeration and sulphur melting purposes.

Raw Material and Product Quality

Raw Materials. In Western Europe the basic raw material is sulphur, although
acid is still produced from alternative sources including SO_2 gas from the
burning of metallic sulphide ores.

In many cases metallurgical type plants are built where the metal is the basic
product and the acid plant merely converts the sulphur dioxide to a useful
by-product or where there is an abundance of sulphide bearing ores such as
pyrites and the acid producer is forced into utilising it rather than
importing sulphur. Since the cost of a metallurgical plant is often more
than twice that of a sulphur-burning plant and since maintenance and downtime
are much higher, it is more economical in most cases to use sulphur even taking
into account an indigenous supply of pyrites and the fluctuating cost of
sulphur.

Sulphur is imported into the U.K. from Poland, France, Canada, Mexico and the
U.S.A. About 80% of the sulphur is in liquid or molten form and of the
'Bright' or low carbon grade. Bright sulphur generally contains 0.03% to

0.05% carbonaceous matter whereas Dark sulphur can contain 0.1% or more. Typical analyses of Bright and Dark sulphur are given in Table 2.

TYPICAL SULPHUR ANALYSES — WT%

	Bright Grade	Dark Grade
SULPHUR	99.8 − 99.9	99.6 − 99.8
BITUMEN	0.03 − 0.05	0.1 − 0.3
ASH	0.01 − 0.02	0.01 − 0.04
ACIDITY	<0.01	<0.01

Table 2

The main impurities in sulphur which have an effect on the design of the plant are ash, which can foul the catalyst bed resulting in increased pressure drops and power consumption, and carbonaceous matter which burns to produce water vapour in the sulphur burner and which combines with SO_3 downstream in the process to produce sulphuric acid mist.

The significance of the ash content of sulphur is more easily realised when one considers that the amount of solid material resulting from the burning of sulphur containing 0.04% ash on a 1000 ton per day plant is more than 40 tons per year. This would be unacceptable as most of it would be trapped in the top catalyst layer of the converter which acts as an excellent gas filter. This build-up of material in the catalyst bed would cause such rapid pressure drop increase that plant operation would be impossible even after a few months. With such high ash contents one would normally opt for installing sulphur filtration equipment which can reduce the ash content of the sulphur to less than 0.002% thus reducing the catalyst fouling problem by more than 95% of the above case. Some acid producers prefer to install a gas filter ahead of the converter as an alternative or as a back-up to sulphur filtration. The amount of ash entering the converter determines the frequency of screening the first catalyst pass. It is normally once every year and seldom longer than once every two years.

The water vapour produced from the combustion of a high carbonaceous sulphur can be more than 200 mg/Nm^3 of burner gas compared to less than 50 mg/Nm^3 in the combustion air entering from the drying tower. It follows, therefore, that in the presence of SO_3 the acid mist formation will be significantly increased. In fact actual acid mist loadings before mist collectors are typically in the range 150 to 200 mg/m^3 of gas with 'bright' sulphur and can be as high as 500 − 1000 mg/m^3 gas with 'dark' sulphur.

Acid Quality.   Acid is normally produced in the U.K. at a commercial grade strength of 96% H_2SO_4 to avoid the risk of freezing.

The impurities which can be present in the acid can be categorised as those which result from impurities in the sulphur and process water raw materials and those which are influenced by the plant design or operating conditions. The latter category includes dissolved SO_2, dissolved iron and nitrogen oxides for which typical values have been illustrated in Table 3.

TYPICAL ACID ANALYSIS AS
PRODUCED FROM BURNING SULPHUR

% H_2SO_4 . . . . 96%	
Fe . . . . LESS THAN 10 PPM	
SO_2 . . . . LESS THAN 20 PPM	
NOX . . . . LESS THAN 10 PPM	

Table 3

Dissolved iron is normally picked up in equipment such as cast-iron piping and can be greatly reduced in modern plants where stainless steel acid coolers are used instead of the old cast-iron coolers.   Typical figures range from 20 to 30 ppm with cast-iron coolers to less than 10 ppm with stainless steel coolers.

The acid circulating over the towers contains dissolved SO_2 in equilibrium with the gas passing through, and in order to meet low SO_2 levels in the product acid it is necessary to export acid from out of the final absorption circuit or from the drying tower.   Levels of 20 ppm or less can be obtained if required.

A small amount of nitrogen oxide is produced in the sulphur burner, and part of this is absorbed in the product acid resulting in a nitre content usually less than 10 ppm as HNO_3.

Plant Design Versus the Energy Crisis

For every tonne of sulphuric acid produced there is a potential recovery of
waste heat of approximately 0.78×10^6 kilocalories from the exothermic
oxidation reactions taking place in the IPA process. This heat is usually
recovered in the process in four different locations; after the sulphur
burner in a fire tube boiler, after the first converter pass in a second fire
tube boiler or an extended surface superheater and before each absorption
tower in two extended surface economisers.

The quantity of heat available for recovery is becoming increasingly important
as world energy costs continue to soar. Thirty years ago when plant
capacities of between 10 and 30 tonnes per day were common, the decision
often facing the acid producer and designer was whether a boiler should be
included in the plant, or alternatively, should the design be based on
dissipating the heat by atmospheric cooling. Quite often the decision was
taken to adopt the latter, as this avoided the complication of preparing
suitable boiler quality water and the problem of what to do with the steam.
Even throughout the '50s and early '60s many plants were designed with the
minimum amount of heat recovery producing steam at about 16 bars, this being
the minimum pressure to avoid corrosion, after which it was often condensed
without extracting any useful work from it.

During the last decade however, there has been an increasing awareness of the
value of the waste heat and more and more plants are being designed not only
to maximise heat recovery but also to generate the steam at high levels of
pressure (50 bars or higher) and temperature ($400-420^{\circ}C$). These conditions
are usually chosen to convert the maximum amount of heat into mechanical or
electrical energy. This is done by passing the steam through turbines driving
machinery such as the main blower and acid pumps or through a turbo-
alternator set to produce electric power.

As the need to maximise waste heat recovery increases with increasing fuel
costs, it also makes good sense to optimise within limits the electrical
energy consumed by the plant. This is not as difficult as it sounds, since
60% to 70% of the power is consumed by the main air blower, and by adjusting
the main vessel sizes, it is possible to design for lower or higher pressure
drops, thus affecting the power consumption.

In order to emphasise the energy recovery potential of an acid plant, some
examples are shown in Table 4 to illustrate in simple terms how steam might
be utilised in four different ways on a 1000 tonne per day IPA plant to offset
electric power costs.

ALTERNATIVE STEAM UTILISATION SYSTEMS

Basic plant is a 1,000 TPD sulphur burning IPA plant producing 98% acid and using Bright sulphur. Water cooling towers/pumps are included.	1 Basic plant with all electric drives producing steam at 16 bars for export directly.	2 Plant with back pressure turbine on blower only utilising 46 bars 410°C steam and producing 16 bars steam for export.	3 Plant with back pressure turbines on five main drives utilising 25 bars 290°C steam and producing 2 bars steam for export.	4 Plant with all electric drives producing steam at 46 bars 410°C for use in T/A set to generate power.
Steam for export after deduction for internal consumption.				
Tonnes/day	1220	1025	1125	–
Power consumption MW	–2.8	–1.2	–0.75	–2.8
Power production MW	–	–	–	+10.0
Net power production MW	–2.8	–1.2	–0.75	+7.2
a) Extra cost for turbines over and above basic plant. (Cost approx. £5,000,000)	–	+ £50,000 (Single Stage) + £100,000 (Multi-Stage)	+£100,000	+£1,000,000
b) Extra cost for raising steam at higher pressure and temp. relative to case I.	–	£90,000	£40,000	£90,000
Total extra cost a) and b)	–	**£140,000 (£190,000)**	**£140,000**	**£1,090,000**
Power cost/year at 1.5p per kWH and 330 days/yr.	–£330,000	–£140,000	–£90,000	+£850,000
Saving in annual power costs relative to Basic Plant.	–	**+£190,000**	**+£240,000**	**+£1,180,000**

Table 4

Column 1 is the basic reference plant having all electric drives and having an annual power cost of approximately £330,000 per year.

Column 2 is for the same plant but producing high pressure superheated steam and having a turbine driven air blower where the inlet and exhaust steam conditions have been chosen to enable 16 bars steam to be exported for further use as well as the surplus HP steam. This plant consumes less power and for an additional capital cost outlay of between £140,000 and £190,000, depending upon the type of turbine, there is an annual saving in power cost of about £190,000 compared to the basic plant. There is also the bonus that all the steam can be exported at a relatively high pressure for use elsewhere in the factory.

Column 3 figures take us a step further to the case where the five main drives on the acid plant, namely the blower, two acid pumps, boiler feed pump and sulphur pump, are all provided with back pressure steam turbines. In this case steam is produced at medium pressure and temperature and the turbines

exhaust at 2 bars. For an extra capital outlay of £140,000 there is a power
saving relative to the basic plant of £240,000 per year.

It should be emphasised here that the first three examples shown are designed
to export virtually all the steam at various pressures and this could amount
to more than £600,000 per year additional credit depending upon the value
attached to the grade of steam.

The fourth and final case in Column 4 illustrates a plant designed to produce
high pressure superheated steam as for the second case but this time the
plant has all electric drives and all the steam is passed through a condensing
Turbo-Alternator set to produce power. Although there is a massive capital
cost outlay of more than £1,000,000 there is also a relative saving in
power costs of more than £1,100,000 per year compared to the basic plant.
There is, however, in this case no steam available for export.

The above examples of how steam can be utilised are all based on actual
designs prepared by Sim-Chem for plants built recently at various locations
around the world. Although the cost of the power can vary considerably
depending on local circumstances a constant value of 1.5p per kWh was chosen
for each case as the intention was not to provide absolutely accurate figures
but to indicate the relative merits of different designs. It should be
stressed that in order to make it more correct one would also have to take
into account the values of different grades of export steam and the
increased cost of maintenance when turbines are used. It does however serve
to illustrate how much careful consideration must be given to the selection
of the steam/power system by both the acid producer and the designer when
a new project is being planned.

Plant Design versus the Environment

Before looking at the levels of sulphur dioxide and sulphuric acid mist
emissions which can be achieved on an IPA plant it is worth noting here that
in the U.K. the sulphur emitted from power station flues is more than the
amount of sulphur burnt to produce sulphuric acid.

Sulphur Dioxide.   As mentioned in the introduction, the proportion of
sulphur emitted from IPA plants in the U.K. must not exceed 0.5% of the
sulphur burnt or in other words a conversion of not less than 99.5%.   The
corresponding figure in the U.S.A. is 99.7%.   These conversions can be met
with some margin by paying careful attention to the following design features:-

A) OPERATING AT THE OPTIMUM OXYGEN/SULPHUR DIOXIDE RATIO AT THE INLET
    TO THE CONVERTER.
B) EMPLOYING SUFFICIENT CATALYST OF SUITABLE CHARACTERISTICS INCLUDING
    HIGH ACTIVITY, HARDNESS ETC. AND CHOOSING THE OPTIMUM CATALYST AND
    TEMPERATURE DISTRIBUTIONS.
C) DESIGNING AN OXIDATION SYSTEM WITH THREE PASSES OR LAYERS OF CATALYST
    BEFORE THE PRIMARY ABSORPTION TOWER AND ONE AFTERWARDS.   IN OTHER
    WORDS THE 3 - 1 IPA PROCESS.
D) PREVENTING GAS FROM BY-PASSING INTO THE FOURTH CONVERTER PASS BY
    EMPLOYING A GAS TIGHT DIVISION PLATE ABOVE IT IN THE CONVERTER AND BY
    AVOIDING CORROSION AND SUBSEQUENT GAS TRANSFER IN THE GAS/GAS HEAT
    EXCHANGERS.

It should be mentioned here that the alternative to the 3 - 1 system is the
2 - 2 system or two passes before primary absorption of the SO_3 and two
passes afterwards.   The merits of both systems have been investigated by
computer simulation and by comparing results from actual operating plants.
The results of this investigation indicate that although there is very little
difference in the theoretical conversion under optimum conditions, the 3 - 1
system gives a significantly better performance under plant upset conditions
such as poor first pass conversion, poor temperature control etc.

It is interesting to note that the SO_2 emission from a modern 1000 TPD IPA
plant designed to achieve 99.7% conversion is the same as that from a 150 TPD
conventional plant.   In some locations in Europe planning permission for an
expansion in acid production is only given on condition that the total new
sulphur dioxide emission is not more than the level on the present plant.
In other words an acid producer could in theory shutdown his old conventional

plant(s) and build a new IPA plant producing five times more acid and still
have a lower SO_2 emission.

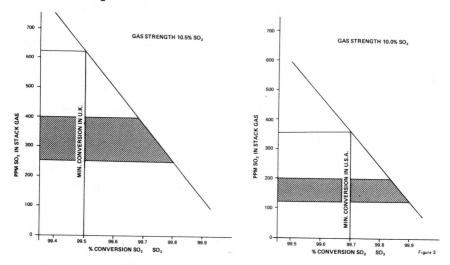

Figure 3 shows the relationship between conversion efficiency and SO_2 content
of the stack gas for different gas strengths.   We can see that based on a
design gas strength of 10.5% SO_2 by volume and a design conversion of 99.5%
the stack gas will contain about 620 ppm of SO_2.   The equilibrium conversion
under this condition is in the region of 125 ppm SO_2 at about 440°C and the
shaded band indicates typical operating results achieved on new plants
designed in this way and operating throughout the world.   Similarly for
99.7% design conversion and a gas strength of 10% SO_2 the stack gas will
contain around 350 ppm of SO_2.   The equilibrium conversion here is in the
region of 100 ppm SO_2 and the band shown represents typical data obtained
from new plants in the U.S.A.

These figures were all obtained by SO_2 monitoring instruments and Reich tests
and even allowing a conservative measuring error it is apparent that very low
SO_2 emissions close to equilibrium values are possible.   On seeing this
excellent performance data you might well ask why the permissible SO_2 limits
should not be reduced even further.   The answer is that although such low
emissions are possible with a well designed plant and new catalyst operated
correctly, a safety margin must be retained to allow for any deterioration
in equipment and catalyst and for any upset in operating conditions.

Sulphuric Acid Mist.   Turning now to the well known problem of mist emission
in sulphuric acid plants let us first spend some time looking at some of the
factors influencing design.

Sulphuric acid mist formation is normally initiated in the plant by a vapour
phase reaction between water vapour, produced from combustion of hydrocarbon
impurities present in the sulphur, and sulphur trioxide, produced in the
converter.   The mist is produced during the subsequent condensation of
sulphuric acid particles from the gas phase as it is cooled in the process
ahead of and inside the absorbing tower.   The quantity of mist produced is
dependent on a combination of factors some of the more important of which
are listed below :-

1) HYDROCARBON CONTENT OF SULPHUR - It is worth noting here that a low
    hydrocarbon (Bright) sulphur can contain less than 0.03% hydrocarbon
    whereas (Dark) sulphur can contain 0.2% or more.

2) TEMPERATURE TO WHICH SO_3 GAS IS COOLED - Normally one designs for a
    higher gas temperature leaving the economiser if dark sulphur is used.

3) WHETHER OLEUM IS PRODUCED AHEAD OF ABSORBING TOWER - It is found that
    the inclusion of an oleum absorber produces a finer mist and more of it.
    This effect increases with increasing oleum strength.

4) MAINTAINING OPTIMUM ABSORPTION CONDITIONS - It is found that too low a
    concentration of acid and/or too low an acid temperature can cause
    excessive mist formation.   This is often particularly true immediately
    following the plant start-up.

Having explained a little of the background to mist formation it is worth
adding here that until recent years phenomenon has probably caused more
problems with the public and the Alkali Inspector than has sulphur dioxide. -
MIST IS VISIBLE, SULPHUR DIOXIDE ISN'T!

The situation was changed dramatically in the late '60's when specially
designed highly efficient mist eliminators particularly the now world famous
"Monsanto Brink" packed fibre bed mist eliminators were developed and tried
out on commercial plants.   From then on the use of the mist eliminators
became more widespread until nowadays all new Sim-Chem and Envirochem Plants
incorporate at least one unit.

The Brink Mist Eliminator ranges from the High Velocity type (H-VBME) which relies on impaction for mist collection to the High Efficiency type (H-EBME) where Brownian movement is the controlling mist collection mechanism.   Both types consist of glass fibre packed between either rectangular or cylindrical stainless steel screens.

Some of the operating characteristics are given in Table 5 for the H-V and H-E types as well as the new H-P or high performance type which has just been announced.   With correct selection of the type and efficiency the acid plant designer and customer can be confident of achieving a virtually invisible stack gas appearance irrespective of the type of sulphur used or whether oleum is produced in the plant.

TYPES OF MONSANTO BRINK MIST ELIMINATORS AND THEIR OPERATING CHARACTERISTICS

Type	High Velocity	High Efficiency	High Performance
Controlling Mist Collection Mechanism	Impaction	Brownian Movement	Impaction
Collection Efficiency on Particles >3 Micron	100%	100%	100%
Collection Efficiency on Particles 0.5 − 1.0 Micron	50 − 85%	94 − 99%	70 − 95%
Pressure Drop (mm w.g.)	150 − 200	150 − 400	200 − 250

Table 5

Plant Design versus Equipment Selection

Having discussed the process, energy and environmental considerations, it is a good time to look at some of the alternatives facing the designer when selecting equipment.

One obvious decision to be made is to fix the waste heat recovery system after consideration of the many factors mentioned previously.   The No. 1 boiler following the sulphur burner is the most important piece of waste heat recovery plant and the cooling duty can be achieved in either a water tube or a fire tube design.   The fire tube boiler is generally favoured as it is less expensive in first cost, requires less maintenance and has a relatively low heat retention capacity.   People have talked for many years of the time when plant capacities and steam pressures will increase such that the fire tube design would no longer be mechanically practicable and the designer would be forced into adopting a water tube design.   Such a change-over point may well exist but as yet has not been reached and it certainly looks as though plants of up to 2000 TPD H_2SO_4 can be equipped with fire tube boilers at pressures of up to 50 bars.   It should be mentioned here that over recent years careful attention has been devoted to protecting the hot

inlet of the fire tube boiler with ceramic ferrules inserted in the tube ends

and set in a castable refractory covering the tube sheet.

The acid circulation cooling system offers the designer perhaps more freedom

of choice than any other section of the plant. This is mainly due to the

fact that the irrigated cast-iron pipe coolers used religiously for years

in sulphuric acid plants have become more and more unacceptable from the

maintenance and environmental standpoints thus encouraging a number of

companies to develop special heat exchangers for cooling hot sulphuric acid.

Some of the main types of coolers have been listed in Table 6 and their

relative merits are compared. The choice between plate coolers and shell

and tube units is often made on economic grounds as it can favour either,

depending on the size of the installation. The air-cooled type however,

being more expensive, is normally considered when there is a shortage of

water and no doubt its popularity will increase if water continues to

become scarce and costly in the U.K.

COMPARISON OF ACID COOLERS

Type of cooler	Normal materials of construction	Typical inlet temp. when operating with 98% H_2SO_4	Utilities per tonne H_2SO_4		Possible environmental concerns	Remarks
			Water make up required with a cooling tower circuit	Extra power consumption incl. water cooling		
PIPE COOLERS Water cooled	SPUN CI	100°C	Up to 1.8 tonnes	6 – 8 kwh	Excessive steam plume and spray from pipe cooler and plume from cooling tower.	1. Requires a lot of maintenance. 2. Requires large plot area. 3. System is not acceptable for most new projects.
SHELL AND TUBE TYPE Water cooled	316 LSS	110°C	Up to 1.8 tonnes	6 – 8 kwh	Plume from cooling tower.	1. Protected by anodic passivation equipment. 2. Can be made in large single shells making them very attractive for larger plants.
PLATE TYPE Water cooled	Hastelloy C	90°C	Up to 1.8 tonnes	6 – 8 kwh	Plume from cooling tower.	1. Easily cleaned. 2. Extra surface can be added. 3. Especially attractive on smaller plants since large heat duties require a multiplicity of units.
FIN-FAN TYPE Air cooled	316 LSS	80°C	None	11 – 13 kwh	Noise from fans.	1. Most expensive. 2. Requires large plot area. 3. Requires no water.

Table 6

Selection of the type of mist eliminators has already been mentioned in the

context of the stack appearance. It is equally important on an IPA plant to

install the correct type of mist eliminator in the primary or interpass

absorbing tower as any acid carryover whether as droplets or acid mist can damage heat exchangers and catalyst downstream. The importance of this cannot be over-stressed as this knowledge including the correct solution was acquired from experience with early design of IPA plants.

Interest is being shown in ways to extend the life of equipment subject to corrosion from acid or scaling from high temperature. Examples include the use of membranes such as 'Teflon' or 'Rhepanol' between the acid-resisting brick and steel shells of tower and pump tanks. In this case the membrane is regarded as insurance against the possibility of corrosion attack on the steel shell resulting from acid leaking through the brick lining. Other examples include the use of 'Alonised' tubing for hot heat exchanger tubes and 'Aluminising' of converters and hot ducting, both of which measures are taken to protect the carbon steel against corrosion-scaling effects in a hot sulphurous atmosphere.

Plant Start-up

More and more attention is being given to minimising the sulphur dioxide and acid mist pollution on plant start-up. A good start-up can be achieved by making sure the following conditions are fulfilled :-

1) SULPHUR BURNER AND OXIDATION UNIT IS ADEQUATELY PRE-HEATED.
2) ABSORBING ACID IS OF THE CORRECT STRENGTH AND TEMPERATURE.
3) PLANT IS STARTED INITIALLY AT REDUCED LOAD WITH A LOW GAS STRENGTH.

The plant is prepared for start-up by burning fuel in air and heating up the sulphur burner and converter by direct contact with combustion gases or by indirectly heated air. The general aim is to raise the temperature of the sulphur burner such that sulphur combustion can be sustained and to raise the temperature in the top of the converter above the minimum reaction or strike temperature range of 400 to $425^{\circ}C$ so as to initiate the exothermic reaction when SO_2 contacts the catalyst.

In the past it was often sufficient to heat only the first catalyst pass of the converter above this reaction or strike temperature and then rely on the reaction heat produced to heat up the remaining passes in turn until all four were converting. The main objection to this is that it takes a number of hours to reach a steady state condition and during this time the conversion of SO_2 to SO_3 is increasing only slowly from an initial 70% or so on one pass to the design conversion of more than 99.5% across four passes. This means that there is a period of a few hours when the sulphur dioxide emission from the stack is many times that of the normal value. This mode of start-up is no longer acceptable and acid producers in the U.K. must now ensure that at least two converter passes are pre-heated to reaction temperature before going on line. For a little extra cost additional gas ducting and valves

can be incorporated to enable three or four passes to be heated up above
strike temperature.   This more thorough start-up procedure does involve
burning extra fuel and a longer heat-up period but offers the advantage of
reducing the time during which the SO_2 emission is above the normal design
value.

Another important requirement of a good start-up is to have the acid in the
absorbing system up to the optimum strength for absorption.   If dilution of
the acid has occurred during oil firing the strength can be raised by
importing oleum into the absorbing acid pump tanks.   It is also a good idea
to raise the acid temperature just prior to start-up by passing the heat-up
gas through the absorbing tower whilst circulating acid.   These two measures
will minimise the mist content of the exit gas when the plant goes on line.

Finally, one should ensure that the plant is initially operated at a low
sulphur input load with a low gas strength so as to maximise the conversion
on those passes which are already preheated.   Low load operation should
continue until the four converter passes are raised above the reaction
temperature and then the load should only be increased gradually at the same
time monitoring the SO_2 emission in the stack gas.   During this initial
period when the plant goes on line the absorbing acid temperature should be
raised quickly by either by-passing the absorbing acid cooler or by regulating
the flow of cooling medium to the cooler.

What of the Future ?

When one considers the tremendous changes in Sulphuric Acid Plants and
equipment design over the last decade, it makes it very difficult to look
into a crystal ball and predict the sort of development we might expect in
the next ten years.

Sulphur dioxide emission levels may well be tightened up even further,
requiring conversions of more than 99.7% to be achieved at all times during
normal operation, as is the case in the U.S.A.   It is expected that the IPA
process will continue to be the "norm" although interest may still be shown
in tail gas scrubbing plants where levels of 100 ppm or lower are demanded
in a local situation.

In the short term, catalyst development work should continue gradually to
improve its main characteristics including activity, hardness and pressure
drop.   Looking further ahead in the catalyst field, it is possible that one
day a new "low temperature" catalyst might be developed, which could
revolutionise plant design by making it possible to achieve the required level
of sulphur dioxide emission from a conventional or single absorption plant.

Development work should continue on plants operating at an elevated gas pressure by which means higher conversions and smaller sizes are possible. These advantages might, however, be more than off-set by extra mechanical problems and higher energy consumption making the idea not quite as attactive.

It is unlikely that tighter controls on mist emissions will be demanded as a stack "free of persistent mist" is called for at the moment. Instead, mist eliminator designs will probably be refined resulting in lower costs and pressure drops.

Start-up controls can be expected to be tightened up, calling for further pre-heating of the catalyst and more attention to be paid to ensuring that the absorbing acid is at the optimum strength and temperature.

The continuing energy crisis will force the designers to look carefully at power consumption and to maximise waste heat recovery. Acid producers can be expected to make more use of the steam produced by employing more turbine drives or generating power in a turbo alternator set.

Smaller conventional acid plants in the U.K. will probably be phased out slowly and in the case of large acid producers will be replaced by IPA plants, having capacities around 1000 tons per day. Major equipment design will continue to be improved, particularly the sulphur burner, converter and absorption towers. Improvements can also be expected within the design of smaller items such as acid valves and instruments for acid measurements and SO_2 monitoring.

Hazard studies will see the introduction of additional safety measures in plant design and operation to safeguard against the possibility of explosions during oil firing and against major SO_2 and acid leaks. More attention may be given to the siting of acid storage tanks.

Noise control will result in more silencing equipment and acoustic barriers as well as the selection of low noise machinery and control valves.

REFERENCES

1.   National Sulphuric Acid Association Limited, London
      - Data for 2nd and 3rd quarter 1976.

2.   110th Annual Report on Alkali, & C. Works, 1973.

Modern Processes for the Production of Ammonia, Nitric Acid and Ammonium Nitrate

By S.P.S. ANDREW

ICI Ltd., Agricultural Division.

Introduction

The fixation of atmospheric nitrogen either in the form of ammonia or of nitric acid is one of the foundations of the modern chemical industry. Combined in the form of ammonium nitrate, urea or ammonium phosphate over 50 million tonnes/year are used worldwide as fertilizers. Other uses of fixed nitrogen amount to some 10 million tons/year, explosives, dyestuffs and resin manufacture predominating. The great quantitative demand for fixed nitrogen, both as fertilizer and for explosives uses, led at the beginning of this century to the development of a new form of chemical industry. Whereas previously the industry had been based on processing gases and solids at substantially atmospheric pressure, the new technology required high pressures. Where previously chemical reactions usually occurred often as the result of the use of high temperatures, the new technology was based on the use of solid catalysts operating at relatively low temperatures for promoting the desired gas phase reactions. The last fifty years have seen a vast expansion of the catalytic processing based on heterogeneous catalysts so that now not only ammonia and nitric acid production but also much of the rest of the chemical and petroleum refining industries employ this method of accelerating chemical reactions.

Great advances have been made during the past fifty years both in the catalysts used and in the engineering of the processes for fixing nitrogen. Nevertheless the basic chemistry of these processes remains very much as it was worked out during the first

World War by Haber and Bosch, in which the fixation of nitrogen

is brought about by the union of nitrogen and hydrogen to form

ammonia. As a fertilizer, ammonia derivatives are,

thermodynamically, an inefficient form of supplying nitrogen to

plants, as the hydrogen content of the ammonia, which represents

a substantial input of energy, is virtually completely wasted,

being oxidised to water by the nitrate forming soil bacteria

before the ammonia is taken up by the plants which must then

expend further energy, derived by photosynthesis, in re-reducing

to ammonia this nitrate. Plant biochemistry does not, however,

permit the rapid absorption of large quantities of ammonia and

commercially viable processes for fixing nitrogen as nitrates do

not yet exist. This basic thermodynamic inefficiency of

fertilizer ammonia use has long been known but no satisfactory

means of overcoming it appears at present to be in sight.

Thermodynamics and Energetics of Nitrogen Fixation

In order to consider in more detail the theoretical heat

requirements of nitrogen fixation it is useful to set down the

heats of the following four reactions for fixing a gram atom of

nitrogen from those readily available species, atmospheric

nitrogen, atmospheric oxygen and liquid water :-

$$\tfrac{1}{2}N_2 + 3/2H_2O \rightarrow NH_3 + \tfrac{3}{4}O_2 + \text{(91.5 k cals absorbed/ g mol)}$$

$$\tfrac{1}{2}N_2 + \tfrac{1}{2}O_2 \rightarrow NO + \text{(21.6 k cals absorbed)}$$

$$\tfrac{1}{2}N_2 + O_2 \rightarrow NO_2 + \text{(8.0 k cals absorbed)}$$

$$\tfrac{1}{2}N_2 + \tfrac{1}{2}H_2O + 1\tfrac{1}{4}O_2 \rightarrow HNO_3 + \text{(7.2 k cals evolved)}$$

From the above it will be seen that fixation of nitrogen as

ammonia is very demanding in energy supply, whereas,

theroetically, its fixation as nitric acid supplies energy.

Correspondingly, at normal pressures and temperatures the

equilibrium of the ammonia forming reaction is extremely

unfavourable whereas that of the nitric acid forming reaction is moderately favourable. Clearly, however, there must be a most severe activation energy barrier to this last reaction otherwise the sea salts would consist predominantly of nitrates! So far no catalysts have been found to reduce this activation energy barrier to a commercially acceptable value. No satisfactory catalyst is even available for the NO_2 forming reaction, which might offer an acceptable energetic alternative. The equilibrium of the NO forming reaction is adverse, except at very high temperatures (> 2400°C), where equilibration is rapid even in the absence of a catalyst. A non catalytic fixation process based on this reaction using regenerative heat exchangers for heating and cooling air to about 2600°C has been proposed and no doubt could be operated as a lower energy consumption process for fixing nitrogen than the ammonia route. The high capital cost of the regenerators, however, under current commercial conditions render this possibility unattractive.

Ammonia Production

Thermodynamics. The ammonia route, though energetically the least satisfactory is currently, and looks as if it will remain for many years, the only one which is commercially viable for industrial nitrogen fixation. As written, the equation implies the splitting of water to give hydrogen for the ammonia and rejecting oxygen. This is the overall mass balance for an ammonia synthesis process using electrolysis of water to supply the hydrogen and a liquid air fractionation unit to supply the nitrogen. The pure product hydrogen and nitrogen from these plants could then be reacted together catalytically to give ammonia.

This last catalytic process is know after its commercial developer - Fritz Haber - who realised that in order to bring

about the synthesis reaction in commercially acceptable yields
at an adequate rate the reactor would have to operate at an
elevated pressure (100-200 atmosphere) and furthermore that it
would even then be necessary to recirculate unconverted nitrogen
and hydrogen to the synthesis reactor, having removed the ammonia
formed. This arrangement, known as 'the synthesis loop' has
remained a feature of all ammonia synthesis plants since Haber's
patent of 1908.

It was early realised that the supply of hydrogen by the
electrolysis of water was expensive except in regions where
electricity was very cheap. An alternative process was therefore
sought. Carl Bosch of BASF, working with Fritz Haber developed
the use of purified water-gas, obtained by the reaction of water
with red hot coke. The reactions involve are:-

$$C + H_2O \rightarrow CO + H_2 + (41.9 \text{ k cals absorbed/g mol})$$

$$C + 2H_2O \rightarrow CO_2 + 2H_2 + (42.6 \text{ k cals absorbed})$$

In order to supply the heat for the reaction it is necessary to
operate the water gas generator in a cyclic manner, alternating
the production of water gas with periods of heating the coke bed
by blowing air through it. At the high temperature of the water
gas generator coke bed, the reaction:- $CO + H_2O \rightleftharpoons CO_2 + H_2 +$
(9.8 k cals produced) known as the water gas shift reaction,
produced a substantial fraction of carbon monoxide, representing
a loss of potential hydrogen. This loss was, however, obviated
by following the water gas generators with a water gas shift
reactor operating at a much lower temperature (c $450^{\circ}C$)
containing an iron oxide–chromium oxide catalyst. The equilibrium
of the shift reaction is moved to the right by operating at a
reduced temperature resulting in a high conversion of the carbon
of the coke to carbon dioxide, which was removed from the hydrogen

containing gas stream, after compression, by a water wash. The
nitrogen for the synthesis reaction was admitted into the system
as air which was passed through continuously operating 'producer
gas' generators where the oxygen was removed by reaction with
red hot coke. The overall stoichiometry for the ideal ammonia
synthesis process for this reaction sequence is thus (simplified):-

$$\text{Air} \qquad \text{Water} \qquad \text{Coke}$$
$$(\tfrac{1}{2}N_2 + \tfrac{1}{8}O_2) + 3/2H_2O + 7/8C$$
$$NH_3 + {}^{\downarrow}7/8CO_2 + (9.2 \text{ k cals absorbed/g mol})$$

Though the overall reaction appears now to be only slightly
endothermic, unfortunately no very effective means had been
developed for recycling heat from the very hot exit gases of the
coke bed to the ingoing steam and air. In consequence
considerable quantities of heat had to be supplied to the coke
bed of the water gas generators during the air blow period.
Furthermore as the gas generators operated at about atmospheric
pressure and the ammonia synthesis reaction occurred at over a
hundred atmospheres, there is a very considerable power
requirement for gas compression, which if supplied by steam
driven machines requires an extensive steam generation plant with
even further coal consumption. Instead therefore of using the
theoretical about one mole of carbon required to produce one mole
of ammonia, in practice over four times this figure was used.
All of the remaining three moles of carbon being burnt to form
heat which was ultimately and expensively rejected, normally to
cooling water.

This, in essence was the process as developed by Haber and Bosch
by 1916. Over the past sixty years as a result of major efforts
the thermal efficiency of ammonia synthesis for air, water and
a carbonaceous feedstock has been improved from 20% to 60% and

the capital cost/ton year of production reduced tenfold. Though
significant advances have been made in the catalytic chemistry,
the metallurgy and the mechanical engineering of ammonia plants
in these years, the major cause of this great improvement in
both efficiency and capital cost is the move from using a solid
fuel, coal and coke, to a liquid hydrocarbon or even natural gas.
The advantage of greatly increased scale of production consequent
on greatly expanding the use of nitrogenous fertilizers is also
very significant.

Ammonia Production from Air, Water and Hydrocarbons

The Steam Reforming Process. In the UK, ammonia currently is
entirely made using methane as the purchased feedstock. The
simplified internal stoichiometry of ammonia production is:-

$(\frac{1}{2}N_2 + \frac{3}{8}O_2) + \frac{5}{8}H_2O + 7/16CH_4 \rightarrow NH_3 + 7/16CO_2 +$ (1.6 k cal evolved/
                                                                                                                    g mol)

Both the simplified theoretical stoichiometrics for methane and
the carbon feedstocks can be written as:-

$(\frac{1}{2}N_2 + \frac{3}{8}O_2)$ + water + 91.4 k cals/g mol $\rightarrow NH_3$ + carbon dioxide

With a coal or coke plant supplying the heat however, the
practical heat input is, even with the best modern plants still
markedly higher than that for modern methane using plants as
the recycling of heat is much more difficult.

The use of methane as a feedstock for ammonia production was
first commercially undertaken in the early 1930's as a result
of the development of the methane-steam reforming process. This
process essentially makes a water-gas by reacting methane with
steam over a nickel catalyst at 700-800°C. The overall reaction,
as with the coke process, is endothermic and the catalyst is
disposed inside narrow tubes through which the reactant gases
pass, these tubes being suspended in a furnace which is maintained

at about 1000°C. Use of this process remained severely
restricted by the supply of natural gas in the UK and Europe
until the discoveries of the last fifteen years. The extension
of the steam reforming of methane process to higher boiling
feedstocks widened its applicability; the major step of catalyst
modification being made in 1960 by ICI whereby naphtha (final
boiling point up to 180°C) could be reformed. Extensive use of
naphtha reforming was made throughout the world during the 1960's
for producing ammonia. The overall internal stoichiometry of the
ideal synthesis process for naphtha (approximate formula C_7H_{14})
requires about 0.084 mols of C_7H_{14} per mol of NH_3 made. Modern
plants use about 0.13 mols of C_7H_{14} total. The fuel required by
the internal stoichiometry, which, like that for using a methane
or even a carbon feed, is almost sufficient in theory to supply
the energy for the complete chemical transformation to produce
gaseous ammonia. The extra fuel, amounting to a further 55% over
the stoichiometric is all used in modern plants for raising the
temperature of the gases leaving the steam reformer to a temperature at which the equilibrium of the reaction:-

$$CH_4 + H_2O \rightleftharpoons CO + 3H_2 \text{ (the methane-steam reaction)}$$
is displaced well to the right.

The Partial Oxidation Process

Whereas both the water gas and the steam reforming processes
use an external supply of heat to raise the carbonaceous fuel-
steam-air mixture to a sufficiently high temperature for
favourable kinetics and equilibria for synthesis gas production,
the partial oxidation process uses internally generated heat.
In effect, additional fuel is burnt inside the process stream to
generate this heat. This fuel cannot, however, be burnt with
air as otherwise the overall internal stoichiometry leading to

the $3N_2:1H_2$ synthesis would fail due to excessive nitrogen.

Therefore the excess fuel must be burnt using pure oxygen. The

partial oxidation process in consequence requires an air

separation plant to supply the process with a N_2-O_2 mixture

enriched in oxygen relative to normal air. The requisite N_2 and

O_2 are not however, usually supplied as a mixture but as pure

oxygen, which is fed to the fuel-oxygen-steam burner, and pure

nitrogen which is added later to the process gas stream just

prior to the ammonia synthesis stage. The surplus nitrogen to

that required is vented from the air separation plant.

Though, at first glance, it might be thought that such a process

using entirely combustion in the process stream to produce a high

temperature in the process gases and hence a favourable

equilibrium of the methane-steam reaction (see above), would be

more thermal efficient than the externally heated catalytic

steam reforming process, in fact this is not so. Even an advanced

partial oxidation process still uses some 10% more fuel to produce

an equivalent amount of ammonia compared with a steam reforming

process. The reason for this lower efficiency lies in the

detailed flowsheeting of the two processes and the relative

abilities to utilise waste heat to drive the air compressors,

gas compressors and refrigeration equipment in these processes.

In order to assess these details it is first necessary to consider

the chemical, catalytic and engineering limitations of the

various parts of these processes.

The Ammonia Synthesis Operation

An understanding of the whole process is best started from the

rear end, the ammonia synthesis reaction, as the conditions under

which this reaction occur define the nature of the gas mixture

which the front part of the process must supply. Two factors are

of crucial importance in the synthesis - firstly, the
equilibrium of the synthesis reaction, secondly, the kinetics
of synthesis over the heterogeneous catalyst. The synthesis
reaction equilibrium:-

$$N_2 + 3H_2 \rightarrow 2NH_3 + (21.9 \text{ kcal evolved/g mol})$$

is markedly temperature and pressure sensitive as can be seen
from the following table which gives equilibrium % ammonia for a
feed of $N_2 + 3H_2$ at various temperatures and pressures:-

	25 atm	50 atm	100 atm	200 atm	400 atm
100°C	91.7	94.5	96.7	98.4	99.4
200°C	63.6	73.5	82.0	89.0	94.6
300°C	27.4	39.6	53.1	66.7	79.7
400°C	8.7	15.4	25.4	38.8	55.4
500°C	2.9	5.6	10.5	18.3	31.9

The kinetics of the synthesis reaction is determined by the
efficacy of the synthesis catalyst. Since the early days of
commercial ammonia synthesis this catalyst has been essentially
metallic iron promoted by KOH and containing a small amount of
mixed refractory oxides such as Al_2O_3, SiO_2 and MgO. The
promoting effect of KOH is considerable (activities being raised
by roughly an order of magnitude per unit surface area of iron).
Other metals, particularly ruthenium are also very active for
ammonia synthesis. They too are greatly promoted by KOH. Other
alkaline hydroxides particularly caesium hydroxide are good
promoters. Appreciable, though lesser activity for ammonia
synthesis is obtained with cobalt, nickel, rhodium, osmium,
iridium and molybdenum. The combination of good activity together
with cheapness of raw materials and low cost of fabrication have
secured a firm position for iron as the catalytic metal.

Fabrication of the catalyst is simply the melting together of the ingredients in their oxide forms and the casting of the molten iron oxide (Fe_3O_4) melt containing dissolved potassium oxide, aluminium and other refractory oxides on a table. The resulting solid sheet is then broken up and pieces sieved to give 5-10 mm chunks of catalyst in the oxidised state. To activate the catalyst it is reduced by ammonia synthesis gas ($N_2 + 3H_2$) inside the ammonia synthesis converter. In its active state the catalyst consists predominantly of iron crystallites of a few hundred Angstrom units in size. These crystallites are separated by amorphous refractory oxides and partially covered by alkali promoter probably in the form of KOH.

Over the fifty years during which this catalyst has been used a number of improvements have been made to its formulation as a result of much empirical experiment. The resultant catalyst is now very significantly more active. An equally large improvement in catalyst activity and life has been secured by eliminating catalyst poisons from the gas entering the ammonia synthesis reactor. Hydrogen sulphide and water vapour are both poisons which in the earlier days severely limited catalyst life and activity. Both are temporary (reversible) poisons and water vapour is, in addition, a potent sintering agent, resulting in a recrystallisation of the catalysts structure giving larger iron crystallites and hence a lower gas/metal interfacial area with consequent loss in catalytic activity. Water vapour can enter the catalyst either as itself or as CO or CO_2, for both of these gases will be hydrogenated to methane and water vapour over the catalyst. Rigorous exclusion of H_2S, H_2O, CO and CO_2 from the catalyst is therefore desirable. Indeed the life of the catalyst (which is nowadays from 2 to 10 years) is very dependent on the ability to effect this exclusion down to levels of less than 0.1 ppm. Methane and argon are inert and merely

act as diluents.

The commercial operating range of ammonia synthesis catalyst

is between $400^{\circ}C$ and $540^{\circ}C$. Below $400^{\circ}C$ the catalyst is not

sufficiently active and above $540^{\circ}C$ it loses surface area by

sintering too rapidly. When well removed from equilibrium, the

synthesis rate is roughly proportional to the synthesis gas

pressure to the 1.5 index and doubles for every $20^{\circ}C$

temperature rise. In order, therefore, to obtain reasonable

conversions per pass through the converter, bearing in mind the

above equilibria, synthesis pressures in modern plants

fall in the range 80-350 atm.

Because of the improvement in equilibrium conversion with falling

temperature and the increase in synthesis rate when remote from

equilibrium with rise in temperature, it is valuable to arrange

for the temperature of the reactant gas mixture passing through

the catalyst bed to fall as the reaction approaches equilibrium.

As the reaction evolves heat this end can only be attained either

by removing heat through the walls of a heat exchanger or by

adding cold gas. The designer of ammonia synthesis converters at

this point has the choice of improving the thermal efficiency of

his process by raising steam using boiler tubes inside the

reactor at the expense of greater engineering complexity, higher

capital cost and a likely lower reliability of the equipment, or

aiming for somewhat less thermodynamic efficiency with lower

capital cost and higher reliability through using a simpler cold

gas addition system.

This choice typifies the problem of the ammonia plant designer,

who, throughout the whole design of the process must balance the cost of complexity against savings in fuel. He must also bear in mind the practical, as opposed to the theoretical, availability of equipment, particularly the availability of suitable gas compressors and turbine drives and of large pressure vessels. In no part of the process are such considerations more significant than in synthesis. The choice of synthesis pressure is little affected by considerations of the total mechanical power requirements for gas compression as, over a wide range of pressures (100-300 atm), the sum of the powers required for gas compression plus gas circulation around the synthesis loop plus power for driving a refrigeration system for condensing liquid ammonia (at temperatures as low as -30°C for the low pressure synthesis) remains virtually constant. The lower make-up gas compression power of the low pressure synthesis system is almost exactly counter-balanced by the higher refrigeration and circulation power. The current trend to lower synthesis pressures has been greatly influenced by the availability of suitable cheap rotary gas compressors capable of handling relatively efficiently the volume of make-up gas entering the synthesis loop. As rotary compressors are not available for the efficient compression of low volume flows, the synthesis pressure may be dropped so that the volumetric flow rises into the region in which commercial compressors are available. Steam turbine drives are usually employed for these compressors, the steam being either partially or wholly generated by heat rejected from the process. An auxiliary boiler is essential in partial oxidation plants using turbine drives as the power requirements of the cold production machinery exceeds that which can be derived from steam generated from waste heat. For steam reforming plants an auxiliary boiler may be necessary for start-up and for part load running where compressor and turbine efficiencies are much lower than at full load.

Synthesis Gas Purification

When a hydrocarbon is used for effecting the production of
synthesis gas, the requirements of the synthesis catalyst as
indicated above for purity of the make-up gas, entail a number
of stages of gas purification. Major quantities of CO, CO_2 and
H_2O must be removed together with lesser amounts of sulphur
containing compounds such as COS and H_2S. If a residual fuel or
coal is used then, in addition, ash must be removed. The chemical
nature of this ash, coupled with the variability of its
composition is, perhaps, the greatest disincentive to the use of
such process feedstock.

Working back from ammonia synthesis, the final stage of purification of the make-up gas is, in modern plants, a liquid scrubbing
of the gas. In steam reforming plants, the liquid used is ammonia
which absorbs residual water and carbon dioxide - unfortunately
carbon monoxide is not absorbed. In partial oxidation plants,
where pure nitrogen is available from the air separation plant,
a liquid nitrogen wash is employed, which also removes this
carbon monoxide, together with inerts such as methane and argon.
These inerts would have had to be purged from the synthesis
loop in a steam reforming process.

Because of the inability of the ammonia wash to remove carbon
monoxide, the steam reforming plants require a further final
purification stage for this component. This is a catalytic
conversion of the monoxide (and also of any dioxide) to methane
by passing the gases over a hydrogenation promoting catalyst.
This catalyst is usually metallic nickel supported on alumina or
on a combination of suitable refractory oxides. It operates at
around 300-350°C and CO levels are reduced to less than 0.1 ppm.

The great volume of carbon dioxide produced from all the carbon that enters the process stream, is removed immediately prior to the above final purification stages. Two types of continuous removal systems are in use. Those which employ a chemical reagent wash of the gas and those which use phsyical solution of the gas in a suitable solvent. The former tend to operate at higher temperatures than the latter, as high temperature enhances the velocity of chemical reaction whereas low temperature enhances physical solubility. Chemical reagents fall into two classes, those which use an aqueous alkaline absorbent and operate by the reaction:-

$$CO_2 + OH^- \rightarrow HCO_3^-$$

and those which use an amine reagent which forms a carbamate:-

$$CO_2 + R-NH_2 \rightarrow R-NHCOO^- + H^+$$

Frequently a mixture of these two types of reagents are used, for instance, aqueous potassium carbonate plus di-ethanolamine. Having absorbed the carbon dioxide by counter-current contacting of the process gas with the wash liquor, the latter is regenerated by stripping out the carbon dioxide using a suitable stripping gas. When the wash liquor is aqueous and hot, the only suitable stripping gas is water vapour and these processes usually operate at the boiling point of the liquor in the stripping column (about $110^\circ C$). This type of wash process is usually employed in steam reformer plants where there is a good supply of heat available for generating stripping steam.

In contrast the physical solvent systems are chosen for partial oxidation plants which are short of heat but are better supplied with cold. A low melting point, but not too volatile oxygenated organic solvent is customary, for instance methanol, and the process may operate at a sub-zero temperature. In addition to removing CO_2 both types of wash process also remove H_2S. Carbon monoxide in the gas leaving either the steam reformer or the partial oxidation burner is the result of only partial completion of the desired gas-making reaction. This partial completion is caused by the water gas shift reaction $CO_2 + H_2 \rightleftharpoons CO + H_2O$ being displaced to the right at the high temperatures ($1100^{\circ}C$) at the end exit of the partial oxidation burner or of the reforming system. In order to complete the hydrogen making reaction it is necessary to drop the temperature in the presence of a catalyst which promoted selectively the water-gas shift reactions. Such catalysts are iron oxide/chromia (active at c $400^{\circ}C$) and copper/zinc oxide/alumina (active at c $200^{\circ}C$). Frequently these catalysts are used in series as a means of reducing the carbon monoxide from about 10% (on a dry basis) of the gases leaving the reformer or oxidiser to about 0.2% exit the second (Cu catalyst) shift converter.

Synthesis Gas Production

So far in this note nothing has been written concerning the pressure of gas passing along the purification system. Very simply this is chosen to be as high as possible consistent with engineering limitations, all of which occur in the gas production stage. To understand these limitations it is necessary to consider the alternative gas production processes of steam reforming and partial oxidation.

The steam reforming process, whether using naphtha or methane as a feedstock is limited in its ability to convert hydrocarbon to hydrogen and carbon monoxide by the position of the methane-steam equilibrium:-

$$CH_4 + H_2O \rightleftharpoons CO + 3H_2$$

High temperatures displace this equilibrium to the right and high
pressures to the left. For practical operation, high pressure
gas making therefore means high temperatures, circa 1000-1100°C
at 30 atmospheres. Because the reaction for making the synthesis
gas mixture from air, steam and hydrocarbon is endothermic (see
above) and the extra heat must be supplied by heat transfer
through a heat exchanger wall suspended in a furnace, then this
wall has to withstand both a high temperature and a high pressure
differential. The duty required of these heat exchanger tubes is
mitigated somewhat by splitting the steam reformer into two
sections, a colder first half (the primary reformer) where only
steam plus hydrocarbon pass through tubes suspended in a furnace
and a hotter second half which operates adiabatically being
heated internally by injection of the necessary air and combustion
of its oxygen to CO and CO_2.

Nevertheless exit temperatures of the process gas from the
furnace of 850°C and tube wall temperatures of 950°C are now
normal. At these temperatures current tubes are limited to
about 35 atmospheres differential pressure in order to retain
an adequate creep life (10,000 hrs). Thus 35-40 atmospheres is
about the highest convenient steam reforming pressure. As
ammonia synthesis equipment becomes very voluminous at
pressures below 80 atmospheres, some make-up gas compression is
therefore almost essential in a steam reformer plant. This
compression usually takes place after the methanation stage of
purification and before ammonia wash which is combined with
product ammonia separation by condensation in the synthesis loop.

Limitations to synthesis pressure are not nearly so severe in
partial oxidation, as the whole process of burning hydrocarbons

with oxygen and steam takes place in an internally insulated vessel. In order to secure the highest possible temperature, the nitrogen required is normally admitted after the oxidation burner, usually in the nitrogen wash purification stage. Care must be taken, particularly at high pressures, that the carbon forming equilibrium $2CO \rightarrow CO_2 + C$ is not crossed excessively in any part of the combustion process otherwise soot formation would be excessive. Pressure of up to 100 atmospheres are commercially possible at present with this type of process, so that single pressure plant for gas production, purification and ammonia synthesis can be constructed.

Hydrocarbon Feedstock Purification

Steam reforming is a catalytic process employing metallic nickel supported on a suitable refractory ceramic. With higher molecular weight hydrocarbons than propane it becomes increasingly necessary to add a further component to the catalyst to ensure that should any carbonaceous material be formed on the catalyst surface by thermal cracking of the hydrocarbon, then this carbon is readily gasified by reaction with steam. A suitable catalyst addition for this purpose is an alkali such as KOH, which, being two-dimensionally mobile at the operating temperature of the process, is capable of climbing onto the surface of any carbon deposits and promoting their steam gasification.

Nickel catalysts are, however, poisoned by sulphur, vanadium and arsenic and compounds of these, especially sulphur compounds must be excluded from entering the primary steam reformer. The hydrocarbon feedstock is therefore first hydrodesulphurised over a cobalt-molybdenum-alumina catalyst and the resulting H_2S formed very effectively separated.

The efficiency of this hydrodesulphurisation plus H_2S
absorption step must be very high, as the steam reforming catalyst
is severely poisoned when sulphur levels exceed about 0.5 ppm.
The ability of the hydrodesulphurisation step to hydrogenate
out sulphur from the complex thiophenes, disulphides and
mercaptans existing in a liquid hydrocarbon feedstock is very
dependent on the final boiling point of this feedstock as the
heavier more complex molecules are much more difficult to
process. In practice this difficulty sets an upper limit to
the boiling point of liquid feedstocks that can commercially be
steam reformed, using a nickel catalyst. Materials with a final
boiling point higher than $200^{\circ}C$, though they could be steam
reformed satisfactorily over a suitable nickel catalyst if they
had been desulphurised to less than 0.5 ppm, cannot readily be
so desulphurised.

The partial oxidation gasification process, being a simple non-
catalytic combustion process, is not limited in feedstock by
the above considerations. It therefore offers a more flexible
choice of feedstocks and the potential of purchasing feedstock
cheaper on a thermal basis than steam reforming. This is,
normally, the strongest argument for choice of partial
oxidation.

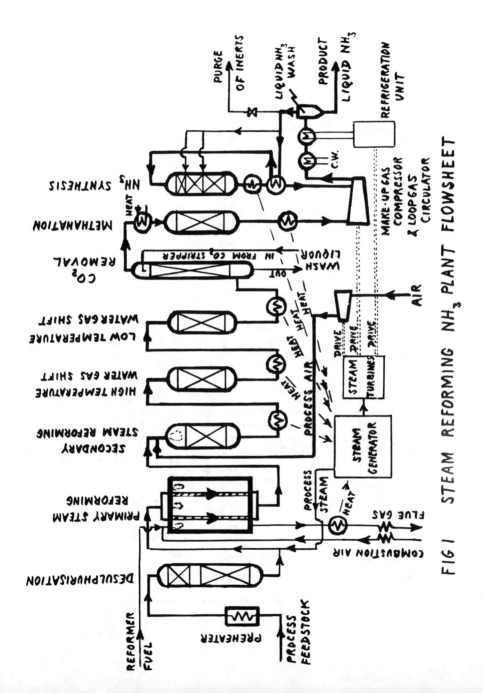

FIG I STEAM REFORMING NH₃ PLANT FLOWSHEET

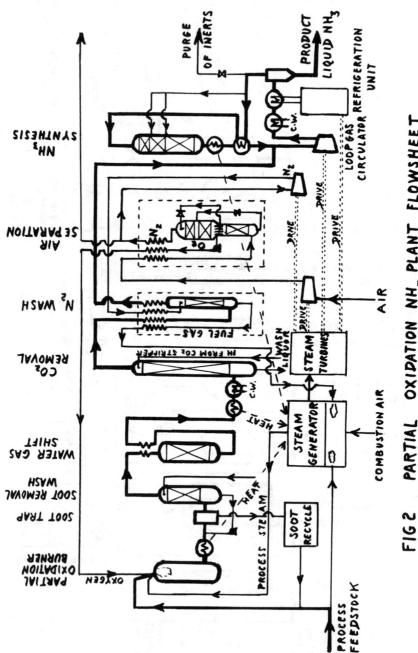

FIG 2   PARTIAL OXIDATION NH₃ PLANT FLOWSHEET

Comparison Between Partial Oxidation and Steam Reforming Processes

Referring to Figs 1 and 2 where much simplified flowsheets of modern representatives of the two types of process are shown it can be seen that on the process gas side the partial oxidation process is apparently simpler and capable of consuming a much wider range of feedstocks than the steam reforming route.

This apparent simplicity is, however, deceptive. The 'cold' system of the air-plant and nitrogen wash are complex and costly and they also require a considerable amount of mechanical power to drive them. Even with the best current flow-sheets where steam turbine drives are employed for all the main machines, the partial oxidation process requires some 10% more total feedstock (heat equivalent) per ton of ammonia than does the steam reforming process, as it is not inherently capable of powering its air compression machinery without the supply of steam from an external boiler. A steam reforming plant is self-powered. The greater total complexity of the cold generating equipment also results in a capital cost which is some 10% greater than that for the steam reforming route. Partial oxidation is therefore only competitive with steam reforming when its feedstock is at least 20% cheaper per unit heating value than steam reforming feedstock.

Nitric Acid Production

Thermodynamics. Having described at some length the sophisticated and highly efficient process for making ammonia it is an anticlimax to now consider nitric acid manufacture, a process wherein the whole of the hydrogen so carefully synthesised and purified is burnt virtually to warm the atmosphere around the nitric plant. The reactions taking place are:-

$$NH_3 + 1\tfrac{1}{4}O_2 \rightarrow NO + 1\tfrac{1}{2}H_2O + (69.9 \; \text{kcal evolved/g mol})$$

$$NO + O_2 \rightarrow NO_2 + (13.6 \; \text{kcal evolved})$$

$$3NO_2 + H_2O \rightarrow 2HNO_3 + NO + (16.7 \; \text{kcal evolved})$$

overall $NH_3 + 2O_2 \rightarrow HNO_3 + H_2O + (98.71 \; \text{kcal evolved/g mol})$

The equilibria of all these reactions are favourable, they all
evolve heat, the first reaction may be catalysed relatively
easily with high selectivity and the other two are non-catalytic
with high yields. Yet because the heat evolved from these
processes is at too low a temperature it cannot effectively be
made use of in ammonia production. This limitation can best be
understood by considering the three reactions of the acid making
process:-

Ammonia Oxidation. Platinum or platinum-rhodium in the form of
a fine gauze has long been known as a very selective and active
catalyst for the ammonia oxidation reaction provided the
oxidation process is run under the optimum conditions. Various
side reactions occur leading to loss of product and must be
avoided. In particular the pre-catalyst reaction of ammonia
oxidising to nitrogen and water on hot surfaces of equipment
prior to the gas contacting the catalyst, the decomposition of
NO to nitrogen and oxygen on the catalyst and the reaction of
unreacted ammonia with NO after the catalyst leading again to
nitrogen formation can all be reduced to low levels by suitable
process design. Nevertheless though the molecular yield of the

process is rendered high (circa 95%) by avoiding these reactions,
the means which are employed to avoid them renders the useful
thermal yield of the process very low. Thus to avoid loss of
ammonia by non-selective oxidation prior to the catalyst on the
reactor walls and in the preheater, it is necessary to ensure
that these walls are not excessively hot. In order to avoid
loss by reaction between NO and NH_3, it is necessary to ensure
that these two species are in as little contact as possible - i.e.
the inlet gas must not contain appreciable NO and the exit gas
from the catalyst must not contain appreciable NH_3, furthermore
the volume of gas hold-up in the catalyst space must be as low
as possible. This last requirement is easily satisfied as the
catalyst bed, some 4 to 30 fine gauzes, is very shallow. A
further restriction on the conditions obtaining in the oxidation
reactor is set by the dynamics of the nitric acid forming
reactions in the nitrous gas absorber where NO is oxidised to
NO_2 by a homogeneous gas phase reaction and the NO_2 so formed
reacts with liquid water to give aqueous nitric acid. Both
these reactions are kinetically second order in nitrous gas
concentration. Furthermore they are slow. In consequence the
stainless steel absorption towers are large and expensive. There
is thus a very considerable commercial advantage in increasing
the concentration of nitrous gas to the maximum as the volume
of these towers is inversely proportional to the square of the
nitrous gas concentration at inlet and hence of the ammonia
percentage in the gas mixture entering the ammonia oxidation
reactor. The use of high ammonia percentages is, therefore, very
advantageous. If all the above requirements are satisfied
then a large temperature rise across the oxidation catalyst is
inevitable and, as this catalyst is normally operated at around
$900^{\circ}C$, a temperature at which the fragile gauze slowly

disintegrates, the inlet temperature must be quite low - say only about $100^{o}C$ as the temperature rise per percent of NH_3 burnt is about $70^{o}C$. In consequence most of the heat of oxidation is liberated at a temperature below which it can be usefully employed for power steam generation. The only practical way of using such heat efficiently would be by utilising the oxidation reactor as the combustor of a gas turbine expanding the hot products of reaction. Unfortunately the kinetics of the oxidation reaction are such that its selectivity falls as reaction pressure increases and the power generated by such a turbine - which might for instance be used for a drive on an integrated ammonia-nitric acid plant would be more than offset by the loss of ammonia by its conversion to nitrogen.

Though in essence an apparently simple process, it can thus be seen that the ammonia oxidation route to nitric acid is hedged about with a great number of chemical kinetic limitations some of which spring from the performance of the ammonia oxidation catalysts and some from our inability to catalyse the subsequent conversion of NO to HNO_3. Despite many attempts over the last fifty years to develop a radically improved ammonia oxidation catalyst, which would have good selectivity (>95%) at high pressures of operation (>10 atm), and catalysts for accelerating the most inconveniently slow absorption process, no significant advance has been made.

The platinum-rhodium gauze reactor operates, when in good mechanical condition, with efficiencies of conversion of NH_3 to NO of 97-98% at 1 atmosphere, 95-96% at 4 atmospheres and 92-93% at 10 atmospheres pressure with the optimum number of gauzes. Too few gauzes result in ammonia slip, too many result in excessive NO decomposition. The reaction, when the gauzes are operating at about 900°C is almost (but not quite entirely)

gas film limited. The oxygen for combustion is always supplied

as air so that the total process has a large nitrogen effluent

to the atmosphere. There is no significant disadvantage in

this use of dilute oxygen in the oxidation reactor as the

percentage of ammonia that could be passed into the burner is

limited by the lower spontaneous inflammability limit of the

mixture to below 14%. It is however, disadvantageous in the

absorption system as the velocity of the NO oxidation reaction

is first order in oxygen partial pressure. Normally further air

is added to increase this partial pressure before the gases

leaving the burner enter the absorption system.

Nitrous Gas Absorption. The oxidation of the NO and its

subsequent absorption in water to give HNO_3 and further NO

takes place after the nitrous gases leaving the burner have been

cooled. Refrigeration of the aqueous nitric acid to around 2°C

is particularly useful if strong (c 65% acid) is to be made as

both the equilibrium of the acid forming reaction and the

kinetics of the absorption and of the NO oxidation are improved

by this means. The homogeneous gas phase NO oxidation reaction

has a negative temperature coefficient. As nitric acid

formation is accompanied by NO evolution requiring further

oxidation and subsequent absorption of the NO_2 formed, the

absorption equipment consists of a large tower filled with

packing or plates operating at quite low linear gas velocities

to give residence time in the gas phase allowing the NO oxidation

reaction to proceed. As was indicated previously, high

absorption pressures are very advantageous in reducing capital

cost of this equipment because of the inverse square dependence

of volume on pressure. Unfortunately, this consideration runs

contrary to the desirability of low pressure oxidation to obtain
good yields. Modern plants therefore tend to employ 1 to 4
atmospheres pressure oxidation followed by gas compression to
up to 10 atmospheres resulting in lower cost and efficient
pressure absorption. Increased environmental pressures against
the discharge of nitrous fumes along with the effluent nitrogen
from the absorption system enhance the need for good but cheap
absorption systems and increasingly result in higher pressures
for absorption.

The inverse square dependence of absorption rate on gas
concentration makes it particularly difficult to remove the final
traces of nitrous gases from the nitrogen and various means have
been sought to accelerate the process in this concentration
region. For instance, the addition of hydrogen peroxide or
ozone is quite effective but much too costly. Commercially
practical means are based on the catalytic reduction of the
nitrous gases using an added reductant such as methane, hydrogen
or ammonia to the gases which are then passed over a catalyst
after heating to some 300 or $400^{\circ}C$. For removal of both NO and
NO_2,without reducing. the residual oxygen within the effluent gas,
it is necessary to employ a selective reduction catalyst such as
a supported vanadium oxide and use ammonia as a reductant.
Methane and hydrogen with platinum catalysts are less selective,
being capable of reducing NO_2 preferentially, compared with
oxygen, but not NO. Reductive processes for removal of nitrous
gases from the effluent are all, clearly, wasteful not only of
the nitrous gas but also of reductant and heat. The obvious
method of removal of nitrous gas, its decomposition to nitrogen
and oxygen is both thermodynamically possible and exothermic.
Unfortunately, no practical catalyst has been found to promote
this reaction at reasonably low temperatures.

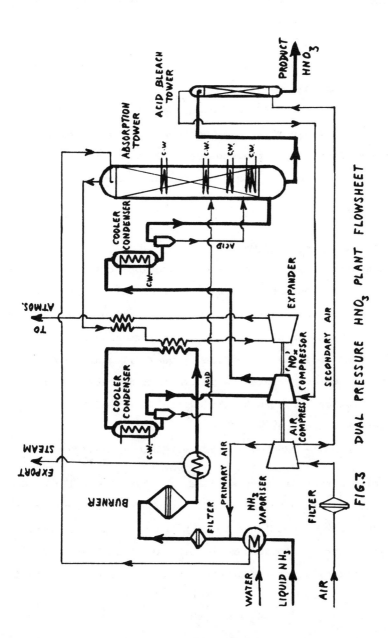

FIG.3   DUAL PRESSURE HNO₃ PLANT FLOWSHEET

Flowsheet. A typical, much simplified, flowsheet for a dual
pressure plant employing atmospheric pressure burning and
pressure absorption is shown in Fig 3. The air and gas
compressors are driven by the gas expander so as to reduce
power import to the process. As is indicated, effluent gas from
absorption is reheated by heat exchange with gas exit the burner
in order to increase its volume and hence its availability to
generate shaft power in the expander. Product nitric acid is
always blown with air to strip off dissolved nitrous gas before
shipping.

Ammonium Nitrate Production

Thermodynamics. Ammonium nitrate production from aqueous nitric
acid and ammonia essentially only involves the neutralisation of
the acid by absorbing gaseous ammonia followed by the
concentration and crystallisation of the ammonium nitrate so
formed. Some care has to be taken, however, to avoid what
could be a hazardous operation as ammonium nitrate is potentially
explosive, as can be seen by the exothermicity of the
decomposition reaction:-

$$NH_4NO_3 \rightarrow 2/5HNO_3 + 4/5N_2 + 9/5H_2O + (25.4 \text{ kcal evolved/g mol})$$

As ammonium nitrate is very soluble in water, the solid product
is strongly hygroscopic and if it is to be used directly as a
fertilizer precautions have to be taken to avoid the ingress
of moisture during storage and to mitigate the effects of any
that does enter, otherwise severe caking of the product would
result.

Neutralisation and Concentration

The heat of neutralisation of the aqueous 55-60% nitric acid is sufficient to evaporate nearly all of the water present provided the neutraliser is uncooled and operates at the boiling point of the mixture. This is the basis of the commercial process for concentration of the ammonium nitrate to a liquid melt containing only a few percent of water. This is a potentially hazardous region of the process and as a result of many investigations into the cause of ammonium nitrate explosions the following precautions are taken:-

1   The mass of concentrated ammonium nitrate melt (<8% H_2O) is kept as low as possible.

2   The melt is not allowed to fall below a pH of 4.5.

3   The neutraliser is well vented so that the pressure cannot build up.

4   Various decomposition catalysts, in particular chloride, metals such as cobalt and readily combustible organics must be excluded.

One arrangement for complying with these requirements is the use of a free boiling atmospheric pressure neutraliser to concentrate up to 10% water followed by a low hold-up evaporator (eg a falling film evaporator) for final water removal; injection of ammonia gas into the evaporator holds the pH above the critical value.

When the ammonium nitrate is to be used as a fertilizer and purity is not important, an additive such as 0.5-1% magnesium ions in the solid ammonium nitrate crystals both represses phase changes in the crystals and also acts as a chemical sink for water which may enter during storage, as the magnesium ions hydrate unlike the ammonium ions.

Solidification

Chilling the melt from the final stage of concentration results
in its freezing, the melting point being about 170°C. A
convenient way of effecting this chilling when the product is to
be used as a fertilizer is to use a 'shot' or prilling tower
down which the melt falls in drops of about 2mm diameter and up
which cold air is drawn. With a suitably high tower (>70m) little
after-cooling is required before bagging. A much simplified
flowsheet of such a plant is shown in Fig 4.

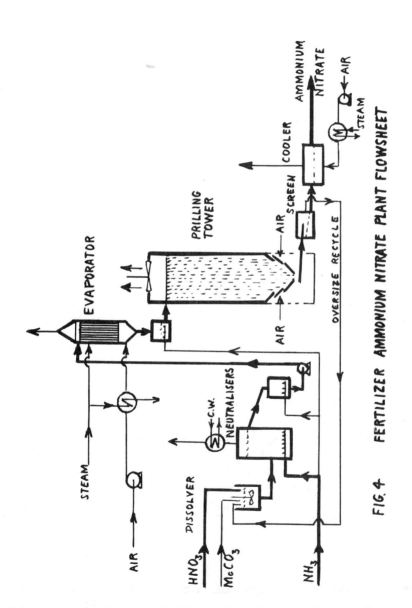

FIG. 4   FERTILIZER AMMONIUM NITRATE PLANT FLOWSHEET

The Manufacture, Properties and Uses of Hydrogen Peroxide
and other Inorganic Peroxy Compounds.

By C.A.CRAMPTON, G.FABER, R.JONES, J.P.LEAVER and S.SCHELLE
Interox Coordination, Solvay & Cie S.A.

1.   Hydrogen Peroxide

1.1.   Introduction

Hydrogen peroxide was discovered in 1818 by the
Frenchman Louis-Jacques Thenard[1] who treated nitric acid with
barium peroxide and obtained a new compound which he first of all
identified as a peroxygenated acid.

It is quite probable that hydrogen peroxide has been
prepared in the laboratory by electrochemical methods prior to
1818 but without the authors realising that they had obtained a
new compound. This applied in the case of the English chemist
Humphrey Davy in 1806[2] and Thenard himself during the course of
work carried out some years later jointly with Gay-Lussac[3].

Despite various problems which he was able to
overcome, he subsequently succeeded in concentrating the product
y subjecting it to slow evaporation over concentrated sulphuric
acid[4,5,6,7,8,9]. He thus obtained a remarkable achievement
considering the means at his disposal - a practically anhydrous
product (95%). He was able to reveal the oxidising power of
hydrogen peroxide as well as the latter's facility for
decomposing catalytically or when subjected to heat. He was
however unable to offer any explanation as regards the reducing
action of hydrogen peroxide.

From this beginning the manufacture of hydrogen
peroxide has developed via several processes so that by 1966 the
world's annual capacity (excluding the USSR) was approximately
200,000 tonnes and this figure increased significantly, to over
500,000 at the end of 1975 (see Table 1).

Table 1 H_2O_2 Capacities

Geographical Area	Capacity (end 1975) as 100 % H_2O_2	Major Producers
Europe	325,000 tonnes	DEGUSSA, INTEROX, MONTEDISON, OXYSYNTHESE
Japan	75,000 tonnes	MITSUBISHI GAS, NIPPON PEROXIDE, TOKAI ELECTRO-CHEMICAL Co
USA	100,000 tonnes	DU PONT, FMC, SHELL
Others	20,000 tonnes	ELECTRO QUIMICA MEXICANA, PEROXIDOS DO BRAZIL
Total	520,000 tonnes	

The main uses of hydrogen peroxide are associated with its oxidizing power as exhibited in the bleaching of textiles, oils and fats, wood pulp and in the manufacture of sodium perborate and percarbonate for inclusion in detergents.

It is anticipated that there will, in the future, also be large scale uses in environmental applications and in the manufacture of organic chemicals.

The relationship between the raw materials for hydrogen peroxide and some of the uses to which it is put are shown in Figure 1.

Figure 1

Uses of H_2O_2

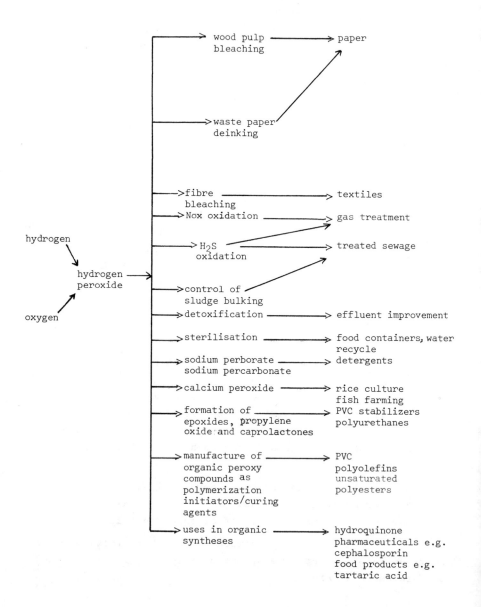

1.2. Manufacturing Processes

There are three industrial processes in use for the production of hydrogen peroxide viz :
- the alkyl-anthraquinone autoxidation or A.O. processes
- the Shell isopropanol autoxidation process providing simultaneously hydrogen peroxide and acetone in equimolecular quantities
- various electrolytic processes, which have in recent years gradually been disappearing owing to the high costs of electric power and capital investment

The A.O. processes dominate the worlds production representing more than 90 % of installed capacity.

1.2.1. **A.O. Processes.** An organic solution of an alkyl-anthraquinone is hydrogenated in the presence of a catalyst to give an alkyl-anthraquinol. The catalyst, which is commonly Raney nickel or palladium on a support, is separated from the solution and the latter then is subjected to oxidation by oxygen or air to give the original alkyl-anthraquinone and to form a dilute solution of hydrogen peroxide which is extracted with water.

The corresponding chemical equations are :

The process is cyclic and theoretically the alkyl-anthraquinone solution (referred to as the working solution) could, following the extraction stage, be recycled for hydrogenation. In fact, as a consequence of certain secondary reactions, degradation products form which must either be removed or reconverted into useful compounds.

The aqueous solution of hydrogen peroxide has to be purified to remove solvent and quinones which may have been co-extracted, an operation which is usually performed by washing with an organic solvent. The hydrogen peroxide solution, analysing at this stage 20 to 40 % by weight[10], is then concentrated by vacuum distillation and stabilised in order to provide a finished product which is available commercially at strengths from 20 % to 85 %.

The whole of the operations are summarised in Figure 2.

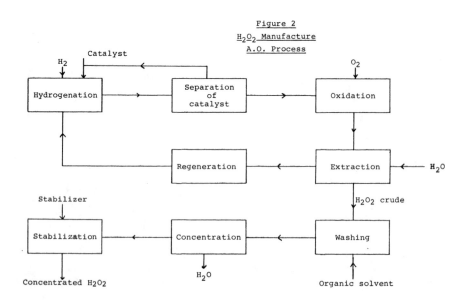

Figure 2
H_2O_2 Manufacture
A.O. Process

In addition, industrial plants frequently incorporate extra stages, which are economically important, viz :
- the recovery of solvents from the excess air used in the oxidation
- the recovery of solvents used for the purification of the hydrogen peroxide
- the preparation of hydrogenation catalyst and regeneration of spent catalyst
- hydrogen generation where this raw material is not available on the production site.

The choice of alkyl-anthraquinone is mainly dictated by its solubility and that of the corresponding alkyl-anthraquinol in the solvent medium involved.

2-Ethyl anthraquinone was initially used by IG Farbenindustrie[11] and is probably still the quinone most widely employed on an industrial scale; in addition the use of 2-t-butylanthraquinone[12], 2-t-amylanthraquinone[13], the mixture of 2-t-amylanthraquinone and 2-s-isoamylanthraquinone[14] are reported. Similarly, it is possible to use a mixture of alkyl- anthraquinones of eutectic composition and thus possessing a solubility greater than that of the individual ingredients[15].

The 2-alkyl tetrahydroanthraquinones can also be used[16,17], viz :

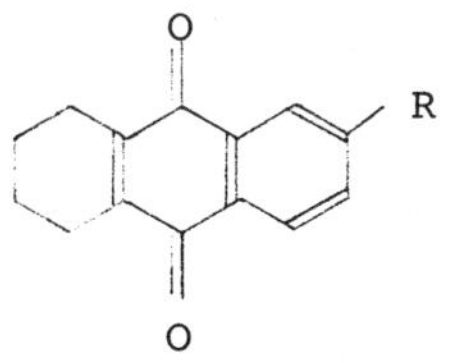

They form moreover during the hydrogenation stage of the process. Depending upon the nature of the alkyl group, the solubility of the alkyl-tetrahydroanthraquinone and the corresponding quinol may be superior or inferior to that of the original alkyl-anthraquinone.

The solvent for its part must possess a number of characteristics, viz :
- it must possess a good dissolving power both for the anthraquinone form and for the anthraquinol form; this is only rarely achieved by a single solvent but rather by a mixture of two or more solvents
- it should offer adequate chemical resistance under the conditions of the various hydrogen peroxide production stages
- it should not inhibit the reactions of hydrogenation and oxidation of the anthraquinones
- it should only be slightly soluble in water and possess for the solvent-water system an H_2O_2 partition coefficient distinctly in favour of the aqueous phase
- both specific gravity and viscosity should be low in order to facilitate H_2O_2 extraction
- it should only be slightly volatile so as to reduce losses in the oxidation residual gases
- it should not produce explosive mixtures with air and hydrogen peroxide under normal conditions or conditions likely to occur in the oxidation section
- it should be available in industrial quantity at a satisfactory price.

Although certain esters are claimed as the only solvent[18,19,20,21,22] a mixture of two or more solvents is more frequently preferred, one of these being the solvent of the quinone form (particularly aromatic solvents) while the other is the solvent of the quinol form (polar solvents).

Benzene, which is too dangerous and too toxic, is no longer used as aromatic solvent. In preference alkylbenzenes are used, particularly as a mixture in the form of petroleum fractions containing mainly trimethylbenzenes[23,24,25,26,27]. Also claimed as solvent for the quinone form are naphthalene,and its alkylated or chlorinated derivatives[28.29], diphenyl and its derivatives[30],and dichloroethane[31].

As polar solvent, for the quinol form, aliphatic alcohols e.g. diisobutyl carbinol or aromatic alcohols[32,21,16], certain aliphatic or aromatic ketones[33,34,35,36], esters e.g. acetates, phthalates, benzoates, phosphates or esters of phosphoric acid[19,20,21,22,24,34,35,37,38,39,40] are claimed.

The choice of the respective proportions of aromatic and polar solvents used in the process depends not only upon the desired solubility for the quinone and quinol forms but also upon the effect of each of the constituents on the rate and selectivity of the reactions at the hydrogenation, oxidation and working solution regeneration stages.

1.2.2. The Shell Isopropanol Process. The basis of the process is the air oxidation of isopropanol giving at the same time hydrogen peroxide and acetone :

$$\underset{\underset{CH_3}{|}}{\overset{\overset{CH_3}{|}}{OH-C-H}} + O_2 \longrightarrow \underset{\underset{CH_3}{|}}{\overset{\overset{CH_3}{|}}{OH-C-OOH}}$$

$$\underset{\underset{CH_3}{|}}{\overset{\overset{CH_3}{|}}{OH-C-OOH}} \longrightarrow \underset{\underset{CH_3}{|}}{\overset{\overset{CH_3}{|}}{C=O}} + H_2O_2$$

Studies by Brodskii, Franchuk et al[41] have shown that all the oxygen present in the hydrogen peroxide comes from the oxygen and not from the isopropanol.

The oxidation reaction can be performed in the vapour phase[42,43] or in the liquid phase. The latter process has been used industrially by Shell in their Norco (USA) plant where a plant operates with a capacity of 15,000 tonnes H_2O_2 100 %/year (see Figure 3) plus 30,000 tonnes of acetone.

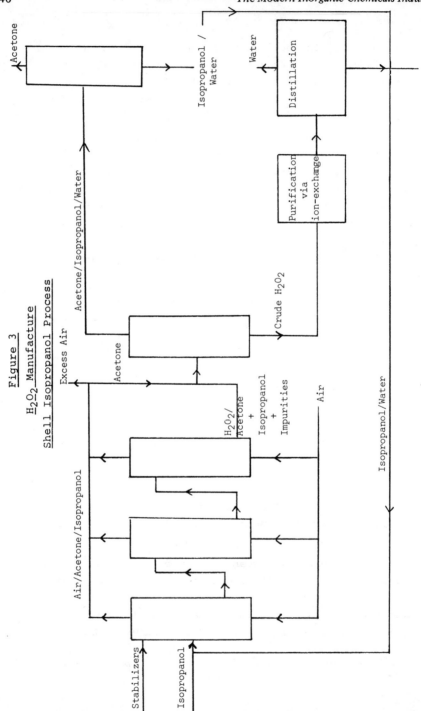

Figure 3
H₂O₂ Manufacture
Shell Isopropanol Process

The oxidation reaction takes place under a pressure of 15 to 20 atm and at a temperature of 90 to 140°C. Several reactors are used in series, and in order to limit secondary reactions resulting in the production of organic peroxides of aldehydes and organic acids the conversion rate is fixed at approximately 30%, with a selectivity of 80%[44].

The feed of isopropanol includes both fresh product and unconverted product which is recycled at the same time as a certain quantity of water. It has been shown that the speed of H_2O_2 formation is directly proportional to the isopropanol concentration in the reaction medium[45].

The reaction does not call for a catalyst but the presence of 0.5 to 1% hydrogen peroxide recycled with iso-propanol avoids induction periods by acting as an initiator.

At the oxidation section outlet, since the reaction mixture contains hydrogen peroxide, the acetone, water and isopropanol which has not been converted is diluted with water and passed into a distillation column which separates the isopropanol-acetone-water mixture at the top and the H_2O_2 solution at the bottom. The water flow at the column inlet is regulated in such a manner that the H_2O_2 strength at the outlet does not exceed 20%.

The column operates under pressure with temperatures between approx. 55°C at the top and 120°C at the boiler. This enables a temperature to be maintained in the column at a level whereby certain organic peroxides present as by-products of the oxidation reaction are reconverted into hydrogen peroxide.

The isopropanol-acetone-water mixture is passed into a rectifier column in which acetone is obtained at the top and the isopropanol-water mixture at the bottom which is recycled to the oxidation stage.

The crude hydrogen peroxide solution is then purified by ion exchange resins[47].

The final stage of the manufacture is a vacuum distillation process.

1.2.3. Electrolytic Processes. The various electrolytic
routes to hydrogen peroxide are largely of historical
value and those that still operate account for a small
proportion of the world's production. The processes are based
upon the anodic coupling of sulphate ions to give peroxydisulphate ions during the electrolysis of sulphuric acid
or bisulphates[48]. Hydrogen peroxide is obtained by subsequent
hydrolysis of the peroxydisulphate ion.

1.3. Properties of Hydrogen Peroxide

1.3.1. Physical Properties. The principal physical
properties of pure hydrogen peroxide are summarized in Table 2
and those of aqueous solutions in Figure 4. However as hydrogen
peroxide is miscible with water in all proportions and it is
the queous solutions which are normally handled, the
properties of the latter are of more interest. A hydrate
$H_2O_2 2H_2O$ appears to be formed in the solid phase and evidence
is against formation of solid solutions; eutectics exist at
45.2 and 61.2 wt % H_2O_2.

Structurally the atoms in hydrogen peroxide are
joined by simple covalent bonds, H - O - O - H. The O - O
distance is 1.453 $\pm$ 0.007 A and O - H is 0.988 $\pm$ 0.005 A

Table 2

Physical Properties of H_2O_2

Melting point $^\circ$C	−0.43
Boiling point $^\circ$C	150.2
Heat of fusion kcal/mol	2.987
Heat of vapourisation kcal/mol at 25°C	12.3
Density at 25°C g/ml	1.4425
Heat of formation at 25°C kcal/mol	−32.52
Surface tension at 20°C dyn/cm	80.5
Electrolytic dissociation at 20°C, K	1.39×10^{-12}
Specific heat, liquid, cal/g $^\circ$C	0.628
Heat of decomposition to water and oxygen kcal/mol at 25°C	23.44

Figure 4 - Physical data for the system H₂O - H₂O₂.

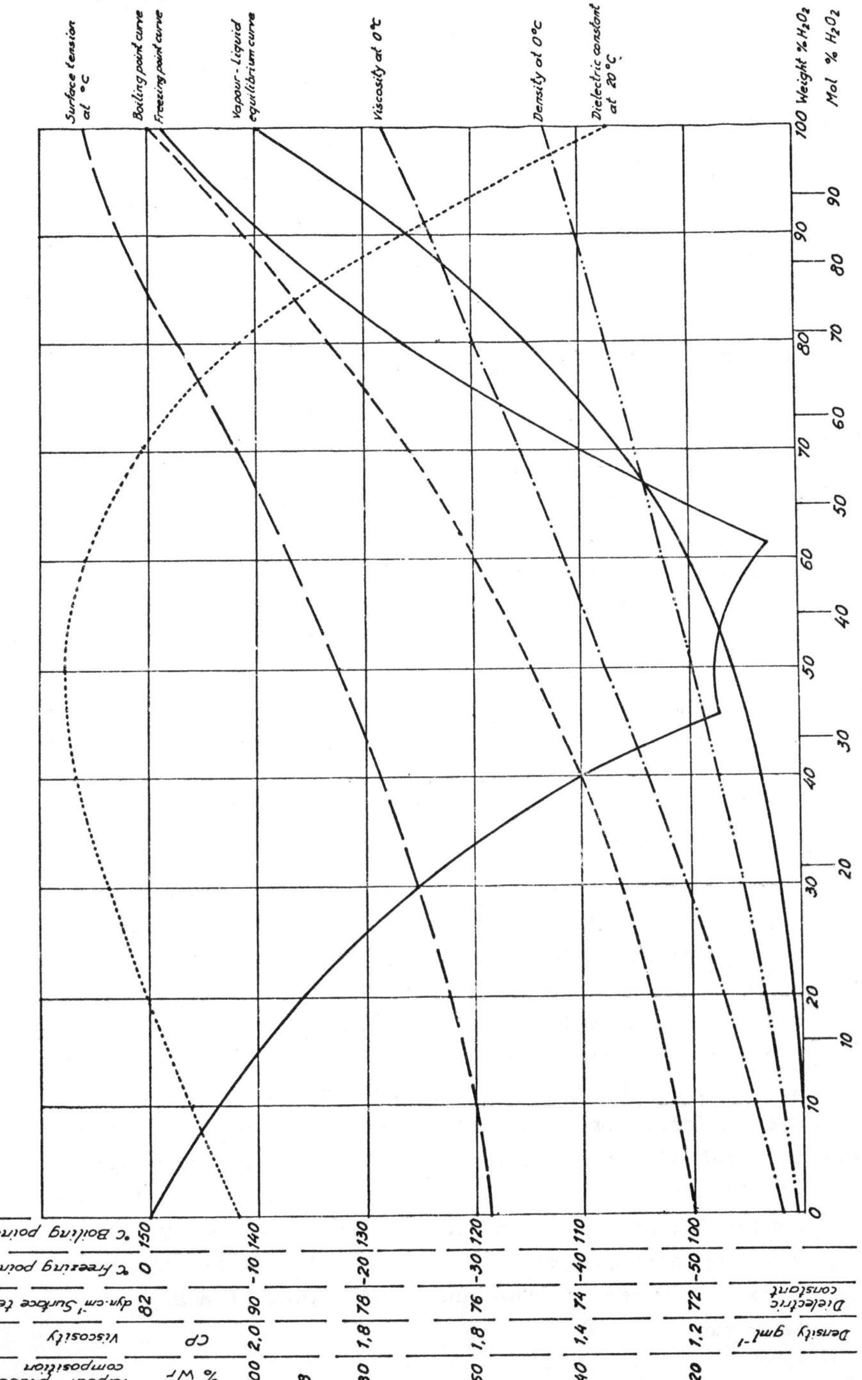

Based on reference 125

1.3.2. <u>Chemical Properties</u>. The reactions of hydrogen
peroxide can be listed as follows :

Decomposition $2 H_2O_2 \longrightarrow 2 H_2O + O_2$

Addition $H_2O_2 + A \longrightarrow A.H_2O_2$

Substitution $H_2O_2 + BC \longrightarrow BOOH + HC$

Oxidation $H_2O_2 + D \longrightarrow DO + H_2O$

Reduction $H_2O_2 + E \longrightarrow EH_2 + O_2$

In its reactions hydrogen peroxide may react
as a molecule, as the ionised species, or through dissociation into free radicals, and catalysis often plays
an important part.

The decomposition of hydrogen peroxide is an
important characteristic which has to be minimised during
manufacture, storage and use, not only for economic
reasons, but because release of oxygen and heat may cause
safety problems. The mechanism of decomposition depends
on many factors that include pH[49], temperature and catalysis. Increasing pH and temperature both increase the
rate of decomposition, and transition metal compounds such
as those of iron, copper and manganese, either in solution
or in the solid state act as catalysts. The decomposition is reduced by purification during manufacture and
by the addition of stabilisers.

Stabilisers act by inactivating traces of
transition metal catalysts that may be present, either
by complex formation as with 8-hydroxyquinoline, or
by colloidal absorption as with sodium stannate.
While these examples are effective under acidic conditions,
ethylene diamine tetra-acetic acid and magnesium silicate
are typical choices for alkaline conditions. Modern
systems often use combinations of stabilisers to reduce
the risk of contamination. A corrosion inhibitor (such
as nitrate with aluminium vessels) is also often included.
Commercial stabilised hydrogen peroxide properly stored
would be expected to lose less than 1 % of its hydrogen
peroxide content in the course of a year at ambient
temperatures.

Hydrogen peroxide adds on to many compounds by hydrogen bonding to form perhydrates that are analogous to the hydrates formed by water. Perhydrates are chiefly formed by oxyacid salts (such as sodium sulphate and carbonate) but also by peroxides themselves. Examples are $Na_2SO_4, 0.5\ H_2O_2$, $Na_2CO_3, 1.5H_2O_2$, $Na_4P_2O_7, 2\ H_2O_2$, $Li_2O_2, H_2O_2, 3\ H_2O$, $Na_2O_2, 2\ H_2O_2$.

Ammonia also forms adducts as do urea and many other amines, including the sodium salt of ethylene diamine tetra-acetic acid.

Typical inorganic substitution reactions are with metal oxides, hydroxides and salts to form peroxides based on O_2^{2-} and hydroperoxides derived from HO_2^-. Although even the alkali metals can form hydroperoxides the tendency increases with reducing basicity through to the acidic oxides whose hydroperoxides are themselves acids e.g. peroxymonosulphuric acid, commonly called permonosulphuric acid or Caro's acid ($HO.SO_2.OOH$).

Furthermore, these peroxy acids can often be substituted at the remaining hydrogen atom attached to the peroxy link by an acid radical to give a peroxy-dicompound such as peroxydisulphuric acid, known as persulphuric acid ($HO.SO_2.OO.SO_2.OH$).

Analogous peroxy derivatives of phosphoric acid exist namely peroxymonophosphoric acid $(HO)_2.PO_2.OOH$ and peroxydiphosphoric acid $(HO)_2.PO_2.OO.PO_2.(OH)_2$. An unstable peroxymononitric acid, $NO_2.OOH$ is known, but the peroxydinitric has not been described.

In addition to these simple derivatives compounds
are known containing oxide and peroxide, hydroperoxide etc
groups together in the same molecule, particularly among
the transition metal derivatives, and catenated and cyclic
structures also occur. Thus a common commercial persalt,
sodium perborate tetrahydrate, the bleaching agent in most
domestic washing powders, has the structure[50]

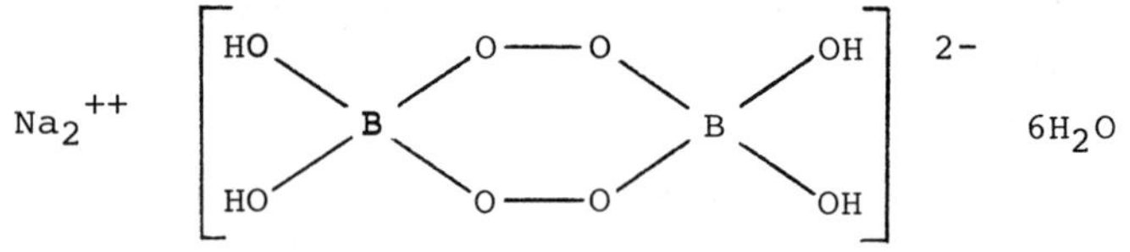

$$Na_2^{++} \left[\begin{array}{c} HO \diagdown \underset{B}{} \diagup O - O \diagdown \underset{B}{} \diagup OH \\ HO \diagup \underset{}{} \diagdown O - O \diagup \underset{}{} \diagdown OH \end{array} \right]^{2-} \quad 6H_2O$$

Hydrogen peroxide is a powerful oxidising agent
and many of its uses and those of its derivatives depend
on this property. The electrolytic couple shown below is
for acidic solutions

$$2H_2O \rightleftharpoons H_2O_2 + 2H^+ + 2e^- \quad E_{298}^\circ = -1.77 \text{ volts}$$

and for alkaline solutions

$$3OH^- \rightleftharpoons HO_2^- + H_2O + 2e^- \quad E_{298}^\circ = 0.87 \text{ volts}$$

A wide variety of inorganic and organic compounds
are oxidised and the mechanisms of these reactions are too
varied for discussion here.

Hydrogen peroxide is a weak reducing agent that
reduces stronger oxidising agents than itself such as ceric
sulphate, permanganates and hypochlorites.

Further details can be found in the standard
reference works on hydrogen peroxide[48,51,52,53,54].

1.4. Uses of Hydrogen Peroxide

Prior to 1939 the principal uses of hydrogen
peroxide were in bleaching and as an antiseptic, with a
total world capacity of 12,000 tons per annum. By 1974
the total usage had grown to approximately 350,000 tons

due to its employment in an expanding and wide range of applications, see Table 3.

Table 3

Analysis of Hydrogen Peroxide Uses (1974)

(excluding USSR)

Application	% of Total Consumption

Chemical uses including
  - manufacture of sodium perborate
    and percarbonate
  - manufacture of epoxy compounds
	61

Bleaching of textiles	16
Bleaching of wood, pulp and paper	14
Miscellaneous uses including environmental applications	9

1.4.1. Bleaching. The biggest single application for hydrogen peroxide is still in the bleaching field, mainly for the provision of a solid stable form of peroxide bleach in domestic detergents. The traditionally favoured compound is sodium perborate but in recent years sodium percarbonate has been introduced.

The use of sodium perborate is more popular in Europe than the USA (see Table 4) and particularly in the UK where the bulk of detergents contain sodium perborate or percarbonate.

Table 4

Usages of Percompounds in Domestic Detergents

Country	Proportion of domestic detergents containing sodium perborate or percarbonate %	Concentration of sodium perborate or sodium percarbonate in detergents %
BELGIUM	68	15-25
FRANCE	82	15-25
GERMANY (WEST)	75	23
ITALY	65	20-30
SPAIN	60	12-18
UK	> 95	15-30
USA	< 10*	≤ 10*

* Used as an external bleaching agent and generally. not incorporated in a detergent.

Textile bleaching had itself increased to approximately 50,000 tons by 1974, with cotton as the prime application, followed by wool and flax. Systems based on hydrogen peroxide have dominated the textile bleaching market with good whiteness increase accompanied by minimum fabric damage, easy handling characteristics and relatively good effluent properties, a factor which has become increasingly important.

Another growth area in bleaching has been in the paper pulp industry, this market was about 50,000 tons in 1974, with approximately half being in Japan.

The majority of the hydrogen peroxide was used to bleach mechanical pulp and other high yield pulps. With the development of thermomechanical and chemimechanical pulps the proportion of high yield pulp used in printing, writing and tissue papers will grow and the use of H_2O_2 will therefore develop futher. This is because hydrogen peroxide is the only chemical which will enable most of these pulps to reach economically a combined high level of brightness (up to 80 °SCAN) and brightness stability.

The peroxide bleaching is normally carried out at high consistency in a tower, but bleaching in the refiner is expected to become an interesting alternative soon. This technique, which enables the same brightness gain to be obtained more rapidly, will reduce the investment cost for bleaching refiner pulps[55].

Hydrogen peroxide has also been used for a long time as a final stage in the bleaching sequence of chemical pulp in order to improve the brightness stability. The introduction of H_2O_2 into the alkaline extraction stages of the bleaching sequences is being developed not only in sulphite pulp[56] but also in sulphate pulp[57,58]. In this case, the use of H_2O_2 is very simple, and as no sodium silicate is required, the chemical cost and effluent pollution can both be reduced ; in addition the pulp quality and the flexibility of the bleaching plant can be improved. In the future, hydrogen peroxide is expected to become even more attractive for chemical pulp as it facilitates, together with oxygen and ozone, bleaching sequences without the use of any chlorinated compounds, that are undesirable environmentally[59].

Hydrogen peroxide is also used in the deinking process for the recycle of waste paper. It is compatible with the deinking process[60] which normally works under alkaline conditions and on materials which should not be delignified and this application will expand further with the increasing use of waste paper in news print, printing, writing and tissue qualities.

As the uses of hydrogen peroxide and peroxyacids in bleaching form such a large section of these applications it is worth considering the reaction mechanisms involved. Although there are many naturally occurring coloured sytems, the colour of organic molecules is associated with electronic excitation that causes absorption in the visible wavelength ; the principal contributor to this state is the double bond. The objective of bleaching is to disrupt this excitation by oxidation or reduction so that electronic transitions move out of the visible region. As hydrogen peroxide bleaches more effectively under alkaline conditions it was at one time considered that the perhydroxyl ion,

HO_2^-, was the effective bleaching species. However
the rate of bleaching does not relate directly to the
perhydroxyl ion concentration so a more sophisticated
explanation is needed. When the coloured substrate is
in solution then the maximum rate of bleaching usually
takes place at the pK value for dissociation of the
hydrogen from the peroxidic oxygen function i.e. when
there is an equal concentration of the dissociated and
undissociated species. The same relationship holds for
the maximum rate of decomposition[54], so it strongly suggests
that an intermediate formed from reaction of dissociated
and undissociated species is responsible either for attack
of the substrate or for decomposition, and in fact, as
bleaching is more efficient when the substrate is more
concentrated a competitive state exists as might be
expected. In general too, the lower the pK value the
more reactive the oxidising species, thus peroxyacetic
acid[61,53] > peroxymonosulphuric acid[62,53] > hydrogen
peroxide[56,49] > peroxymonophosphoric acid[58]. When
the coloured constituent is attached to, or absorbed by a
surface then there are additional complicating physical
factors to be taken into account such as the penetration
of the oxidising species, and the degree of dissociation
of the coloured substrate from the surface.

1.4.2. Environmental Applications. The current pre-
occupation with the need to improve the environment
including ways of treating domestic and industrial effluents
and the purification of drinking water has resulted in an
increasing interest in the use of hydrogen peroxide. This
can be illustrated by the fact that some 300 patents on
the use of percompounds in the environmental or related
biological fields have been granted in the last 5 years.

An example of the use of hydrogen peroxide is
in the treatment of raw sewage which becomes anaerobic
and generates odorous hydrogen sulphide if it is kept for
some hours in absence of air, e.g. in pipe-lines or collection points. Hydrogen peroxide has become a popular means

of both destroying the sulphide and restoring aerobic conditions. If the hydrogen peroxide is injected at a point shortly before the odour is liberated into the atmosphere then the peroxide is relatively selective in sulphide oxidation and the usage is little more than theoretical[63]. Other features of sewage treatment that have received attention recently have been hydrogen peroxide as an oxygen source in the secondary stage of biological treatment units[64], as an aid to nitrification[65], to improve bulking conditions[66] and as a flotation assistant[67]. It is also proposed to improve the operation of tertiary treatment sand beds by use of hydrogen peroxide in the feed at the back wash[67a] and the sterilisation of sewage effluent by hydrogen peroxide and peracids is being studied.

An established sterilisation application is in pre-treatment of milk cartons, and the treatment of swimming pools in conjunction with an algicide is also proposed[68]. Peracetic acid is an effective sporicide against anthrax and is also being considered for action against dermatitis in intensified chicken rearing[69]. Hydrogen peroxide also finds use as a slimicide in recirculating water systems[70].

A major area for hydrogen peroxide application is detoxification of liquid and gaseous effluents. Effluents containing cyanides are treated with hydrogen peroxide if the initial cyanide concentration is sufficiently high and a moderate residual content is acceptable, otherwise a peroxy acid such as permonosulphuric is required. Heat treatment salts[71,72] and acrylonitrile effluents are treated by the first procedure and steel plant effluents by the second[73] and both are effective unless there are present substantial quantities of oxidisable impurities that reduce efficiencies. Procedures for removal of chromates and other metals[74] with hydrogen peroxide have been described. Toxic gaseous effluents such as waste NO_x gases from nitric acid manufacture are removed by reaction with hydrogen peroxide[75]. The treatment of hydrogen sulphide[76], sulphur dioxide[77] and automobile exhausts has also been described[78]. Other treatments aimed at effluent improvement are BOD and COD reduction[79],

colour[80] and odour[81] removal from fluids (typically
dyehouse and animal waste products respectively), oxidation of specific obnoxious organics such as phenols and
aldehydes[82] and dechlorination[83].

1.4.3. Chemical Uses. An important and expanding sector
of hydrogen peroxide use is in organic chemical manufacture,
of which the largest part depends on electrophilic epoxidation and hydroxylation of double bonds, normally employing
a peroxycarboxylic acid for the purpose. Soya bean oil
is epoxidised by performic or peracetic acids to provide
a stabiliser for PVC, and allyl alcohol is epoxidised
during an intermediate stage of glycerol manufacture.
Other significant applications are in manufacture of long
chain amine oxides, for shampoos, thiourea dioxide for
textile processing, hydroquinone[84] and caprolactone[85].
A development that uses a different epoxidation route is
the nucleophilic tungstate catalysed epoxidation of maleic
anhydride by hydrogen peroxide, and of particular interest
is the breeding of suitable enzymes to hydrolyse the resulting epoxy succinic acid quantitatively to d-tartaric acid
rather than the racemic mixture formed by chemical
hydrolyses[86].

An important use of hydrogen peroxide is in the
manufacture of organic peroxides[62]. This class of compounds
includes peroxides based upon ketones e.g. methyl ethyl
ketone or cyclohexanone, alkyl hydroperoxides such as tert-
butyl hydroperoxide, peresters e.g. tert-butyl perbenzoate
and peroxides containing alkyl, and acyl groups e.g.
di-tert-butyl peroxide and dibenzoyl peroxide. Organic
peroxides are principally used as catalysts or initiators
via the generation of free radicals for the polymerization
of ethylene, vinyl chloride, acrylonitrile, styrene, unsaturated polyesters and other olefins.

Looking to the future there have been significant
developments[87,88] in the potential use of hydrogen peroxide
for the manufacture of propylene oxide using peroxy acids
or hydrogen peroxide and arsenic catalysts. Should

these applications result in the erection of large commercial scale plants then the market for hydrogen peroxide will increase dramatically and new plants will need to be erected.

1.4.4. Energy Use. Although more powerful and cheaper propellants have to a great extent displaced hydrogen peroxide, large quantities were previously used for rocket propulsion.

2. Hydrogen Peroxide Derivatives

2.1. Peroxides and Hydroperoxides

Hydrogen peroxide ionises first to give hydroperoxide ion OOH^- and secondly to the peroxide ion O_2^{2-}, analogous to hydroxide and oxide ions derived from water. Well characterised derivatives are formed with the alkali metals, but more complex and often less well defined compounds are formed with transition metals. In extreme cases, such as Cr, Mo and W as many as four peroxy groups can be bound to a single metal atom. With the more electronegative elements the hydroperoxides and peroxides become acidic in character and their salts are also formed. As in the difficult cases of distinguishing true hydroxides from hydrous oxides and hydrated oxides and hydroxides, there exists the same experience with the hydroperoxides, peroxides and perhydrates with the added complication of mixtures of the water and peroxide derivatives. Several reviews are available[48,89,90,91,92].

Sodium peroxide (Na_2O_2) has been made for many years based on a process developed by Castner. Sodium is oxidised in two stages by dry air first to the monoxide and then to the peroxide at high efficiency. Its principal use has always been in textiles and paper pulp bleaching.

Small quantities of magnesium, calcium, zinc and
barium peroxides are also made for specialist uses where
a slow release of hydrogen peroxide is desired via
hydrolysis with water i.e.

$CaO_2 + 2 H_2O \longrightarrow Ca(OH)_2 + H_2O_2$.

More recently calcium peroxide has attracted interest to
vulcanise thiokol rubbers and as an assistant to seed
germination and growth, particularly under conditions of
limited oxygen supply such as rice germination from seeds
planted under water.

2.2. Peroxyacids and Peroxyhydrates

In the peroxyacids the substitution of one or
both hydrogen atoms of hydrogen peroxide by an acid radical
modifies the properties of the peroxy link in its
chemical reactions and therefore offers the opportunity of
improving the properties of hydrogen peroxide where
(1) a greater oxidation potential may be needed
(2) a modification in polarity of the molecule and its ions
   is beneficial
(3) susceptibility to catalytic decomposition of hydrogen
   peroxide reduces reaction efficiency
(4) where physical properties e.g. solubility in organic
   media are inadequate.

A particular example of this change in properties
is found in the alkyl peroxycarboxylic acids, e.g. peracetic
acid, which are finding widespread and increasing use in
organic reactions. It is likely that the simplest inorganic
counterpart, peroxymonosulphuric acid (Caro's acid) H_2SO_5
will be developed similarly, although present uses are confined to shrinkproofing of wool and certain organic oxidation reactions. One such new use under consideration is its
application in the hydrometallurgy of zinc extraction[93], to
precipitate low valence metal impurities e.g. Fe^{II} and Mn^{II}
as the less soluble high valence hydrous oxides. Permonosulphuric acid is made by reaction of hydrogen peroxide and
sulphuric acid and minimum water content of the
reaction favours conversion efficiency. The salts

tend to be unstable hygroscopic solids and only the
potassium salt is sufficiently stable in air to be
marketable, as the composition $2 \text{ KHSO}_5, \text{KHSO}_4, \text{K}_2\text{SO}_4$ which
is used for specialised laundry bleach products[94].

2.2.1. Sodium Perborate and Percarbonate. These
compounds, which should be called sodium peroxy borate
$\text{NaBO}_3, 4\text{H}_2\text{O}$ and sodium carbonate perhydrate $(\text{Na}_2\text{CO}_3,$
$1.5 \text{ H}_2\text{O}_2)$ are the most important derivatives of hydrogen
peroxide.

The former is the most commonly used bleaching
component of domestic detergents, thus in Europe the
percentage of sodium perborate in detergents can reach
30 %. The large scale use of sodium percarbonate in detergents is of more recent origin. It has the advantages of
a more rapid solution rate, a high active oxygen content,
indigenous raw material, and with certain grades, more
attrition resistant properties, to set against a greater
tendency to loss of active oxygen in the presence of other
detergent raw materials.

2.2.2. - Manufacturing Processes - Sodium Perborate.
The principal process for the production of sodium perborate is the conversion of a boron ore into a solution of
sodium metaborate followed by reaction with hydrogen
peroxide.

The sources of boron most commonly used are
Borax $(\text{Na}_2\text{B}_4\text{O}_7, 10\text{H}_2\text{O})$ or Rasorite ®
Tincal or native borax $(\text{Na}_2\text{B}_4\text{O}_7, 10\text{H}_2\text{O})$
Ulexite $(\text{Na}_2\text{O}, 2\text{CaO}, 5\text{B}_2\text{O}_3, 16\text{H}_2\text{O})$
Colemanite $(2 \text{ CaO}, 3\text{B}_2\text{O}_3, 5\text{H}_2\text{O})$
Boric Acid (H_3BO_3)

The preparation of the sodium metaborate solution
is achieved by addition of caustic soda solution to the boron
mineral. At this stage clarifying agents are often
used to assist production of clear metaborate solution.

When a calciferous mineral is used it is necessary to
attack the ore with a blend of caustic soda and sodium
carbonate.

$$Na_2B_4O_7 + 2NaOH \longrightarrow 4NaBO_2 + H_2O$$

$$NaBO_2 + H_2O_2 + 3 H_2O \longrightarrow NaBO_3.4H_2O$$

The reaction between sodium metaborate solution
and hydrogen peroxide is carried out in large vessels in
the presence of stabilisers such as magnesium silicate.
The reaction mixture, which is continuously agitated, is
cooled and the sodium perborate formed is separated by
filtration or centrifuging. The product obtained, which
contains 3-10 % moisture, is dried by warm air at a
temperature below 60 °C to avoid the perborate melting
in its own water of crystallisation. The mother liquors
may be recycled.

As in all processes calling for a crystallisation
step, the conditions of precipitation, temperature,
agitation and seeding, have a very important influence
on the physical properties of the crystals, in particular
their particle size, distribution, the mean size and
the specific weight.

2.2.3. Manufacturing Processes - Sodium Percarbonate.
The manufacture of sodium percarbonate $Na_2CO_3.1.5$ H_2O_2
may be accomplished by either a "dry" or a "wet" process.

In the dry process, hydrogen peroxide droplets
are added to anhydrous sodium carbonate intermittently.
Each peroxide addition is followed by drying with hot
air, the function of the latter being to limit decomposition and to keep the solid mass sufficiently mobile[95,96].

A variation of this process consists in effecting
the drying in a continuous fluid bed in presence of percarbonate particles that have a smaller particle size
than those that eventually emerge from the drier[97].

In the dry process stabilisers (silicate, phosphate, etc)
are generally introduced in the peroxide solution or by
dispersion on the solid mass.

A recent "dry" process developed by Interox
consists of injecting hydrogen peroxide solution and
sodium carbonate solution into a fluid bed fed by warm
air[98]. The fluid bed serves as a combined reactor, granulator and drier. The percarbonate leaving the fluid bed
is sieved and the fraction inferior to the granulometry
desired is recycled to the base of the fluid bed, and the
remainder is ground. This process has the advantage of
avoiding the generation of an effluent and of yielding
a product of improved stability whose abrasion resistance
is particularly good.

The "wet" process is very similar to that used
for manufacture of sodium perborate. Solutions of hydrogen
peroxide and of sodium carbonate react together
in large vessels in proportions close to molecular
stoichiometry of the end product. The carbonate solution
has a tendency to supersaturate and contains solid carbonate
in suspension. As in the dry process stabilisers are used.
The precipitation is made by cooling and sometimes by
vacuum evaporation. The process may be made continuous
with recycle of mother liquor. Precipitation can also
be encouraged by salting out with sodium chloride.

In a variation disclosed by Degussa[99,100] the
reactants are sodium bicarbonate, hydrogen peroxide and
sodium peroxide.

$$Na_2O_2 + 2\ H_2O_2 + 2\ NaHCO_3 \longrightarrow 2\ Na_2CO_3 \cdot 1,5\ H_2O_2$$

To improve the stability of sodium percarbonate
in the presence of washing powder constituents, stabilisation can be achieved by coating with a range of products
amongst which are certain resins[101], sodium sesquicarbonate
or mixtures of sodium carbonate and sulphate, sometimes
in a mixture with small quantities of silicate[102,103,104].

This coating operation, which is applicable to products
from both wet and dry processes, can be carried out in a
mixer, but is best achieved in a fluid bed by evaporating
the solvent from a solution of the coating agent applied.

2.2.4. Properties of Sodium Perborate and Percarbonate.
The structural formula of sodium perborate has been shown
to be[50]

and by careful heating of sodium perborate in a dry air
stream or under vacuum three molecules of water are removed
to give the so called sodium perborate monohydrate without
basic alteration of the structure and the chemical properties of the compound. However reference to the above
structure illustrates that further dehydration by heating
disrupts the molecule[105]. The peroxygen link is retained
although its chemical behaviour now approximates to that
of a superoxide e.g. decomposition with water to liberate
gaseous oxygen. The amorphous nature of anhydrous sodium
perborate has not allowed elucidation of its structure.

Both sodium perborate and sodium percarbonate
dissolve readily in water and under the conditions of
domestic bleaching the effective species is a mildly alkaline
solution of hydrogen peroxide.

3. Peroxydisulphates

3.1. Introduction

Peroxydisulphuric acid was discovered by Berthelot[106] in 1878 when he electrolysed a solution of dilute sulphuric acid, but the establishment of the formula, $H_2S_2O_8$ was due to Caro[107] and Baeyer and Villiger[108]. By 1909 the first commercial plant, based upon this reaction, was commissioned at Weissenstein - Austria. The acid produced was a precursor in the manufacture of hydrogen peroxide i.e.

$$S_2O_8^{2-} + H_2O \xrightarrow{H^+} HSO_4^- + HSO_5^-$$

$$HSO_5^- + H_2O \underset{\xleftarrow{\hspace{1cm}}}{\xrightarrow{H^+}} HSO_4^- + H_2O_2$$

The replacement of this route to hydrogen peroxide led to the conversion of some units to the manufacture of solid peroxydisulphates and by 1975 the world's annual production, excluding Eastern Europe was about 40,000 tonnes with the ammonium salt predominating, followed by the potassium and sodium salts.

The main applications of the salts are as polymerization initiators and as metal etching compounds particularly in the fabrication of printed circuits.

3.2. Manufacturing Processes

There are two electrolytic routes to the manufacture of ammonium peroxydisulphate and the corresponding sodium and potassium salts :

- The classical process, which was developed from the earlier production of hydrogen peroxide. In this process with the use of a diaphragm cell, ammonium sulphate/ sulphuric acid solutions are electrolysed to produce ammonium peroxydisulphate[109,110,111,112,113,114].

The two other technically important peroxydisulphates of
potassium and sodium are produced from the ammonium salt
by double decomposition with potassium sulphate, or sodium
hydroxide respectively[115,116,117,118,119,120,121,122].

- The new processes, which produce the individual peroxydisulphates directly from the corresponding bisulphate
solutions, use a cell without a diaphragm[123]. During the
process the peroxydisulphates deposit from the electrolyte
and are continuously removed.

In the classic electrolytic cells the diaphragm
has to prevent the peroxydisulphate ions produced at the
anode from approaching the cathode, and being subsequently
reduced. To separate the cathode compartments from
those of the anode, ceramic or porcelain diaphragms are
used. If the anodes and cathodes are in the same electrolytic compartment the cathodes are protected by bands of
asbestos or synthetic material. As cathode materials,
graphite and lead are normally used but sometimes titanium,
zirconium and certain types of refined steel in bar,
cylinder or plate form.

As anode material only platinum is used, in the
form of wire or foil, held on supporting materials, such
as tantalum, titanium and zirconium and also iron or aluminium coated with hard rubber or polyethylene.

The theoretical potential for the formation of
peroxydisulphate is about 2.1 V but industrially 5-7 V are
needed, resulting from several contributions which are
themselves partly dependent on the cell construction : for
example, loss of voltage across the diaphragm, at the anode
and cathode (according to the material but adding to 0.5 V)
and in the electrolytic solution (varying with concentration
but < 0.5 V). The anodic overvoltage which suppresses the
formation of oxygen at the anode, and is therefore absolutely necessary, is primarily dependent on the current density
and amounts to $\sim$ 1.1 - 1.3 V at 0.5 - 1 A/cm^2.

In addition to the cell voltage the current
efficiency is an important factor for investment and running
costs (the cost of energy amounts to approximately one
third of production cost). The optimisation of the electrolytic formation of peroxydisulphates is a difficult undertaking
because of the conflicting influence of most of the parameters on voltage and current efficiency. Increasing the
permeability of the diaphragm decreases not only the
voltage, but also the current efficiency. An increase
of the ammonium sulphate concentration increases the
current efficiency but also the voltage needed as does
increasing current density. Higher sulphuric acid concentrations and temperatures decrease the cell voltage, but
on the other hand the rate of hydrolysis of the peroxydisulphate to peroxymonosulphate is increased, and the latter
has a strongly negative influence on the current efficiency.

During the process the electrolyte solution,
when it passes through the electrolytic cell, is enriched
by 1-1.5 equivalents/l of peroxydisulphate and depleted by
the same amount of ammonium sulphate and sulphuric acid.
After the electrolysis the solution is cooled until the
ammonium peroxydisulphate deposits. The crystals are
separated, washed and dried. A high purity product (> 99 %)
is obtained. The mother liquor obtained is recycled to electrolysis after the addition of sulphuric acid and ammonium
sulphate.

To produce potassium peroxydisulphate the ammonium
salt is treated with potassium bisulphate i.e.

$$(NH_4)_2S_2O_8 + 2 KHSO_4 \longrightarrow K_2S_2O_8 + 2 NH_4HSO_4$$

The sparingly soluble potassium peroxydisulphate
is deposited and the ammonium bisulphate is recycled to
electrolysis. This procedure is ineffective with sodium
peroxydisulphate as it is as soluble as ammonium peroxydisulphate. Instead a solution of caustic soda or sodium
carbonate must be used to displace the ammonia.

$$(NH_4)_2S_2O_8 + 2 NaOH \longrightarrow Na_2S_2O_8 + 2 NH_3 + 2 H_2O$$

The reaction is critical since the ammonia
which is produced can form an explosive mixture with oxygen
and so great care must be taken to ensure that the
operation excludes air and is below the decomposition
temperature of the peroxydisulphates.

The new processes to produce potassium and sodium
peroxydisulphates directly electrolyse the corresponding
bisulphate solutions without diaphragms. This requires a
completely different cell construction and process. Almost
saturated solutions of bisulphates are chosen to keep the
concentration of the peroxydisulphates as low as possible in
order to minimise their cathodic reduction. The peroxydisulphate precipitates during the electrolysis and is removed
in a subsequent step.

The electrodes consist of two cylinders placed
inside each other, or adjacent plates spaced only a
few mm apart, through which the electrolyte is pumped.
Suitable anodes are platinum with titanium, zirconium or
tantalum as supporting material and cathodes are titanium, zirconium, graphite and highly alloyed refined steel.

The new electrolytic process for production of
potassium peroxydisulphate clearly has advantages, since with
the same electrical energy being used, the displacement
of ammonium peroxydisulphate is circumvented. For the
sodium salt more energy is required than for the classical
procedures ($\sim$ 20 %), but the dangerous reaction stage
of ammonium peroxydisulphate plus caustic soda is avoided.
In principle also, ammonium peroxydisulphate can be
produced in a cell without diaphragms.

A decisive point in favour of the new procedures
is that a multi-purpose plant can be used to produce all
three peroxydisulphates by merely changing the electrolyte
solution.

The Interox Hydrogen peroxide plant at Jemeppe, Belgium

3.3. Properties and Uses of Peroxydisulphates

3.3.1. Physical and Chemical Properties. Pure peroxydisulphuric acid forms unstable and very hygroscopic crystals that melt with decomposition at 65 °C. It is a strong dibasic acid, almost completely dissociated at both stages and the acid salts are not known in the solid state.

The peroxydisulphates of ammonium, sodium, potassium, rubidium and caesium are, if pure and dry, very stable and only slightly hygroscopic. The salts of calcium and strontium exist only as hydrates. The peroxydisulphates of lithium, magnesium, zinc and thallium can be obtained only as very unstable and mostly hygroscopic mixtures with the corresponding sulphates.

The solid commercial peroxydisulphates are of almost analytical grade quality, are stable, and decompose at high temperatures. In the case of ammonium peroxydisulphate a propagating decomposition can occur once it has been locally initiated by a small fire. Such propagation is difficult to control. Inflammable materials such as paper and wood wool can catch fire in presence of decomposing peroxydisulphates which also release large quantities of suffocating acid vapour that make extinction difficult.

In common with other peroxy compounds the peroxydisulphate ion is thermodynamically unstable and may decompose into compounds of lower energy values. This decomposition, which is scarcely measurable for the pure dry salts takes place rather quickly in aqueous solutions even at moderate temperatures. During this decomposition the peroxydisulphate ion exhibits considerable oxidising properties. In aqueous solutions the following reaction takes place under weakly acid to alkaline conditions.

$$S_2O_8^{2-} + H_2O \longrightarrow 2\ HSO_4^- + 1/2\ O_2$$

In strong acid solution i.e. above 2 N, Caro's acid is chiefly formed and from this in turn hydrogen peroxide.

$$S_2O_8^{2-} + H_2O \longrightarrow HSO_5^- + HSO_4^-$$

$$HSO_5^- + H_2O \longrightarrow HSO_4^- + H_2O_2$$

As the HSO_5^- decomposes to $HSO_4^- + H_2O_2$, the H_2O_2 decomposes to $H_2O + 1/2\ O_2$. Thus the overall reaction for the complete decomposition of $S_2O_8^{2-}$ ions, under strong acid conditions also follows the first equation. In an as yet unexplained subsidiary reaction some ozone is formed in the strong acid region. Moreover the hydrogen peroxide formed reacts with the peroxydisulphate ions according to the reaction

$$H_2O_2 + S_2O_8^{2-} \longrightarrow 2\ HSO_4^- + O_2$$

3.3.2. Uses of Peroxydisulphates. Many of the uses of the peroxydisulphates depend upon their oxidising properties. The redox potential under standard conditions for the reaction

$$2\ SO_4^{2-}\ (aq) \longrightarrow S_2O_8^{2-}\ (aq) + 2\ e^-$$

is 2.1 volts. The peroxydisulphate ion is therefore a powerful oxidising agent. The proof of this is in the oxidation of Mn^{2+} to MnO_4^-. This reaction, which only takes place with measurable speed in the presence of silver ions, also demonstrates that the oxidative power of the peroxydisulphate anion can often only be developed fully in the presence of heavy metal ions or other additives.

Considerable quantities of the freely soluble ammonium peroxydisulphate, increasingly replaced by sodium peroxydisulphate, are used to etch metal surfaces, and particularly to remove thin layers of copper in the production of printed circuits. Here the peroxydisulphate ion, absorbed on the metal surface, most probably oxidises a copper atom by two electron transfer to a cupric ion, with two sulphate ion residues. By using additives the speed and effect of the etching can be regulated.

The major use of peroxydisulphates is however as
a source of radicals to initiate the polymerization of
certain olefins and the procedure is limited to those
monomers which are sufficiently soluble in water (polymerization with precipitation of polyacrylonitrile for fibre production), or in emulsion polymerisation of e.g. styrene-buta-
diène, vinyl chloride, vinyl acetate and acrylic esters.

The polymerisation rate depends on the quantity
of radicals generated per unit of time and volume i.e.
on the decomposition rate of peroxydisulphate ion. As
this rate can be considerably accelerated by heavy metal
ions and sulphites, these reducing agents are added to the
polymerization process or batch. By such a redox system a
high polymerisation rate is obtained even at moderate temperature, and although relatively small quantities of peroxydisulphates are required (0.5-2 % relative to the monomer), the
bulk of the potassium and sodium peroxydisulphates produced,
and a large amount of ammonium salt, are used for this purpose.

4. Explosive and Health Hazards

As with many chemicals hydrogen peroxide is
handled safely in large quantities, but certain precautions
have to be observed.

Care should be taken to avoid contact with
the skin and particularly the eyes and to avoid ingestion
or inhalation. As is to be expected physiological
effects become more pronounced the greater the concentration.
Hydrogen peroxide from spillage should be washed away with
excess water. Dilute hydrogen peroxide is in fact used as
a mouth wash and zinc peroxide in skin dressings.

The explosive properties of liquid hydrogen
peroxide alone have been carefully studied. Only at concentrations approaching 100 % is there any evidence that
hydrogen peroxide can be exploded and that with a heavy

detonating charge. However concentrations above 26 mole %
are explosible in the vapour phase at atmospheric pressure
(see Figure 4 for compositions generated at boiling point).

Mixtures of hydrogen peroxide with reducing
compounds and organic chemicals may generate heat, increase
decomposition and vapour concentration and may even give
violent explosion. The explosive limits of hydrogen peroxide
with a range of organic compounds have been determined[124].
Thus although commercial hydrogen peroxide is stabilised
it is important to avoid contamination, to handle in clean
vented containers (glass, aluminium and certain grades of
stainless steel are prefered) and during reactions to control
reactant concentrations and cooling.

References

1 - J.L. Thenard, <u>Ann. Chim. Phys.</u>, 1818, <u>8</u>, 306

2 - J. Davy, <u>The Collected Works of Sir Humphrey Davy</u>
    (Smith, Elder and Co, London, 1840), Vol. 5

3 - J.L. Gay-Lussac and L.J. Thenard, <u>Recherches Physico-</u>
    <u>Chimiques</u> (Deterville, Paris, 1811), Vol.1, pp. 159, 169

4 - J.L. Thenard, <u>Ann. Chim. Phys.</u> Série <u>2</u>, 1818, <u>9</u>, 51,
    94, 314, 441 ; 1819, <u>10</u>, 114, 335 ; 1819, <u>11</u>, 85, 208 ;
    1831, <u>48</u>, 79 ; 1832, <u>50</u>, 80

5 - J.L. Thenard, <u>Comptes Rendus</u>, 1855, <u>41</u>, 341

6 - J.L. Thenard, <u>Mém. Acad. Sci.</u>, 1818, <u>3</u>, 345

7 - J.L. Thenard, <u>Amer. J. Sci.</u>, 1833, <u>23</u>, 382

8 - J.L. Thenard, <u>Traité de Chimie</u> (Crochard, Paris, 3e Ed.
    1821), Vol. 1, pp. 562-601

9 - J.L. Thenard et P. Thenard, <u>Ann. Chim. Phys.</u> Série 3,
    1856, <u>47</u>, 173

10 - Anonym., <u>Chemical Age</u>, 1965, April, 507

11 - I.G. Farbenindustrie, Ger. P. 671 318 (1937)

12 - Du Pont, U.S. P. 2 689 169 (1949)

13 - Solvay, Belg. P. 750 704 (1970)

14 - Edogawa, Jap. P. 260 362 (1957)

15 - F.M.C., U.S. P. 2 966 397 (1956)

16 - I.G. Farbenindustrie, Brit. P. 465 070 (1935)

17 - Allied Chemical, U.S. P. 2 995 424 (1958)

18 - Buffalo Electrochemical Co, U.S. P. 2 455 238 (1946)

19 - Laporte, Brit. P. 734 204 (1952) and
     777 138 (1957)

20 - Etat Français, Ugine et Société des Produits Peroxydés,
     Fr. P. 1 055 424 (1952)

21 - Degussa, Ger. P. 1 039 500 (1956)

22 - Montecatini, It. P. 601 202 (1957)

23 - Laporte, Brit. P. 1 132 693 (1968)

24 - Laporte, Brit. P. 695 779 (1951)

26 - Edogawa, Jap. P. 261 120 (1957)

27 - Pittsburg Plate Glass, Brit. P. 991 338 (1965)

28 - F.M.C., U.S. P. 2 768 065 (1951)

29 - Solvay, Belg. P. 531 403 (1954)

30 - F.M.C., U.S. P. 2 768 066 (1952)

31 - Montecatini, It. P. 570 456 (1956)

32 - Du Pont, U.S. P. 2 668 753 (1949)

33 - Etat Français, Ugine, L'Air Liquide, Fr. P. 1 092 393 (1955)

34 - Degussa, Ger.P. 1037 433 (1956)

35 - Allied Chemical Corp., U.S. P. 2 890 105 (1954)

36 - Columbia Southern Chemical Corp., Brit. P. 865 113 (1957)

37 - Solvay et Cie, Belg. P. 531 402 (1954)

38 - Nihon Kagaku Kogyo, Jap. P. 252 684 (1957)

39 - Buffalo Electrochemical Co, U.S. P. 2 537 516 (1950)

40 - Buffalo Electrochemical Co, U.S. P. 2 537 655 (1950)

41 - A.E. Brodskii, V.I. Franchuk, M.M. Alexankin and V.A. Lunenok-Burmakina, Dokl. Akad. Nauk SSSR, 1953, 123, N° 1, 117

42 - M. Baccaredda et C. Pedrazzini, Rivista dei Combustibili, 1954, VIII, N° 6, 417

43 - A.R. Burgess, J. Appl. Chem., 1961, 11, July, 235

44 - N.V. de Bataafsche Petroleum Mij, Brit. P. 758 907 (1956)

45 - T. Kunugi, T. Matsuura and S. Oguni, Hydroc. Processing, 1965, 44, N° 7 116-122

46 - N.V. de Bataafsche Petroleum Mij, Brit. P. 758 967 (1954)

47 - N.V. de Bataafsche Petroleum Mij, DAS 1 106 738 (1959)

48 - W.C. Schumb, C.N. Satterfield and R.C. Wentworth Hydrogen Peroxide (Rheinhold, New York - 1955)

49 - F.R. Duke and T.W. Haas, J. Phys. Chem. 1961, 65, 304

50 - A. Hansson, Acta Chem. scand., 1961, 15, 934

51 - A.F. Chadwick and G.L.K. Hoh in Encyclopedia of Chemical Technology (Interscience, New York - 1966) Vol. 11, p. 391

52 - P.A. Giguere in Compléments du nouveau traité de Chimie Minérale (Masson, Paris - 1975) Vol. 4

53 - Gmelin, Handbuch der Anorganische Chemie (Verlag, Weinheim 1966) Sauerstoff, System 3, Lieferung 7

54 - J.O. Edwards, Peroxide Reaction Mechanisms (Wiley, New York 1967)

55 - V. Lorås et al, A.T.I.P., to be published

56 - R. De Vooght et al, to be published

57 - M.C. Delattre and G. Papageorges, 6th Annual Convention of Brazilian Technical Association of Cellulose and Paper, Sao Paulo, November, 1973

58 - P.K. Paasonen, Paperi Ja Puu, 1974, 9, 696

59 - S. Rothenburg et al, Tappi, 1975, 58, 182

60 - G. Papageorges and J. Deceuster, Indian Chemical Age, 1976, 27, 451

61 - A.G. Davies, Organic Peroxides (Butterworth, London - 1961)

62 - D. Swern, Organic Peroxides (Wiley, New York - 1970)

63 - J. Stephend and M.F. Hobbs, Water and Sewage Works, 1973, 8, 67

64 - Sigmund, Ger. P. 2211890 (1972)

65 - W.R. Muller and I. Sekoulov, Conference on nitrogen as a water pollutant, Copenhagen, 1975

66 - P.J. Keller and C.A. Cole, Water and Waste Engineering,
     1973, E4

67 - N. Wolters et al, Ger. P. 2446511 (1976)

67a - Kurita Industries, Jap. P. 37455 (1974)

68 - Degussa, Belg. P. 802466, 1972

69 - S.N. Hussaini, Veterinary Record, 1976, 98, 257

70 - Chemical Research Laboratories, U.S. P. 3082 146

71 - Interox booklet, Decomposition of Heat Treatment Salts,
     1975

72 - Degussa, Ger. P. Appli. 12352 856, 1973

73 - Traitement des surfaces, 1976, 146, 37

74 - Kurita Industries, Jap. P. 367374 (1969)

75 - Ugine Kuhlman, Belg. P. 825 241 (1975)

76 - Institut Français Pétrole, Fr. P. 1583 732 (1967)

77 - L'Air Liquide, Belg. P. 773, 172 (1971)

78 - Tilhaud, Fr. P. 2123735 (1971)

79 - Minnesota Mining, U.S. P. 4721624 (1971)

80 - Cellulose Attisalz, Swiss P. 573 874, (1973)

81 - F.M.C.,Ger. P. 2328532 (1973)

82 - L'Air Liquide, Fr. P., 2181551 (1971)

83 - F.M.C., U.S. P. 3878208 (1973)

84 - Rhône-Poulenc, Ger. P. 2 064 497 (1970)

85 - Friess et al J. Am. Chem. Soc., 1951, 73, 3768

86 - Japanese Chemical Week, 20 Nov. 1975, 4

87 - Interox, Belg. P. 838068 (1976)

88 - Union Carbide, Belg. P. 838953 (1976)

89 - I. Volnov, Peroxides, Superoxides and Ozonides of Alkaline
     Earth Metals (Plenum, New York, 1966)

90 - E.S. Shanley and J.O. Edwards, in Encyclopedia of Chemical
     Technology (Interscience, New York, 1966) Vol. 14, p. 746

91 - J.A. Cannon et al in Advances in Inorganic Chemistry and
     Radio Chemistry (Academic Press, New York, 1964) Vol. 6

92 - Gmelin, Handbuch der Anorganische Chemie (Verlag, Weinheim 1960) Sulphur, System 9, Lieferung 2, Teil B, p. 798

93 - Hudson Bay Mining and Smelting Co, Can. P. 550346 (1956)

94 - Du Pont, U.S. P. 2802722

95 - Osterr. Chem. Werke GmbH, Ger. P. 608 830 (1932)

96 - Osterr. Chem. Werke GmbH, Ger. P. 610 611 (1932)

97 - Solvay, Fr. P. 2 076 430 (1970)

98 - Solvay, Fr. P. 2 160 251 (1971)

99 - B.I.O.S. Final Report N° 854 - Item N° 22

100 - Degussa, Ger. P. 732 501 (1937)

101 - Solvay, Belg. P. 810 289 (1974)

102 - Solvay, Lux. P. 67 482 (1973)

103 - Solvay, Lux. P. 68 831 (1973)

104 - Interox, Belg. P. 842 014 (1976)

105 - J.O. Edwards et al J.A.C.S., 1969, 91, 1095

106 - Berthelot, Compt. Rend., 1878, 86, 20

107 - Caro, Z. Anorg. Chem., 1898, 845

108 - A. Baeyer and V. Villiger, Ber. 1900, 33, 124 and Ber. 1901, 34, 853

109 - Du Pont, U.S. P. 1937 621 (1931)

110 - Du Pont, U.S. P. 2,094, 384 (1934)

111 - Du Pont, Brit. P. 508045 (1936)

112 - Henkel, Ger. P. 659923 (1935)

113 - H. Schmidt, Ger. P. 2,018, 216 (1970)

114 - Du Pont, Fr. P. 836834 (1937)

115 - Laporte, Brit. P. 503625 (1932)

116 - F.M.C., Brit. P. 785,660 (1955)

117 - F.M.C., Brit. P. 816,125 (1956)

118 - A. Pietzsch, Ger. P. 243,366 (1909)

119 - Degussa, Ger. P. 724945 (1940)

120 - Ugine, Fr. P. 953041 (1944)

121 - L'Air Liquide, Fr. P. 1493, 723 (1966)

122 - F.M.C., U.S. P. 2899272 (1957)

123 - Peroxid-Chemie, Ger. P. 2,346, 945 (1973)

124 - E.S. Shanley and F.P. Greenspan Ind. Eng. Chem., 1947, 39, 1536

125 - H. Pistor in Chemische Technologie, Band I, Anorganische Technologie I, (Carl Hanser, Muenchen) 1970, 517

Industrial Gases

W J Grant
S L Redfearn
BOC Limited

Industrial Gases is a vague but convenient title used here to cover some gases regularly manufactured and distributed to industry for diverse uses.

The description originally seems to have meant any gas used in industry, other than town gas and other than gases produced and consumed in integrated processes without isolation as commercial entities, but developments in recent times have even further blurred this already vague definition, and at present almost any gas could be embraced by the title.

Industrial gases considered here are some of the commercially well-established inorganic gases – the atmospheric gases; helium; and hydrogen – others being subjects of other chapters.

Atmospheric Gases

Air is the source of industrial gases Oxygen, Nitrogen, Argon, Neon, Krypton and Xenon. It is convenient to group these gases together for description because of their common raw material (air) and their primary processing in an air separation unit.

The composition of air at low altitude varies slightly regionally and temporally. The main variable component is water vapour, ranging between about 4% v/v, in a tropical jungle to a very low value in cold dry climates. Other variations are localized and due to unusual geological circumstances (eg high sulphurous gas content in volcanic regions) or to human activities (eg chimney exhausts). The major invariant part of air has the following composition –

Nitrogen	–	78.03% by volume
Oxygen	–	20.99%
Argon	–	0.93%

Carbon dioxide	–	0.03%
Neon	–	0.0015%
Hydrogen	–	0.0010%
Helium	–	0.0005%
Krypton	–	0.0001%
Xenon	–	0.000008%

– and this mixture may be considered to be the raw material from which atmospheric industrial gases are separated.

The total amount of Earth's atmosphere is estimated to be about 4.5×10^{15} tonnes, so that for practical purposes the supply is unlimited. The seemingly very large scale extraction of oxygen (of the order of 1×10^8 tonnes/year on a worldwide scale) is insignificant in comparison with the total resource and has no perceptible effect on air composition. Proportions of the constituents oxygen and carbon dioxide are in dynamic equilibrium, oxygen being continually consumed by animal metabolism, fires, and other oxidative reactions, and replenished by plant photosynthesis, while carbon dioxide is generated by animals, fires, etc., and consumed by plants.

Separation of air into its components can be done by various chemical and physical means, but the most generally advantageous method on an industrial scale is fractional distillation of liquid air. Modern large-scale manufacture of atmospheric gases stems from the invention at the beginning of this century of practical economic processes for air liquefaction, following the work of Hampson (Britain), Linde (Germany), and Claude (France). Hampson and Linde used the Joule-Thomson cooling effect to liquefy air, by the process shown diagrammatically in Fig.1. In this process, air is compressed isothermally and then expanded adiabatically at a throttle to obtain Joule-Thomson cooling. The cooled air is recycled through a heat-exchanger to cool the compressed air before expansion, thereby accumulating the cooling effect until the temperature is lowered sufficiently to liquefy part of the cycling air.

The original Hampson-Linde process is not thermodynamically efficient and requires a lot of air to be highly compressed to give the necessary refrigeration. Claude's great contribution to air liquefaction was to find practicable means for carrying

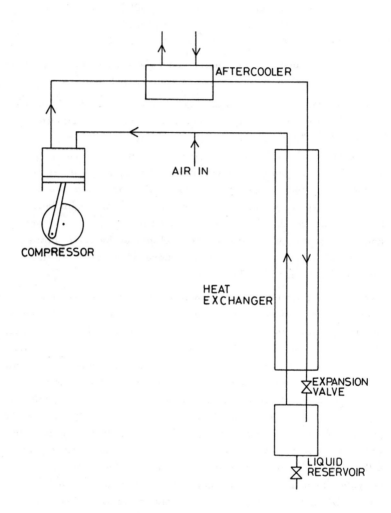

FIG. 1. LINDE - HAMPSON PROCESS

out part of the expansion isentropically in an expansion engine from
which mechanical work is obtained, thus producing a much greater
cooling effect than the Joule-Thomson effect alone. The Claude
process is outlined in Fig.2, and some form of it is now used in
all large air liquefaction plants.

Fractional distillation of liquid air is based on the same principles
as fractionation of other mixed liquids such as petroleum. An air
still however necessarily operates at very low temperature, and
special arrangements are required to recondense vapour for reflux
at low temperature, conventional coolants such as water or ambient
air not being usable. This problem is solved by using the oxygen
product itself at a lower pressure (and hence lower boiling point)
to condense vapour for reflux at a higher pressure. The most
widely used embodiment of this principle is the double-column air
still shown in Fig.3. In this, liquid air at a higher pressure
enters the lower column and is partly separated therein, the more
volatile nitrogen refluxing at the head and less volatile enriched
oxygen concentrating at the base. The oxygen-enriched mixture
at the base of the lower column is transferred to the upper column
at a lower pressure, where it is finally separated into oxygen and
nitrogen. Liquid oxygen boiling at the base of the upper column
is cold enough at the operating pressure to condense nitrogen for
reflux in the lower column.

The most volatile air constituents - helium, hydrogen, and neon -
are not condensed and accumulate as a gaseous mixture with
nitrogen at the top of the lower column. This uncondensed gas
may be processed to recover neon. Argon has a volatility between
those of oxygen and nitrogen and concentrates in the upper column
in a region some way above the base, whence it may be withdrawn
(mixed with oxygen and nitrogen) for further processing and
purification in a separate column (not shown in Fig.3). The least
volatile air constituents - krypton and xenon - accumulate in the
oxygen boiler of the upper column. For recovery of these gases,
a sidestream of liquid oxygen enriched in these elements is withdrawn from the upper column boiler and separately processed.

Complete air separation units each comprising an air liquefaction
and liquid air fractionation section as outlined above, may vary

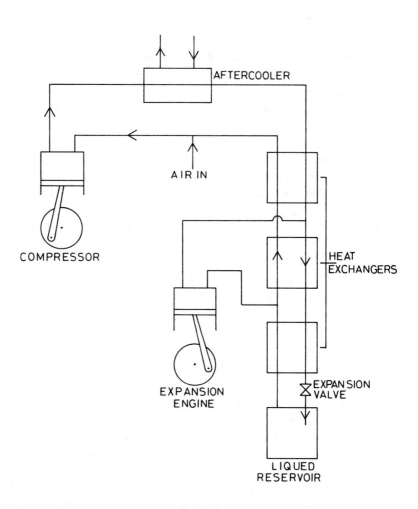

FIG. 2   CLAUDE PROCESS

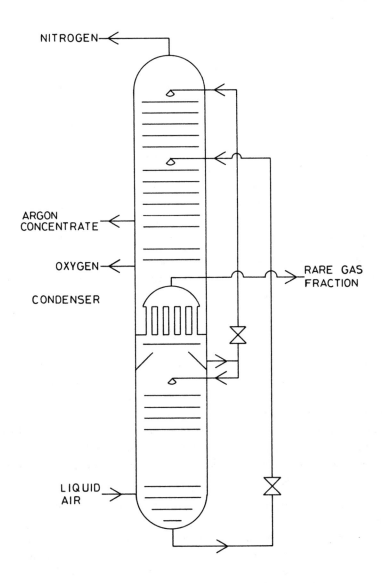

NITROGEN

ARGON CONCENTRATE

OXYGEN

CONDENSER

RARE GAS FRACTION

LIQUID AIR

FIG. 3. DOUBLE COLUMN AIR STILL

considerably in detail design and interconnection. There is no
universally best design, variations in detail being required according
to the scale of the plant, to the ratio of oxygen and nitrogen products
required, and the form in which these products are to be produced
(ie as liquid or as gas at high or low pressure). A line diagram
for a typical modern large-scale air separation unit for production
of oxygen and nitrogen is shown in Fig.4. Notable features of
such plants are the use of reversing heat exchangers for cooling
incoming air with cold waste gas, and the low operating pressure
(about 6 bars) with use of turbines for compression and expansion.
Also note the adsorber for removal of hydrocarbons from the liquid
oxygen. An air separation plant such as that of Fig.4 would form
the basis of an industrial gas complex which might also include
product compressors, nitrogen or oxygen liquefiers, an argon
purification unit, and rare gas processing units.

Atmospheric gas plants on their present optimum economic scale
of many hundreds of tonnes per day of oxygen and nitrogen products
are large energy consumers. The theoretical minimum energy
required to separate 1 normal cubic metre of oxygen from air (all
at 1 bar pressure) is 0.074 kWh. However, in actual plants
separation energy is several times this minimum because of many
thermodynamically irreversible features unavoidable in a practical
design. Also the energy required for compression or liquefaction
of separated products is high, the theoretical minimum energy for
compression of nitrogen to cylinder pressure (170 bars) for
example, being 0.14 kWh per cubic metre, or for liquefaction 0.25 kWh
per normal cubic metre (gas). A large air separation plant
producing, say, 700 tonnes/day products may have a power requirement of the order of 10 MW.

Large resources are needed to set up and operate a nationwide or
wide regional network of atmospheric gas plants on the scale required
by modern industry. This has led to the industry being concentrated
in any country into a very small number of very large organizations.
In Britain the firms concerned with atmospheric gases manufacture
are Air Products Ltd and BOC Ltd.

Many systems for air separation other than liquid air distillation
have been considered or tried, and some are still under study and
development. Older processes rendered obsolete by air distillation

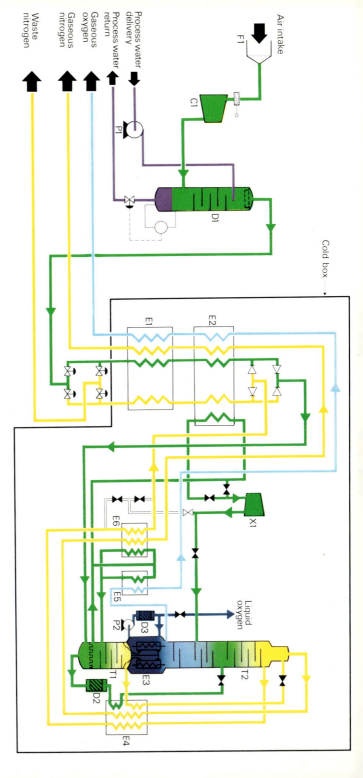

Plant items

C1   Air turbo compressor
D1   Direct cooler
D2   Adsorber (rich liquid)
D3   Adsorber (liquid oxygen)
E1–2 Reversing heat exchangers

E3   Condenser
E4   Poor liquid-rich liquid subcooler
E5   Oxygen heater
E6   Nitrogen heater
F1   Air filter

P1   Direct cooler water pump
P2   Liquid oxygen pump
T1   Lower column
T2   Upper column
X1   Expansion turbine

Waste nitrogen

Gaseous nitrogen

Gaseous oxygen

Gaseous oxygen

Process water return

Process water delivery

Air intake

F1

C1

P1

D1

Cold box

E1

E2

E6

E5

E4

X1

D2

D3

T1

T2

E3

P2

Liquid oxygen

were (i) Brin's process, and (ii) Mallet's process. In the Brin
process barium oxide was heated in air at 700°C and 2 bars pressure
to form the peroxide, from which pure oxygen and reusable barium
oxide were recovered by reducing pressure to 0.05 bar. Mallet's
process depended on the greater solubility of oxygen than nitrogen
in water: by contacting air with water under pressure an oxygen-
enriched dissolved gas can be recovered on reducing pressure, and
by repeating this cycle oxygen of any desired concentration may be
produced. A now obsolete process used on a pilot scale in the
1940/1950 decade was based on the use of organo-cobalt complexes
which form loose oxygen compounds (in a manner recalling
oxyhaemoglobin formation in blood), from which oxygen may be
recovered by altered pressure/temperature conditions.

Recent and current research on new air separation processes is
mainly concerned with (i) differential adsorption, (ii) differential
permeation, and (iii) oxygen ionization. Of these, differential
adsorption seems most promising and has reached a commercial
stage in development. The pressure-swing adsorption (PSA)
process is used for production of concentrated (rather than pure)
oxygen or nitrogen, and depends on the use of special adsorbants
with strong nitrogen or oxygen adsorbing power at room temperature.
Adsorption takes place at a higher pressure and desorption at a
lower pressure, the pressure ratio being about 5 to 1. Adsorbants
have been developed which can regularly produce oxygen of about
92% strength or nitrogen of about 99% strength from air in one
stage. More than one stage is generally uneconomic.

Differential permeation processes depend on the different rates at
which oxygen and nitrogen permeate certain membranes, for
example, silicone rubber films. Present development work is
concerned with producing membranes in optimum form (probably
as thin-walled capillaries) for efficient practical air separation.

Oxygen ionization is the basis of the Allis-Chalmers air separation
process in which atmospheric oxygen is dissolved (as OH^- or
equivalent ions) in electrolyte at a fuel cell type of electrode, and
discharged at a second similar electrode in a cell. In theory
the power required for this process is very low, as both cell electrodes are oxygen-electrodes and should have no potential difference

other than that due to ohmic electrolyte resistance to flow of ions.
In the present state of development, irreversible secondary electrode
processes give rise to a PD of about 1 volt per cell, requiring a
large expenditure of power which renders the process uneconomic.

Oxygen
Production: Oxygen was first produced commercially towards
the end of the last century by the now obsolete Brin process. Brins
Oxygen Co. later became the British Oxygen Co. (the present BOC
Ltd).

Minor amounts of industrial oxygen are currently produced by the
PSA process already referred to, and as a co-product in the electrolysis of water for hydrogen production.   Very small amounts of
breathable oxygen for use in emergencies (eg failure of normal
supply in aircraft or submarines) are made by decomposition of
sodium chlorate in 'oxygen candles'.

Air separation was revolutionized by processes for liquefaction and
distillation of air, as already outlined, and nearly all industrial
oxygen is produced by this process.

Oxygen is manufactured on a very large scale, outputs from a single
modern plant usually being several hundred tonnes per day. Some
production estimates showing the order of scale are in Table 1.

Table 1 : Oxygen Production Estimates,
000's tonnes/yr

	1970	1975
UK	1800	2200
USA	13000	15000
World	45000	80000

Commercial oxygen is distributed to users as piped gas; as liquid
oxygen in road tankers; or as compressed gas in cylinders,
according to scale of use.   In Britain there are at present (1977)
about 50 miles of main oxygen pipeline serving large users.

In relation to the fairly low ex-factory cost of oxygen, distribution
costs are high, and works are usually set up near to centres of

high demand to serve limited surrounding regions. The price of oxygen to a user varies greatly according to the method and scale of distribution to him, reflecting the high distribution costs. Currently in Britain, the order of oxygen prices may range from about £2 per 100 cubic metres delivered on a large scale by pipeline direct from works, through about £7 per 100 cu.m as liquid d/d in large tanker loads, to £20 (or more) per 100 cu.m for small amounts delivered to user in cylinders.

Technology: Oxygen is the most abundant element in the parts of Earth accessible to man, forming about 23% by weight of the atmosphere, 89% of water, and about 50% of the mixed oxides which form the rocks of the Earth's crust.

The oxygen atom exists in three stable isotopic forms 0^{16}, 0^{17}, 0^{18}, which occur in nature in the ratios 1000/3.7/20 respectively. Several unstable radioactive isotopes (0^{14}, 0^{15}, 0^{19}, 0^{20}) have been prepared artificially but these have short half-lives and are not found in natural oxygen. The 0^{18} isotope can be separated in concentrated molecular form and is available commercially for use as a tracer in studies of chemical reaction mechanisms. Oxygen atoms can combine in two allotropic molecular forms, 0_2 (ordinary molecular oxygen) and 0_3 (ozone). Industrial oxygen separated from air consists of a mixture of predominantly 0_2^{16} with some $0^{16}0^{17}$, 0_2^{17}, etc. etc.

Some properties of oxygen are shown in Table 4. The gas condenses to a pale blue liquid at $-183.1^{\circ}C$, 1 bar, and on further cooling below $-217^{\circ}C$ freezes to a solid (γ) form which undergoes transitions to other solid forms at still lower temperatures. Oxygen in the gaseous, liquid, and solid states is paramagnetic, due to unpaired electrons in the atom.

Oxygen is an extremely reactive gas, combining with many substances in ordinary ambient conditions, and with many more at elevated pressure and temperature, this reactivity being the basis of oxygen uses. Familiar examples of oxygen reactions are the combustion of fuels; rusting of iron; souring of beer and wine; rancidification of oils and fats; polymerization of paint oils; denaturation or discolouration of substances such as perfumery ingredients, photographic developers; and biochemical oxidation in animals.

The reactivity and unusual properties of oxygen make possible many
basic reference methods for its analysis. The most generally used
chemical methods are based on absorption of oxygen in alkaline
pyrogallol or in ammoniacal cuprous chloride. Physical methods
based on the unusual paramagnetism of oxygen, or on its reactivity
as an electric cell depolarizer, are also in wide use. Commercial
oxygen is a highly purified product, typically containing less than
1 vpm nitrogen and other air components.

Many metals, notably steel, burn vigorously in pure oxygen if raised
to ignition temperature. This is utilized in oxygen-cutting, but
is also the reason for the special precautions used in conveying
oxygen in steel pipes and equipment. Oxygen equipment must be
clean and free from oily or combustible matter which might start
a metal fire. The linear speed of oxygen in pipes should be limited
(to 25 m/sec at 40 bars) to minimize the hazard of impacts between
particles of pipescale carried along by the gas and internal walls
at pipe bends, valves, or fittings; energetic impacts of this kind
having been known to initiate fires. Any material combustible in
air will burn rapidly in oxygen, and clothing, asphalt roads, wooden
furniture, and so on, ordinarily regarded as low fire risks, become
hazardous in the presence of pure oxygen or oxygen-enriched air.

Oxygen in the diluted form of air is essential to human and animal
life, but in pure form it has a definite toxicity. Pure oxygen is
often beneficially administered under medical supervision at
controlled dosage in a variety of respiratory disorders, but long-term
breathing of pure 100% oxygen is hazardous.

Uses: The largest single use for oxygen is in converting iron to
steel by modifications of the Bessemer process. Use of oxygen
instead of air greatly hastens the process and increases productivity
of the steel plant. This use of oxygen commenced on a large scale
in the late 1950s and has accounted for most of the very large
increase in oxygen consumption during the last two decades.
Oxygen processes are now almost universally used in steelmaking.
Current use of oxygen for this purpose in Britain is of the order
of 2 million tonnes/yr, or 85% of all oxygen produced.

There is a growing use in iron blast furnaces where oxygen enrichment of the blast makes it possible to use heavy fuel oil to replace

some of the more expensive metallurgical coke.

Coke reduction has been an important incentive for iron founders to use oxygen enrichment in iron melting cupolas. The economic advantages achieved are due to increased melting zone temperature and improved combustion efficiency. Oxygen is used on a large scale in other furnace applications such as ferrous and non ferrous melting and in glass manufacture where higher temperatures, greater productivity, reduced power consumption and/or longer furnace life are important benefits.

The chemical industry is a large consumer of oxygen, which can often confer benefits in comparison with air as oxidant by giving faster reaction and greater reactor productivity, or improved yield or energy-efficiency by avoiding dilution of reaction gases with nitrogen. These benefits may lead to lower capital cost plant, lower energy consumption, and environmental improvements due to significant reduction in the amount of purge gas.

Oxygen is used on a large scale in the direct oxidation of ethylene to ethylene oxide; in production of titanium dioxide by the chloride route; in the manufacture of vinyl chloride; vinyl acetate; synthesis gas (hydrogen and carbon monoxide); propylene oxide and chlorohydrocarbons. Oxygen-using processes exist for manufacture of acetylene; terephthalic acid; cyclohexanone; nitric acid; nitrous oxide, sulphuric acid; acetaldehyde; acetic acid and other chemicals.

Medium and smaller scale oxygen uses are very diverse and are briefly summarised in the list below –

Steel cutting	River revival
Oxy-gas welding	Fish farming and transport
Oxygen lancing (drilling	Diving and respiration
concrete)	gases
Paper pulp bleaching	Medical uses
Sewage treatment	

Nitrogen

Production: Nitrogen is produced as a co-product with oxygen in
air separation by fractional distillation.   It has been available on
a large scale in pure form only since the 1950/1960 decade,
following the great expansion of oxygen production for steelmaking.
Industrial nitrogen is now made and used in large quantities, as
shown in the estimated figures in Table 2 –

Table 2 : Nitrogen Production Estimates
000's tonnes/yr

	1970	1975
UK	700	1000
USA	5000	8000
World	20000	30000

Commercial nitrogen is distributed as piped gas; as liquid nitrogen
delivered in road or rail tankers; or as compressed gas in cylinders;
according to scale of use.   As with oxygen, distribution costs are
high in relation to the ex-factory cost of nitrogen, and the price
of the gas to a user can vary considerably depending on the method
and scale of distribution used.   Currently (1977) in Britain the order
of nitrogen prices may range from about £1 per 100 cubic metres
delivered by pipeline direct from works, through about £6 per
100 cu.m for large tanker loads of liquid nitrogen, to £20 (or more)
per 100 cu.m for small amounts of nitrogen in cylinders.

Large amounts of low-purity nitrogen (mixed with 10-15% CO_2 and
other minor impurities) are generated and used on location in some
industries by burning gas or oil with airflow controlled so as to
minimize (circa 1%) excess oxygen or incomplete combustion
products.   The total amount of such low-grade nitrogen made and
used is probably comparable with the total of commercial high-
grade nitrogen.   Combustion gas 'nitrogen' can of course be
further purified by removing CO_2 with a regenerable alkaline scrubbing
medium, and some is so treated, but more usually the unpurified
combustion gas finds application in cases where a crude inerting
gas is sufficient, such as inerting or fire fighting in ship's holds.

Nitrogen can also be made economically by the pressure-swing
adsorption air separation process.

Minor amounts of nitrogen, or more usually nitrogen/hydrogen mixtures, are also made on location in some industries by cracking ammonia and burning some or all of the hydrogen in the cracked gas with air.

<u>Technology</u>:   Nitrogen forms about 76% by weight of the atmosphere and is the most abundant uncombined element accessible. Atmospheric nitrogen consists mainly of N^{14} isotope (99.63%) with a little N^{15} (0.37%). Other short-lived radioactive isotopes are known. Some properties of nitrogen are shown in Table 4.

Nitrogen is a chemically inert gas, not reacting with any substances in ordinary ambient conditions, and only undergoing limited reactions at higher temperatures. This inertness is the basis of, or an important feature in, the industrial applications of nitrogen.

Liquid nitrogen has wide use as a refrigerant, its inertness making it very suitable for direct contact with the materials to be cooled. Originally the use of liquid nitrogen as a refrigerant tended to be limited to applications in which its very low temperature was essential. More recently however, its use has been extended to applications where only a moderately low temperature is required, as in food freezing. Liquid nitrogen can be economically competitive in some such cases with mechanical refrigeration and other refrigerants such as ice or dry-ice by virtue of its advantages in being in liquid form for direct contact with refrigerated goods by immersion or spray, and its large temperature difference, giving rapid cooling in simple equipment. The total cooling effect of liquid nitrogen may be regarded as being made up of its latent heat of vaporization, and the cooling due to the cold evaporated gas. The total refrigeration available at various initial liquid nitrogen tank pressures and final vented gas temperatures is shown in Fig.5.

The inertness of nitrogen poses a problem in devising basic reference methods of analysis. Its reaction with hot calcium has been mostly used, but more often a physical method for analysis, or simply estimation by difference after analysis of the more reactive components of a mixture, are employed. Commercial nitrogen is a highly purified product, typically containing less than 10 vpm oxygen. Specially purified 'oxygen-free' nitrogen, containing less than

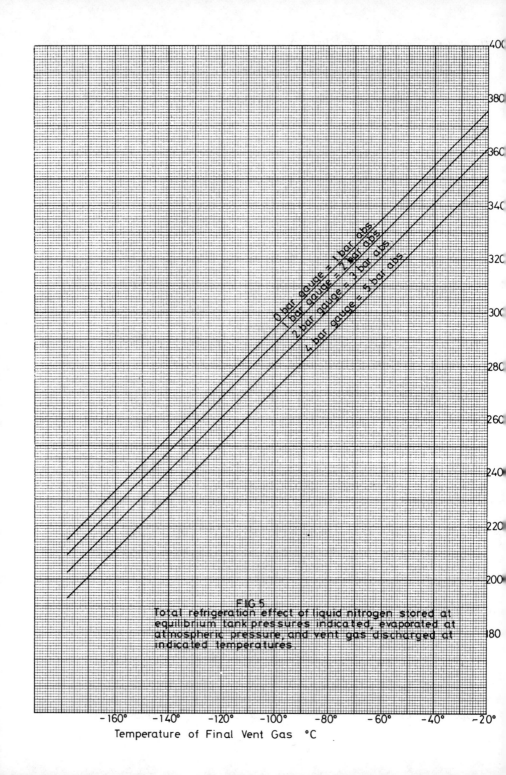

FIG 5
Total refrigeration effect of liquid nitrogen stored at equilibrium tank pressures indicated, evaporated at atmospheric pressure, and vent gas discharged at indicated temperatures.

0 bar gauge = 1 bar abs
1 bar gauge = 2 bar abs
2 bar gauge = 3 bar abs
4 bar gauge = 5 bar abs

Temperature of Final Vent Gas °C

2 vpm O_2 is regularly available on a smaller scale for special purposes.

Nitrogen at atmospheric pressure is non-toxic but asphyxiating. When breathed at higher pressure (as in deep-water diving) it has a definite toxic effect.

Uses: The uses of nitrogen considered here are those in which it enters and leaves the process unchanged in its elemental state. This excludes the very large use of nitrogen as a chemical feed-stock for ammonia manufacture, this nitrogen usually not being isolated externally to the ammonia plant.

Industrial nitrogen has two basic types of use, the larger being as an inert atmosphere, and the other being as liquid nitrogen for refrigeration. In these two kinds of use nitrogen finds application in a great diversity of industries.

Nitrogen is used as an inert atmosphere in many metallurgical, chemical, and other processes where the presence of air would involve fire or explosion hazards, or would give rise to undesir-able oxidation reactions. For example, large amounts of nitrogen are used in petrochemical processing to purge vessels of air before admitting inflammable liquid or gas, or conversely to purge inflammable vapours before admitting air, to avoid forming poten-tially dangerous air/vapour mixtures. It is used to dilute reactants, to transfer or mix liquids and to blanket reactions inhibited by oxygen. An interesting inerting development is the use of nitrogen as the gas in a foam. This is used to make vessels that have contained inflammable liquids safe for cutting or welding operations. Nitrogen atmospheres are also used on a large scale in metallur-gical operations such as heat-treatment to prevent formation of metal oxide scale on the surface of goods being treated and as the base gas in reactive heat-treatment processes. A related large consumer of nitrogen is the float-glass process, in which a nitrogen atmosphere prevents oxidation of the molten tin bath on which the glass floats. A new and growing application is in blast furnaces.

Medium and smaller scale nitrogen uses as an inert gas are in packaging of processed foods, pharmaceuticals, and other sub-stances, to prevent spoilage by air; in storage of vegetable oils,

paint, and perfumery products; for the gas filling in hydraulic
accumulators, hot water systems, aircraft and vehicle tyres, and
other engineering systems where pressurized gas is in contact
with oil, rubber, or oxidizable materials.

Liquid nitrogen is widely used as a direct-contact expendable
refrigerant, the material to be cooled being immersed in, or
sprayed with the liquid. Heat is rapidly transferred directly
from material to refrigerant without passage through an inter-
vening metal wall as in conventional mechanical refrigeration
with recycled refrigerant. Applications in which the very low
temperature of liquid nitrogen is the primary requirement are –

> Freeze grinding, for pulverizing normally soft
> or rubbery materials;
> Shrink-fitting, for assembling engineering com-
> ponents;
> Deflashing, for removing mould flash from rubber
> mouldings;
> Low temperature machining of rubbery materials;
> Fragmentation of steel scrap for remelting;
> Preservation of biological matter such as blood,
> semen, etc.

Refrigeration applications in which a very low temperature is not
essential, but for which nitrogen is regularly used for its convenience or rapidity of action –

> Food freezing;
> Soil freezing, for consolidating unstable ground in
> tunnelling or excavation;
> Environmental testing;
> Freeze-branding, for marking cattle;
> Pipe-freezing, for stopping flow in absence of
> valves;
> In-transit refrigeration.

Argon

Production: Argon is obtained from the concentrated argon fraction
from air separation units, or from ammonia plant tail-gas, the

former being the only source presently used in Britain.

The concentrated argon fraction from an air still typically contains about 10-12% Ar, the rest being mainly oxygen. This mixture is fractionally distilled in a column for which low temperature cooling for reflux condensation is obtained from the main unit. The base product from the argon column – oxygen – is returned to the main unit, and an overhead product ('crude argon') is obtained consisting typically of about 98% Ar with 2% O_2. The crude argon is normally purified by adding hydrogen for catalytic reaction with the oxygen to form water. After drying and recondensing, a highly purified argon product is obtained.

Argon in air consumed in ammonia synthesis remains in concentrated form (typically 8-10%) in the purge gas finally discharged, mixed with nitrogen and hydrogen and other minor impurities. Argon is obtained in pure form from this mixture by fractional distillation.

Estimated production statistics for argon are in Table 3 –

Table 3 : Argon Production Estimates, 000's tonnes/yr

	1970	1975
UK	12	30
USA	130	240
World	400	700

Argon is distributed to industry in liquid form or as compressed gas in cylinders. Current prices in Britain range between about £50 per 100 cubic metres for liquid to £150 (or more) per 100 cu.m for compressed gas.

Technology: Argon forms about 1% by weight of the atmosphere. A large air separation unit or ammonia plant taking in many hundreds of tonnes per day of air has a potential capacity of several tonnes per day argon, so that it is by no means a rare gas, and there is in fact more argon potentially available from these sources than is required by industry.

Argon exists in three stable isotopic forms Ar^{36}, Ar^{38}, Ar^{40}, with natural abundance ratios respectively 0.337%, 0.063%, and 99.60%. Several long-lived radioactive isotopes are also known. Some properties of argon are shown in Table 4.

Argon is chemically inert and forms no compounds existing in normal conditions, this inertness being the basis of its industrial uses. Argon does however form a chelate compound with hydroquinone which has been used in purifying argon.

The total inertness of argon precludes chemical reference methods for analysis, and physical methods must be used. Commercial argon is a highly refined product, typically containing less than 4 vpm O_2 and less than 36 vpm other impurities.

Argon is non-toxic but asphyxiating in pure form.

Uses: The principal use of argon is as an inert gas in situations where the cheaper gases nitrogen and carbon dioxide are reactive. This usually implies the use of argon in high-temperature metallurgical processes. For example, argon is used in gas-stirring and stream shrouding molten steel; and in spectrographic analysis. Recent developments in manufacture of stainless steel will significantly increase argon consumption. Argon is widely used as a shielding gas in electric arc welding; as an inert atmosphere in powder metallurgy and in the processing of atomic energy fuels; and for blanketing hot sodium used as heat-transfer medium in fast breeder nuclear reactors.

Smaller scale but important uses of argon are in filling electric lamps, both filament and gas-discharge types, and other electric discharge devices.

Neon

Neon is recovered from a mixed gas fraction from an air separation unit. The crude gas typically consists of about 50% N_2, 25% Ne, 17% H_2, 8% He. Nitrogen is first largely removed by condensing it, and the residual gas is then separated by fractional adsorption on carbon at low temperature, neon being more readily adsorbed than hydrogen or helium.

Only a relatively small part of the neon potentially recoverable from air separation plants is required by industry. Current consumption of Neon in UK is about 50,000 litres/year.

Neon is sold in sealed glass bulbs containing 1 litre at about atmospheric pressure, or in larger amounts in cylinders. Current (1977) price is around 75p per litre according to quantity taken.

A typical analysis of commercial neon is –

Ne	–	99.998%
He	–	5 vpm
N_2	–	5 vpm
O_2	–	2 vpm
H_2O	–	2 vpm

Table 4 shows some properties of neon.

The principal use of neon is for filling electric discharge tubes, more particularly light tubes in which it gives its characteristic red colour. Neon has also been used on a limited scale as an inert refrigerant in low temperature physics research for situations where liquid hydrogen (which has about the same temperature as liquid neon) would be too hazardous. Neon has also been proposed as an inert constituent of deep-water diving gas, as its blood solubility is low, and as it does not affect speech as badly as the more generally used helium.

Krypton and Xenon

Krypton and Xenon are recovered from a stream of liquid oxygen concentrated in the upper column sump of an air still, the concentration in the liquid sidestream withdrawn being typically about 0.18% Kr and 0.15% Xe. Oxygen is first removed by burning with hydrocarbon, and after absorption of the CO_2 produced, and drying, a residual gas is obtained consisting mainly of krypton and xenon, which are separated by fractional adsorption.

Current (1977) UK consumption of these gases is about 100,000 litres/year Krypton and 5000 litres/year Xenon.

Krypton and Xenon are mostly sold in sealed glass bulbs containing 1 litre gas at about atmospheric pressure or in metal cylinders for larger amounts.   Current (1977) prices are –

Krypton around £1 per litre
Xenon   around £5 per litre

according to amounts taken.

Typical analyses of the commercial gases are –

Krypton: Kr 99.97%

Xenon: Xe 99.997%

Principal uses of krypton and xenon are for filling electric lamps and discharge tubes of various kinds.   Krypton is superior to argon as a filling for filament lamps in giving reduced filament evaporation, allowing longer life and/or greater efficiency. Xenon is used in discharge tubes to give high-intensity white light for such purposes as photographic flash, laser pumping, and searchlights.

Xenon has high blood solubility and acts as an inhalation anaesthetic rather like nitrous oxide, but its high cost and rarity preclude its use as an anaesthetic except for research purposes.

Stable chemical compounds of xenon and krypton with fluorine and oxygen are known, but so far these have not found application in industry.

A radioactive xenon isotope Xe^{133} originating in nuclear reactors is used as a diagnostic aid in medicine.

Helium

Production: Helium is recovered from natural gas, mainly in the USA and USSR.   Algerian natural gas is also a potentially large source.   The helium content of natural gas varies greatly in different gas fields, ranging from 7% to less than 0.01% v/v, and being typically around 1% in the gases regularly used as commercial helium sources.   Helium in North Sea gas wells is only 0.02 – 0.07% which in present circumstances is too low for economic extraction.

Natural gas and helium derived from it are expendable resources, but for the near future there is ample helium to meet foreseeable requirements. In the 1960/70 decade helium consumption had a very high growth rate and there was some concern about future supplies. The USA Government operated a helium conservation programme which involved extraction of helium from all richer natural gas produced, and storage underground of surplus helium. However, consumption eventually levelled off, and this together with fresh discoveries made the conservation programme less compelling.

Helium is separated from natural gas by liquefaction of the methane and other components, leaving an uncondensed gas consisting mainly of helium, which is finally purified if necessary by adsorption of impurities. There is no standard design of plant for helium extraction, details varying according to natural gas composition, pressure, and quantity to be processed. A typical plant would include a nitrogen liquefier to provide refrigeration for condensing the natural gas, and heat exchange equipment to give optimum energy efficiency.

Helium is present in the atmosphere to the extent of 1 volume per 200 000 volumes, representing a vast quantity of helium which is however too dilute for economic separation. Atmospheric helium is concentrated in the rare gas fraction vented from air separation plants, and at one time this concentrate was a source of helium in Britain. This is no longer economic in competition with imported natural gas helium but a Helium/Neon mixture may be recovered for use as such.

Helium is also concentrated in the tail gas from ammonia plants using helium-bearing natural gas feedstock, and this is a source of commercial helium in Holland.

Helium production in the USA is currently about 20 million cubic metres per annum, while the USSR countries (notably Poland) produce about 6 million cu.m. All helium consumed in Britain is imported, currently (1977) at the rate of about 2.5 million cu.m annually. A small fraction of helium used in Britain is recovered after use for purification and re-use, as it is a fairly expensive gas with present price around £5 per cubic metre for compressed gas, and £3.5 per litre for liquid helium.

<u>Technology</u>:   Helium exists as two isotopes He^3 and He^4, the
natural abundance ratio being 1.3/1 000 000.   The small proportion
of helium-3 can be separated by low temperature chromatography.
Properties of natural helium (ie virtually Helium-4) are in Table 4.

Helium is completely inert chemically, and has low solubility in
other liquids.   The gas is very light, only hydrogen having a
lower density.   Liquid helium has a lower boiling point than any
other liquid, only 4^oC above absolute zero at 1 bar.   These
properties lead to many diverse applications for helium.

Liquid helium has a low latent heat of evaporation and is easily
vaporized.   Its industrial handling and use in liquid form has
been possible only since the development of vacuum superinsulation
on an engineering scale for helium vessels.   For example, liquid
helium is regularly shipped from the USA to Europe and Japan in
special vacuum insulated tanks, and sustains only a small commercially acceptable vaporization loss en route.   Liquid helium is
distributed internally in Britain and other countries in small containers of capacities 5 litres and upwards.

Helium is also distributed to users as compressed gas in cylinders.
A typical analysis of commercial helium is Helium 99.98%.
Helium is non-toxic but asphyxiating in pure form.

<u>Uses</u>:   The chemical inertness of helium leads to its application
as an inert atmosphere in situations where the cheaper gases
nitrogen and carbon dioxide would be too reactive.   This generally
implies use in high temperature metallurgical processes such as
shielded arc welding, molten metal treatment, atomic energy
fuel processing, and so on.   These applications overlap those of
argon, and choice between helium and argon for these uses is made
on the basis of relative price when the gases are technically equal,
or on the basis of desired secondary properties of either gas
additional to the primary requirement of inertness.   In the USA
where helium is relatively cheaper, more is used for applications
such as welding where in Britain argon would be employed.

The inertness and favourable thermal conductivity of Helium make
it especially useful as a carrier gas in chromatography, this being

one of the largest applications of Helium in the UK.

The low density and inertness of helium make it an ideal balloon
filling gas. The lifting power of helium in air at sea level is
approximately 1 kg per cubic metre. The same combination of
properties – lightness and inertness – also make helium specially
suitable (mixed with 20% oxygen) as an easy-breathing gas to
replace air for patients with certain kinds of respiratory difficulties.

The low solubility of helium in blood favours its use as an inert
diluent for oxygen in breathing-gas mixtures for deep water diving,
where the solubility of nitrogen at the high pressures encountered
renders air unsuitable, and this has been an important helium use
in Britain in recent times in connection with North Sea gas and
oil exploitation. Low solubility, low boiling point, and inertness
combine to make helium an ideal pressurizing gas for liquefied gas
fuels and oxidants in rockets, and large quantities have been used
for this purpose in the USA.

Liquid helium is the coldest liquid refrigerant available, and is
essential in the study and application of very low temperature phenomena
such as superconductivity. Large recycled helium refrigerators
have been built for operation of superconducting magnets in particle
accelerators and electric motors. For smaller scale applications
and general research in low temperature physics, liquid helium
is usually purchased and used as an expendable refrigerant.

Helium finds some small scale application as a filling gas for
electric discharge tubes and devices.

Hydrogen

Production: Hydrogen is produced on an industrial scale (i) by
reaction of steam and hydrocarbon, or oxygen, steam, and hydrocarbon, (ii) as a by-product in cracking paraffin hydrocarbons,
(iii) as a co-product with chlorine and soda by brine electrolysis,
and (iv) together with oxygen by electrolysis of water.

The steam/hydrocarbon route is generally used to produce large
quantities of hydrogen for chemical feedstock in nearby integrated
processes (eg ammonia; petroleum hydrogenation, etc.) which
consume all the output of the hydrogen units. Similarly hydrogen

from petroleum cracking is normally all used in integrated petrochemical processes. Hydrogen from these sources requires
special purification to make it suitable for the industrial gas market,
usually by a pressure swing adsorption process or by palladium
diffusion.

In Britain, hydrogen available in the general gases market comes
from the steam/hydrocarbon process, from brine electrolysis,
and from water electrolysis. The amount of hydrogen sold in the
general market is only a small fraction of the total hydrogen
generated and used in integrated chemical processes.

Brine electrolysis is carried out primarily to produce chlorine and
caustic soda, the hydrogen being regarded as a by-product to be
disposed of to best advantage in the prevailing circumstances.
In Britain, part of this by-product hydrogen is used by the organizations producing it, and part is sold to industry for a variety of
purposes.

The ratio of hydrogen, chlorine and soda produced by brine electrolysis is of course fixed, and is 0.028 tonne H_2 (= 337 cu.m at 15°C,
1 bar) per tonne chlorine (that is, equal volumes of hydrogen and
chlorine). Total UK chlorine capacity is about 1 million tonnes/yr,
corresponding to 28 000 tonnes/yr by-product hydrogen, far more
than required by the industrial gases market. Chlorine/soda
manufacture is carried on in Britain by ICI, BPC, Associated Octel,
and Staveley Chemicals.

The chlorine/soda process is a large consumer of energy, about
3700 kWh being required per tonne chlorine + 1.125 tonne soda +
0.028 tonne hydrogen.

Industrial hydrogen is also produced by electrolysis of water. The
need for this, in spite of the large amount produced by brine
electrolysis, arises because of the high cost of distribution.
Hydrogen available from chlorine/alkali plants is in the northern
half of England. To serve users in the south, it may be more
economic to generate hydrogen locally than to haul it from the
north.

The electrolyte used in hydrogen manufacture is potassium
hydroxide solution, as pure water is non-conducting. Cell elec-

trodes are made of steel. Cells may be of low-pressure type
(Knowles cells) producing gas at only a little above atmospheric
pressure, or high-pressure type (Lonza cells) producing gas at
up to 30 bars gauge. Theoretically, 2 faradays of current should
produce 1 mole hydrogen and $\frac{1}{2}$ mole oxygen, or that is, 223 000
amp-hours per 100 cu.m hydrogen at 15°C, 1 bar. The theoretical
cell voltage is 1.226 v, leading to an ideal power consumption of
274 kWh per 100 cu.m hydrogen. In practice, with the electrode
current densities normally used the cell voltage is about 2 v, and
this together with transformer and rectifier losses gives a
practical power consumption of about 500 kWh per 100 cu.m hydrogen.

Hydrogen is distributed in Britain by pipeline or as compressed gas
in ordinary or in very large cylinders. Liquid hydrogen has been
distributed for special uses on a small scale in Britain, and the
technology of liquefaction and handling is well advanced if it should
be required. Liquid hydrogen has been manufactured and handled
on a large scale in the USA for use as a rocket fuel.

Current (1977) price for hydrogen (compressed gas) is around £30
per 100 cu.m depending on amount and location.

Technology: Hydrogen exists in three isotopic forms – the ordinary
most abundant form H^1, a stable more rare form H_2 occurring in
nature to the extent of about 1 part in 5550 parts H^1, and a radioactive form H^3 which is exceedingly rare naturally but is produced
in atomic energy reactors. As would be expected with isotopes
varying so greatly in atomic mass ratios, the hydrogen isotopes
are more distinct in properties than are isotopes of heavy atoms,
and merit separate names and symbols to distinguish them –

$$
\begin{array}{lll}
H^1 & H & = \text{hydrogen} \\
H^2 = & D & = \text{deuterium} \\
H^3 = & T & = \text{tritium}
\end{array}
$$

Ordinary industrial hydrogen under discussion here is mainly the
molecular form of hydrogen-1, that is H_2^1. Deuterium and tritium
are both available commercially and have special technologies and
uses which are outside the scope of this article.

Brine electrolysis in Britain is done mainly in mercury cells, and
the crude by-product hydrogen from such cells contains mercury

vapour, which is removed by adsorption. Typical impurity levels
in commercial hydrogen are –

O_2 4 vpm
N_2 100 vpm
H_2O 6 vpm

Hydrogen is unusual in being able to diffuse through some metals,
notably hot palladium, and this is used to prepare extremely pure
hydrogen.

Molecular hydrogen H_2 exists in two forms – ortho and para hydrogen –
differing in neutron spin symmetry. At room temperature the
equilibrium concentrations are 25% para and 75% ortho, but at the
normal boiling point of liquid hydrogen at $-253^{\circ}C$ the ratio changes
to 99.8% para. The transition from ortho to para is an exothermic
process. In order to avoid evolution of heat by ortho-para transition in liquid hydrogen and consequent loss of liquid by evaporation,
it is necessary to carry out the conversion in the gas phase prior
to liquefaction. This can be done by assisting the transition in the
cooled gas by contact with iron gel catalyst .

Hydrogen burns in air and has exceptionally wide concentration
limits for combustion – from 4% to 74% . This together with the
ability of hydrogen flames to pass through exceptionally small
holes makes the use of hydrogen more hazardous than with many
other common combustible gases.

Hydrogen is non-toxic but asphyxiating in pure form .

Uses: The largest uses for hydrogen are as chemical feedstock
for ammonia; methanol; petroleum hydrogenation; cyclohexane
and cyclohexanol; long-chain aliphatic alcohols; hydrogen peroxide;
amines from nitriles, and other chemical intermediates. Hydrogen
for these large-scale uses is usually generated on site.

Applications of marketed hydrogen tend to be on a smaller scale for
which on-site generation would be uneconomic . These applications
are very diverse, some using hydrogen in chemical reactions, some
using it as a reducing atmosphere in metallurgical operations,
and others using miscellaneous properties of hydrogen for special
purposes.

Chemical medium and small scale uses of industrial hydrogen include manufacture of furfuryl alcohol; tetra- and deca-hydronaphthalene; hydrogen sulphide; hardened edible oils; and many fine-chemical, perfumery, and pharmaceutical products.

Metallurgical uses are in reduction of metal oxide to metal (eg tungsten), and in high temperature processes in which small but unavoidable amounts of air or water vapour in the 'inert' atmosphere used would, in the absence of hydrogen, oxidize and spoil the surface of the metal being treated. Addition of a suitable concentration of hydrogen to give a reducing atmosphere in such processes prevents oxidation and keeps the metal bright. The amount of hydrogen necessary ranges from about 3% in bright annealing plain steel, up to 100% for brazing some kinds of stainless steel or tungsten carbide components.

Miscellaneous uses of hydrogen are (i) for filling balloons, in which the low density is exploited, (ii) as a heat transfer medium, where its combination of high thermal conductivity and other properties favour it. The widest use in this field is in cooling large electric generators, (iii) as a fuel gas. Oxy-hydrogen burners and torches are employed where a high temperature is required with the facility for using an oxidizing or reducing flame without presence of carbon or nitrogen. Oxy-hydrogen flames are used, for example, in fabricating precious metals, jewellery, synthetic ruby, etc. A special instance of hydrogen as fuel is its use in liquid form in rocketry. Hydrogen is potentially a large-scale fuel for use in internal combustion engines and fuel cells if the notional 'hydrogen economy' is developed, but at present relatively little pure hydrogen is used as fuel. (iv) in liquid form as a refrigerant and as a bubble chamber fluid, (v) as a carrier gas for deposition substances in the manufacture of solid-state devices.

Table 4 : Properties of Industrial Gases

	Oxygen	Nitrogen	Argon	Neon	Krypton	Xenon	Helium	Hydrogen
Formula	O_2	N_2	Ar	Ne	Kr	Xe	He	H_2
Mol Wt	32.00	28.02	39.94	20.18	83.80	131.3	4.003	2.016
Dens$_g$, 1 b, 15°C	1.337 kg m^{-3}	1.169	1.670	0.842	3.505	5.510	0.1670	0.0840
Dens$_g$, 1 atm, 0°C	1.429 "	1.251	1.784	0.8999	3.749	5.897	0.1785	0.0898
Dens$_l$, 1 b, BPt	1140 "	808.1	1399	1207	3449	3053	124.9	70.81
BPt, 1 b	-183.1 C	-196	-186	-246	-153.6	-108.1	-269	-253
Crit Temp	-118.4	-146.9	-122.3	-228.7	-63.6	16.3	-267.9	-239.9
Crit Press	50.8 bars	33.90	48.95	27.80	55.08	58.42	2.29	12.96
Crit Dens	430 kg m^{-3}	311.0	536	484	908	1105	69.3	31.0
Ht Vapn, BPt	213.4 J g^{-1}	199.5	163.2	85.83	107.7	96.24	20.28	448.3
Sp Ht (g), 25°C	0.915 J g^{-1}K^{-1}	1.015	0.522	1.031	0.248	0.158	5.19	14.3
Sp Ht ratio, 25°C	1.401	1.404	1.667	1.642	1.689	1.666	1.630	1.410
Sp Ht (l), 25°C	1.695 "						3.60	
Acoustic Veloc, 25°C	330 m s^{-1}	334	307.9	435	213	168	970	1284
Visc$_g$, 1 b 0	0.0196 mPa s	0.0168	0.0225	0.0314	0.0249	0.0227	0.0196	0.0087
25	0.0208	0.0179						
100	0.0252	0.0213						
Visc$_l$, BPt	0.189 "							

Table 4 : Properties of Industrial Gases (cont'd)

		Oxygen	Nitrogen	Argon	Neon	Krypton	Xenon	Helium	Hydrogen
Therm Cond (g) 1 b	0	0.0245 $W\,m^{-1}K^{-1}$	0.0242	0.0164	0.0459	0.00873	0.00506	0.142	0.173
	25	0.0260	0.0255						
	100	0.0320	0.0309						
Refr Index (g) n_D		1.0002639	1.000297	1.000284	1.000067	1.000427	1.000702	1.000035	1.000132
Dielect Const (g)		1.000523	1.00058	1.00051	1.000123	1.000768	1.00124	1.000064	1.000264
Ioniz Potl, 1st		12.5 eV	15.6	15.7	21.6	14.0	12.1	24.6	15.4
Mag Susc		$+3449 \times 10^{-6}$	-3.98×10^{-6}	-19.4×10^{-6}	-7.2×10^{-6}	-28×10^{-6}	-43×10^{-6}	1.9×10^{-6}	-4×10^{-6}
Diff Const, 1 b, 0 self diff		0.187 $cm^2\,s^{-1}$	0.172	0.157	0.452	0.0795	0.048	1.48	1.285
O_2/air		0.178							
O_2CO_2		0.136	0.0524						
O_2/water		0.000021	0.000019						
Sol in water Ostwald coeff, 0		0.0493	0.0239	0.0330	0.0105	0.0594	0.1081	0.00861	0.0195
25		0.0310	0.0159						
50		0.0246							

Production and Uses of Inorganic Boron Compounds

By R. THOMPSON

Borax Consolidated Ltd.

Despite many years of intensive research into the preparation and applications
of a wide and interesting range of inorganic boron compounds, those which
make up the boron products industry in 1977 are the chemically unspectacular
substances boric acid, sodium tetraborate (borax) and their simple variants.
With total world-wide production in excess of two million tons annually, they
are firmly in the class of bulk chemicals.   Like many other simple industrial
inorganic compounds whose manufacture is described in this book, production
methods from naturally-occurring salts may be explained by a few simple
equations.   Similarly, the art of manufacture lies in optimising the
conditions for the primary reactions with regard to yield, the elimination of
impurities and the avoidance or control of environmentally unacceptable by-
products;   and in doing so on a very large scale, nowadays with attention to
energy as well as financial economy.

Description of this important sector of the inorganic chemicals industry is
therefore devoted largely to mining and refining processes leading to boric
acid and sodium borate in various states of hydration, and to their applications;   with lesser attention to other inorganic borates, and least to the
chemically more interesting compounds which are usually fully described in
standard inorganic chemistry textbooks.   It is hoped in this way to
introduce an industrial perspective to the subject[1].   The inter-relationship
of the bulk chemicals and their ores is shown schematically in Figure 1.   One
other boron compound might debatably find a place in the diagram, and that is
sodium perborate.   It accounts for a considerable proportion of the total
output of borax, but for a variety of technical, commercial and geographical
reasons its manufacture has traditionally been the preserve of the hydrogen
peroxide manufacturing industry (described elsewhere in this book).   This and
other applications for borax, and uses of borates in other forms, are discussed
later in the present chapter.   However, it is perhaps appropriate to mention
at this point that these simple boron compounds can be used interchangeably
in many industries, subject to economic constraint (the more highly refined
or less hydrated materials are obviously the more expensive);   and also that
the producer and consumer industries deal on the basis of contained B_2O_3.

OCCURRENCE AND MINING

The Western world's borax requirements are derived mainly from native sodium borate minerals (tincal, $Na_2B_4O_7, 10H_2O$ and kernite, $Na_2B_4O_7, 4H_2O$) and lake brines. Deposits occur in sufficiently rich form for economic mining in California, Turkey and Argentina; the brine lake is also in the same region of California. All North American production of boric acid is also from these materials but in almost all other areas, including Europe, it is derived from the calcium ore, colemanite ($Ca_2B_6O_{11}, 5H_2O$) mined for this purpose mainly in Turkey. Colemanite is also won in parts of California, but is currently only used in upgraded form as a glass ingredient. In a few other places (Italy, Eastern Europe) this calcium mineral, again of Turkish origin, is subjected to decomposition by sodium sesquicarbonate to produce borax. Particular local circumstances render this route economically (or politically) viable, but it was abandoned in California when the rich sodium borate reserves were discovered half-a-century ago.

The most important sodium borate deposit now being worked is the open pit mine of the U.S. Borax & Chemical Corporation, which is situated near the town of Boron in the Mojave desert of Southern California. There is also a similar, though much smaller, mine operated by Boroquimica Limitada at Tincalayu in Argentina. Both are wholly-owned subsidiaries of R.T.Z. Borax, a British company. At each location the orebodies are at sufficiently shallow depths to permit opencast mining, which in the Kramer deposit at Boron was made possible by the removal of some 400 feet thickness of over-burden over a pit area which is now about 3500 feet across. This mine with its adjacent refinery is shown in Plate 1. From its discovery until 1957 the deposit was worked by conventional underground mining methods and evidence of the original workings may still be seen.

Tincal predominates at both locations, but the Californian mine also contains substantial amounts of the tetrahydate, kernite (in some textbooks erroneously called Rasorite after its discoverer, but this is now a registered trade name for borax concentrates). Mining at the Boron open pit is by drilling and blasting selectively in benches. Electrically powered shovels transfer the ore via dump trucks to a primary crusher situated at the foot of the conveyor belt (see photograph) which then hauls it to the surface stockpile. From here it receives a secondary crushing before chemical processing.

The sodium borate deposit in Turkey is of more recent discovery and develop-ment. Owned by an agency of the Turkish government, not much has as yet been published about the operation but it can differ but little from the already worked borax deposits.

PRODUCTION OF BORAX FROM SOLID ORES

As implied above, the overall chemical flow-sheet of borax production from its
ore is simple: the crushed feed is dissolved in hot water, the siliceous
gangue which accompanies it is removed by thickening and filtration, after
which the product is cropped by crystallisation and dried. In practice the
chemistry is complicated by the combined facts that for borax yield, solubility
and water conservation reasons the process is operated in a cyclic manner, thus
leading to a build-up of, and the need to eliminate, certain impurities some
of which are also water-soluble. The ore is in fact dissolved in recycled
weak hot mother-liquor, the solution and suspended solids being passed through
vibrating screens to remove the larger particles before the liquor is piped to
the main gangue separation stage, a battery of six thickeners. Hot dissolution
is necessary because of the moderately low solubility in cold water (5g/100g
water at 20°C, rising to 191g at 100°C.). In order to avoid premature
crystallisation in pipe-lines and equipment the liquors are maintained at
around 100°C during processing. This applies equally to the gangue separation
step.   Approximately 230 feet in diameter, when installed some 20 years ago
the thickeners were (and possibly still are) the world's largest covered type.
They are operated in sequence, countercurrent, and thickener pulp or underflow
is piped to sealed tailings ponds. The clarified liquor overflow is
filtered and the borax recovered as either the decahydrate or pentahydrate in
Struthers Wells vacuum crystallisers. Crystal is separated from mother-
liquor by centrifuging, after which the product is dried in rotary dryers.

"Borax" by common usage implies sodium tetraborate decahydràte, $Na_2B_4O_7,10H_2O$.
This contains 47% of water. Transport and storage charges are therefore
potentially doubled compared with the anhydrous salt, but these costs have to
be set against the energy expense of removing the water. From the 1950's
the tendency was towards a greater proportion of total production being in the
anhydrous form, but as fuel became both more expensive and at times scarce,
the early 1970's saw a reversal of this trend. Fortunately, the existence
of a pentahydrate enables a compromise to be made, and by cropping the crystals
at above the transition point of 60.8ºC half of the hydrate water is removed
from the product at substantially zero energy cost. Water consumption is
theoretically reduced also, but in all such processes some liquor bleed is
necessary in order to eliminate impurities. Careful routing of make-up
water from all points of introduction to final discharge via the thickener
system maximises utilisation and reduces consumption. In spite of the
increasing trend towards use of "5-mol" borax (also known by the trade name
NEOBOR[R]) a considerable tonnage of the decahydrate is still required by U.S.
Borax and its customers, and the need to make varying quantities of the two

hydrates according to market demand requires skilled production planning and
control.   (U.S. Borax uses much of the decahydrate production for the
manufacture of its BORAXO[(R)] powdered hand soap and BORATEEM[(R)] laundry additive).

PRODUCTION OF BORAX FROM LAKE BRINES

A minor but significant proportion of North American borax production is from
the brine of Searle's Lake, also situated in the Mojave desert.   This apparently
dry lake of about 34 square miles in area has saturated brine (circa 35% solids)
to within a few inches of the surface, filling channels and voids which occupy
40-50% of the cross-section, effectively about 120 feet thick but in two main
layers.   The brine is rich in soda, potash, sulphate and chloride but
contains about 1% of B_2O_3 (equivalent to 3% $Na_2B_4O_7,10H_2O$).   In the main process,
operated by Kerr-McGee (formerly American Potash) at a location named Trona,
upper-layer brine is pumped from wells and concentrated by evaporation.   Rapid
cooling precipitates potassium chloride, itself a valuable product.   This
leaves a liquor supersaturated with respect to borax, which is recovered above
the transition temperature as the pentahydrate and then recrystallised to the
commercial product.

An alternative process, operated originally in 1925 at another part of the
lake in the plant of the West End Chemical Company (formerly a division of
Stauffer Chemical Company and now also part of Kerr-McGee) but later adopted
by American Potash, involves acidification of the alkaline (and rich in borax)
lower-layer brine with kiln or flue-gas carbon dioxide.   Sodium carbonate is
thereby converted to sodium bicarbonate, which precipitates because of its
lower water solubility.   At the same time the $Na_2O:B_2O_3$ ratio of the liquors
is altered from the 1:2 of borax (N.B. this ratio, even though borax is
historically called the tetraborate) to the 1:5 of sodium pentaborate, at which
the water solubility with respect to B_2O_3 is much greater.   When further
quantities of lake brine are added, the pH and the ratio are restored to their
original values after which the less-soluble borax is crystallised out upon
refrigerated cooling.

Compositional and phase studies of the lake brine were classics of their time
and were published for Teeple as an ACS monograph in 1929[2].

MANUFACTURE OF BORIC ACID

Boric acid to both producers and consumers always means orthoboric acid, H_3BO_3.
Metaboric acid, HBO_2, is not manufactured on a significant scale and when needed
by industry is usually made in situ (see below).   There are now three important
sources of the acid: solid sodium borates, lake brines and colemanite.
Although boric acid occurs naturally in the hot springs or fumaroles of

Tuscany, an historical source, recovery directly therefrom is no longer
practised.

Whether borax or colemanite is the raw material, boric acid is liberated by
decomposition with a stronger acid, which implies any mineral acid.   For
economic reasons sulphuric acid is almost invariably used.   This also has the
production convenience with colemanite that the precipitated calcium sulphate
by-product of the metathetic reaction is insoluble and readily removed by
filtration.   Because of the relatively low solubility of boric acid in cold
water (4.95 g per 100 g at $20^{\circ}C$) the process is carried out at above $80^{\circ}C$
(solubility 23.6 g).   This means that a large crop of boric acid crystals is
readily obtained by cooling the liquor, from which it may readily be recovered
on centrifugal hydro-extractor or similar equipment.   A technical grade can
be obtained directly, by merely introducing a washing cycle to the hydro
operation before discharge, but for higher grades of product the options are
total recrystallisation or the removal of dissolved impurities (both anionic
and cationic) by passing the liquors through ion-exchange columns.   Until
recently crystallisation was effected batch-wise in batteries of air-cooled
or water-cooled granulators, but vacuum crystallisers are now being used
increasingly.   After centrifuging,the moist crop is dried in continuous rotary
dryers (or similar) to free-flowing granules (or, after grinding and sieving,
powder).

Production from borax involves a similar double-decomposition reaction with
sulphuric acid.   The water content of the process is conveniently reduced
(and provision thereby made for later introduction of water for crystal washing)
by starting with the pentahydrate.   For solubility reasons this process is
also operated hot.   Boric acid is the first product to separate on cooling,
and is removed on hydro-extractors.   Further cooling precipitates the sodium
sulphate by-product, which is anhydrous ("salt cake") if separated at above
$33^{\circ}C$.   A process advantage of making boric acid from refined borax is that
the minor impurities have already been removed.

Building upon their expertise in brine processing, American Potash later
developed a supplementary process in which soluble borate in either lake brine
or weak plant end-liquors is removed as an organic complex[3].   This is a solvent-
extraction process in which the alkaline brine is first treated with
1,8-naphthalenediol or an aliphatic or aromatic polyol, alone or dissolved in
a carrier such as kerosene or octanol.   A borate complex is formed and may
precipitate, in which case it is separated by filtration, dissolved in water
and treated with sulphuric acid in order to precipitate the diol.   Boric acid
is then recovered by an evaporation and crystallisation cycle from the aqueous
filtrate (which also contains sodium sulphate).   With some extractants the

borate complex does not precipitate but remains dissolved in the immiscible
organic phase, from which after separation it is stripped with sulphuric acid,
the boric acid recovered as before from the resultant aqueous phase and the
organic phase recycled. The preferred polyol solution is said to be
4-isooctyl-6-chlorosaligenin in kerosene.

ANHYDROUS BULK PRODUCTS

Freedom from combined water is a prerequisite for the borates used in certain
sectors of the glass, pyrometallurgical and chemical industries. Both borax
and boric acid are readily dehydrated on an industrial scale and, although as
stated above the practice is declining somewhat in the case of the former,
these two "fused products" are important items of commerce.

Dehydration of borax is usually effected in two stages: "calcination" at
relatively low temperatures to the approximate dihydrate, followed by fusion
and heating to about 850°C to eliminate remaining amounts of water. In the
process currently operated by U.S. Borax the first stage is accomplished in
oil- or gas-fired rotary furnaces and the second by feeding the powdery intermediate into large tank furnaces similar to the ones used in the glass industry.
Anhydrous borax is in fact a simple glass and the fully dehydrated product
is tapped continuously from the furnace as a vitreous stream, cooled on chill
rolls and finally crushed and sieved. American Potash have described their
somewhat different technique which employs a furnace of circular design,
shaped like a giant tundish with a domed roof. Gas burners are located in
the dome and the partially dehydrated (calcined) feed enters through an
annular space where the dome and tundish meet. The design provides that the
borax is finally dehydrated and fused on the crust of a bed of unfused feed,
so avoiding direct contact with refractory brick[4]. Other designs require
the use of special refractories; molten borax is a powerful solvent for many
refractory oxides, hence its extensive use in the glass industry.

Dehydration of boric acid to boric oxide, B_2O_3, is accomplished at lower
temperatures. The reverberatory furnace equipment is again of generally
similar design to the glass industry's melting furnaces, although much smaller
for market reasons than the plant used for borax dehydration. Discharge,
cooling, crushing and sieving are done similarly. Both the lower fusion
temperature and the lower corrosivity of boric oxide lessen the refractories
problems. U.S. Borax adopted a rather different approach in the 1960's and
developed a process for the production of a technical grade of boric oxide in
bulk direct from borax[5]. It takes advantage of the wide immiscibility gap in
the B_2O_3-Na_2SO_4 phase system. Hydrated borax and concentrated sulphuric acid
are mixed in equimolar amounts to a dry paste, which soon becomes a free-

flowing granular material in which the metathetic reaction has taken place.
This is fed to a furnace where it dehydrates, melts and segregates into two
layers, the upper of which is boric oxide.   Layer separation is effected by
use of a dam or weir, enabling the streams to be discharged individually for
cooling and collection.

OTHER INORGANIC BORATES

Although several other borates are produced commercially, none is made on a
scale greater than a few thousand tons per annum and all are thus minor in
comparison with borax and boric acid.   Nevertheless, they play useful roles
in industry.

Other Sodium Borates.The ready ability of the boron atom to form trigonal or
tetrahedral bonds with oxygen renders possible a wide variety of stoichiometric
combination ratios of B_2O_3, a metal oxide and H_2O.   This is displayed even
with monovalent atoms like sodium, and the metaborate ($NaBO_2$, or Na_2O,B_2O_3)
and pentaborate ($Na_2O,5B_2O_3$) are well-known in addition to borax itself (the
1:2 compound).   The existence in some cases of a number of hydrates and
anhydrous crystalline or glassy products imposes a discipline on producer and
consumer industries to restrict the commercial range.   In practice, a 1:4
borate and the 1:1 sodium metaborate ($Na_2B_2O_4,4H_2O$ or $8H_2O$) are produced
industrially at annual rates of several thousand tons.

Reference was made in the lake brine context to the increased solubility of
the 1:5 borate compared with borax.   Solubility is at about its highest with
the 1:4 $Na_2O:B_2O_3$ ratio, more than five times as much B_2O_3 entering solution
in this form at 30°C compared with the dissolution of borax.   Moreover, the
resultant solution is of essentially neutral pH (7.3 at 15%), an advantage in
many applications.   There is no true compound of this formula, and hence a
homogeneous solid cannot be made by crystallisation.   However, U.S. Borax
developed a process whereby a concentrated solution of the 1:4 ratio is spray-
dried to a stable granular tetrahydrate, and this product has been sold for
many years under the trade-name POLYBOR[R].

Sodium metaborate is made by the interaction of borax and sodium hydroxide,
the product being crystallised readily from solution.   It is more soluble
than borax, as well as being more alkaline (a 1% solution at 20°C has pH = 11.0).

Potassium Borates.   Both the tetraborate ($K_2B_4O_7,4H_2O$) and pentaborate
($K_2B_{10}O_{16},8H_2O$) are in commercial production.   Each can be made either by
the addition of the appropriate amount of potassium hydroxide to boric acid,
or by the interaction of borax and potassium chloride, with or without
carbonation, e.g.:

$$4KCl + 5Na_2B_4O_7,10H_2O + 6CO_2 \longrightarrow 2K_2B_{10}O_{16},8H_2O + 6NaHCO_3 + 4NaCl + 31H_2O$$

U.S. BORAX'S OPEN PIT MINE AND ADJACENT REFINERY NEAR BORON, CALIFORNIA

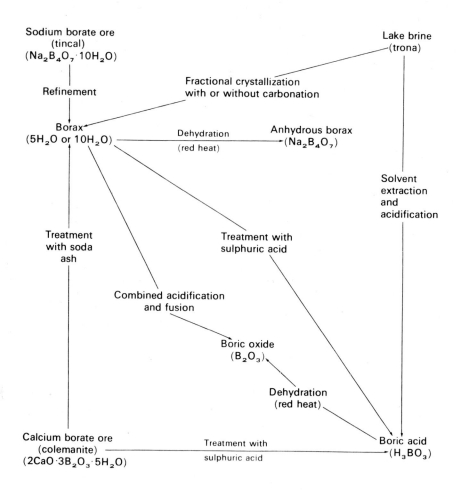

Figure 1

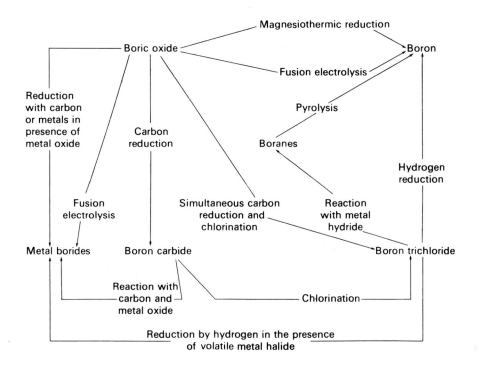

Figure 2

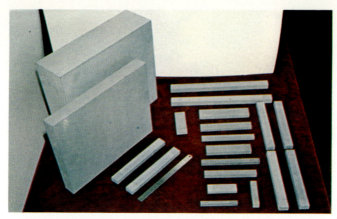

HOT-PRESSED BLANKS AND FINISHED EVAPORATOR BOATS OF TiB$_2$—BN COMPOSITE

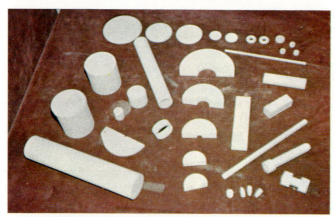

MISCELLANEOUS OBJECTS OF HEXAGONAL BN MACHINED FROM HOT-PRESSED BLOCKS

HOT-PRESSED SHOT-BLAST NOZZLES AND OTHER OBJECTS MADE FROM B$_4$C

<u>Ammonium Borates.</u>   The 1:2 (usually referred to in this case as the <u>bi</u>borate) and 1:5 ratio compounds are made industrially, from ammonia and boric acid. The pentaborate is much the more stable, with no odour of ammonia.   Its principal use is as an electrolyte component in electrolytic condensers.

<u>Calcium Borates.</u>   Hydrated synthetic calcium metaborate has been marketed, but not in significant quantities.   Of far greater importance is the mineral colemanite ($2CaO,3B_2O_3,5H_2O$) in dressed form.   The main use is in glass fibre production (see below).

<u>Barium Borate.</u>   Barium metaborate, $BaB_2O_4,4H_2O$, is produced on an industrial scale by the reaction of barium sulphide (derived from the carbothermic reduction of barytes) with borax.   It is used as a water-borne paint pigment, primarily because of fungicidal properties although it does have corrosion inhibiting qualities.   When prepared in a particular way, coated with silica or in dispersion with anatase or rutile, it is capable of controlled chalking, helping exterior paints to maintain a continually cleaned surface in climates of intermittent rainfall but which are sufficiently warm and humid to encourage mildew and fungal growth[6].

<u>Zinc Borate.</u>   Several grades of hydrated zinc borate are now on the market and used as a fire retardant pigment for plastics[7].   The precise ratio of zinc to boron is less important than the overall chemical properties, notably the amount of hydrate water and the temperature at which it is evolved.   This temperature must be sufficiently high not to cause "blowing" during fabrication or cure of the plastic article, but be released endothermically at the appropriate temperature during the build-up of the combustion process.   The residual zinc borate then controls after-glow.   A compound which has been found to be particularly suitable for incorporation in a wide range of plastics, notably halogenated polyesters and PVC, has the composition $2ZnO,3B_2O_3,3.5H_2O$ and is known in the trade as ZB 2335 or FIREBRAKE ZB(R).   It is made under carefully controlled conditions from zinc oxide and boric acid[8].

<u>Boron Phosphate.</u>   This material is no longer in regular production by the borate industry, users in the petrochemical field preferring to make it in-house.   Its major application is as a heterogeneous acid catalyst, for which purpose proprietary variations are made in $B_2O_3:P_2O_5$ ratio, surface area and the presence of other solids in admixture.   It is especially suitable for the hydration of alkenes and the dehydration of amides to nitriles.

Boron phosphate exists as both low-temperature and high-temperature forms, although there is no sharp distinction.   It is most readily made by

intimately mixing one mol of concentrated phosphoric acid with one mol of fine
boric acid powder and heating the resultant creamy slurry at a low temperature
until an amorphous dry solid is obtained.   The product at this stage is
approximately $BPO_4.H_2O$.   Heating to higher temperatures (preferably after
pulverising) progressively eliminates water, until at about $400^{o}C$ all has been
lost.   At this stage it is still amorphous, but more slowly soluble in
water.   If finally heated to above $1100^{o}C$ it transforms to a white crystalline, insoluble, refractory product, BPO_4, which is isostructural with silica
(cristobalite).   This form had limited use as a ceramic flux in conjunction
with pegmatite, and as an opacifier for glass utensils.

MAJOR APPLICATIONS OF B_2O_3 CONTAINING MATERIALS
_____

The bulk-chemical boron compounds are employed both in fusion processes and
in aqueous solution, making use of the specific qualities of the contained
B_2O_3 in the former case and the overall chemical properties of the particular
compound in the latter.   The glass and related industries are the main
consumers in the one, and the general detergency area in the other.   But there
are few industries, including process industries, which do not use boric acid
or borates in some form.

Glass and the Vitreous Industries.   Borax and boric oxide are powerful
solvents (or fluxes) for many metal oxides, including those commonly used in
glass manufacture.   They enable glass to be fused at lower temperatures,
with consequent economy in fuel.   Most important, they improve the quality
of the product in a number of directions. B_2O_3, like SiO_2, is a "network
former" which readily enters into the main structure of the glass and alters
its properties.   Particularly important are the reduction of thermal
expansion and the increase in brilliance.   These effects are evidenced in
borosilicate glasses, e.g. Pyrex, in craze-free pottery and tile glazes.
In the vitreous industries (glass, glazes, enamels) it is often immaterial
which borate is used, consistent with the qualitative and quantitative
compatability of the accompanying cation with the formulation as a whole.
Price per unit of contained B_2O_3, with due allowance for purity and water-
content, is the primary factor influencing choice when such chemical inter-
changeability is possible.   Sodium tetraborate, including the sodium borate
mineral concentrate RASORITE[R], is used for the production of Pyrex and the
sodium borosilicate glasses, including glass wool for thermal insulation
which is a major consumer.   However, glass fibre (known as "E"-glass) for
plastic reinforcement or textiles must be soda-free, in which case boric oxide
or, more usually nowadays, specially upgraded colemanite ore is used.

<u>Metallurgical Fluxes.</u>   The abilities of molten boric oxide and alkali metal
borates to dissolve other metal oxides are exploited in a wide range of
fluxing applications in the metallurgical and engineering industries. They
are valuable cover fluxes in metal melting, especially of copper alloys.
Borax is used as a flux in the assaying and refining of gold.   In the steel
industry it finds application for scale-free reheating of billets.   Potassium
pentaborate is principally used for welding and brazing stainless steel.
Trimethyl borate, being volatile, is the active constituent of gas welding
fluxes.

<u>Bleaching and Detergency.</u>   A very large proportion of borax production,
and of RASORITE$^{(R)}$, is used in making sodium perborate (see the chapter by
Crampton et al, this book).   Regular bulk shipments are made from California
to Europe and other parts of the world for this purpose.   Established
practice in European households has been to "boil" clothes, or at least to
heat to well above hand-washing temperatures in order to whiten domestic and
personal linen.   Sodium perborate is an increasingly effective bleach at
temperatures above about 55°C.   As average washing temperatures have fallen
towards this over recent years, so has the sodium perborate content of the
packaged washing powder increased to the present level of about 25% (the other
major ingredients being sodium tripolyphosphate and synthetic sulphonated
detergents).   Annual sodium perborate consumption for this purpose in Europe
is now over 500,000 tons.

It has not been the tradition in the U.S.A. to "boil" household laundry, and
in fact far fewer domestic washing machines are fitted with water heaters in
that country.   Consequently, chlorine-based bleaches (hypochlorites, chloro-
cyanurates, etc.) are often used, a practice not generally favoured by most
European housewives.   Nevertheless, it has been customary in the U.S.A. over
several generations for borax to be added to the wash along with the proprietary
detergent powder, and very large tonnages of the salt are used in this way.
Thus in the wash waters of both continents there is a presence of dissolved
sodium borate which acts as a detergent builder of acceptably mild alkalinity
(to hands and clothes).   Borax solution is also used extensively as a home
laundry pre-soak, especially for diapers and the like where its chemical
buffering action and protein solubilising properties are particularly valuable.

<u>Biological Applications.</u>   Boric acid and its salts have low mammalian toxicity
and are not surprisingly rather weak bacteriostats by modern standards. Formerly
employed as food preservatives and medicinally (boric acid is still a favoured
constituent of eye lotions), they have generally been supplanted by other
materials in these areas.   No role for boron has been identified in animal
metabolism.   Soluble inorganic borates are however toxic to certain forms of

insect life, and for example they are specific stomach poisons for cockroaches.
The compounds are simply administered by mixing either borax or boric acid
with a bait such as sugar or starch and applied as powder to infested areas.
The mixtures are harmless to domestic animals.   Ant baits are formulated
similarly.   Borates are also toxic to many wood-boring insects and the water-
soluble 1:4 sodium borate (called TIMBOR[R] for this application) is widely used
in timber preservation.   The process is operated by momentary immersion of
newly-sawn or unseasoned timber in the hot and concentrated aqueous borate
liquor and allowing it to drain;   followed by diffusion of the residual solution
on its surface into the depth of the wood during several weeks of storage and
drying, leaving it preserved throughout (uniquely amongst commercial preservative
systems) in a colourless and odourless manner, harmless to higher life forms.
An ester of boric acid (a dioxaborinane) known as BIOBOR[R]JF is used to control
fungus and other microbial life in diesel and aviation jet fuels.

Boron plays an interesting and unique role in plant life. Above certain levels
(which vary widely according to the plant) it is phytotoxic and borax, often
in admixture with sodium chlorate, was used extensively for many years as a
non-selective weed killer.   More powerful herbicides have been developed,
but nothing can replace boron as a trace nutrient.   Apart from the citrus
family, most other fruit, fodder, oil and beverage crops require boron as an
essential element.   Without it, apple and pear develop cork, yield of sugar
beet is reduced, peanut quality is impaired and alfalfa growth is stunted.
The boron may be introduced along with the fertilizer, in soluble or slow-
release form, or as a foliar feed.   The highly water soluble 1:4 sodium borate,
which is also of substantially neutral pH, is sold for this use under the name
of SOLUBOR[R].

Unless grossly misused, borates are thus environmentally safe materials.

Uses of Boric Acid.   The most important single use for boric acid is now in
the nylon industry.   It is employed to modify the course and selectivity of
the liquid phase oxidation of cyclohexane by molecular oxygen to cyclohexanol
and cyclohexanone, thus enabling a better yield of the co-products required for
nylon 66 manufacture to be achieved.   The function of boric acid in this
industrial extension of the Bashkirov reaction[9] is as a stoichiometric particip-
ant and not merely as a catalyst.   The acid, dehydrated in situ at 160°C to
metaboric, esterifies cyclohexanol (which also helps in separating it from the
cyclohexanone) and is partially recovered during the subsequent hydrolysis of
the ester.   A similar controlled oxidation of n-paraffins can be used to
produce a detergent alkylate for making biodegradable "soft" syndets.

Miscellaneous Applications of Borax and Boric Acid. Both borax and boric acid have been used for many years in flameproofing treatments for combustible building materials and textiles, particularly where the substrate is of a cellulosic nature. The optimum $Na_2O:B_2O_3$ ratio for fire retardancy is close to the maximum water solubility within the system, at approximately 1:4, enabling an appropriately high loading of borate to be introduced into soft wood fibre insulation boards during manufacture. The spray dried product, POLYBOR[R], is frequently used for this purpose. Such solutions also diffuse readily into veneers used in plywood manufacture. Wood particle board (sometimes called chipboard) is best flameproofed with boric acid alone, which is mixed with urea-formaldehyde resin and sprayed onto the wood chips before they are compacted in a heated platten press or between heated rolls. The boric acid, which complexes with hydroxyl groups on the cellulose, also acts as the catalyst in heat-setting the resin. Wood particle board flameproofed with boric acid has the added advantage of producing very little smoke under fire conditions, and "after-glow" is prevented. Matchsticks are also glow-proofed in this way. The largest single outlet for boric acid in this area however is in flameproofing cotton mattresses (mandatory for public institutions in parts of the U.S.A.). For this application the cotton linters and dry boric acid powder are milled intimately together. A mixture of shredded waste newsprint and borax or boric acid is often used as a cavity wall insulation material in the U.S.A.

Borax, with its mild alkalinity and high buffering capacity, is used extensively as a corrosion inhibitor for ferrous metals. Its high solubility in ethylene glycol combines to make it especially useful in car antifreeze formulations (e.g. British Standard 3152). Aqueous solutions of borax have replaced chromates in railroad and other diesel coolants. In the dry drawing of steel wire, borax is used to neutralise residual acid from the pickle and the deposit of the salt remaining on the wire is valuable as a lubricant carrier.

Boric acid and its soluble salts and esters find major or minor applications in textile dyeing, dyestuffs manufacture, leather tanning, the photographic industry, laundry starches, starch-based adhesives for corrugated cardboard manufacture, anodising, electrolytic capacitor production, nuclear shielding, plaster and concrete set retardants, paper sizing and in many other industries.

OTHER BORON COMPOUNDS USED INDUSTRIALLY

The non-oxygenated inorganic boron compounds for which commercial application
has been found may conveniently be divided into two classes: those having
boron-halogen and boron-hydrogen bonds; and the binary interstitial solids of
a hard or refractory nature, to which class elemental boron itself belongs.

Boron Trifluoride. BF_3, a colourless gas b.p. $-100.4^{\circ}C$., is produced in
tonnage quantities by relatively conventional chemical process methods.
It is the only boron halide possessing any hydrolytic stability and may be made
simply by the action of sulphuric acid on a mixture of borax and fluorspar:

$$Na_2B_4O_7 + 6CaF_2 + 8H_2SO_4 \longrightarrow 4BF_3 + 2NaHSO_4 + 6CaSO_4 + 7H_2O$$

Sulphuric acid in excess of the stoichiometric amount acts as a dessicant. The
evolved gas, separated from spray, may be collected and stored in cylinders as
the pure compound; or more usually is absorbed in media such as ether, phenol
or acetic acid with which, being a strongly electrophilic molecule, it readily
forms addition complexes.   Other processes involve the reaction of boric acid
or borax with fluorine sources such as HF, NH_4HF_2 or HSO_3F; each has relative
merits in BF_3 purity and, compared with the fluorspar route, economy in
sulphuric acid usage. BF_3 is used either alone or in the form of one of the
above or similar adducts as a catalyst for a wide variety of alkylation,
polymerisation and other reactions in the petrochemical industry.   Its
dimethyl etherate finds some use as the boron source for the vapour fractionation method of ^{10}B isotope enrichment or separation.

Fluoboric Acid and Metal Fluoborates. HBF_4 is made either by dissolving
boric acid in hydrofluoric acid, or by treating boric acid and fluorspar with
sulphuric acid.   It is sold as an approx. 50% aqueous solution, but production
is not large and uses are limited to metal cleaning and as an organic reaction
catalyst.   Sodium, potassium and ammonium fluoborates are all available
commercially and can be made by neutralising fluoboric acid with the hydroxide
or carbonate of the desired cation.   Alternatively, HF (e.g. waste from other
processes) can be absorbed in borax solution to produce $NaBF_4$ directly; KBF_4
can then be made from this by the addition of KOH.   Fluoborate uses are in
the electroplating industry, with the sodium and potassium salts also having
an important outlet as grain refining agents for aluminium (along with
fluotitanates they react in the molten metal to form titanium diboride nuclei).

Other Boron Halides. BCl_3, BBr_3 and BI_3 are all very readily hydrolysed.
They are prepared under high-temperature anhydrous conditions, usually from
boron carbide and the halogen.   While the tribromide and triiodide are made
only in laboratory amounts, for research or electronics applications, boron
trichloride is made in tonnage quantities.   The preferred process is the
exothermic chlorination of boron carbide at above $700^{\circ}C$ (Figure 2).

Production is accomplished by passing chlorine upwards through a column packed with B_4C granules. Reaction is initiated by heating those at the base of the column to redness, after which it is necessary to control the flow of chlorine such that the reaction zone rises through the bed at a suitable rate. The boron trichloride gas (b.p. $12^{\circ}C$) is condensed, along with the halides of any impurity elements (Al, Si, Fe) which may be present in the boron carbide, and purified by fractional distillation. The process is usually run batch-wise, but continuous operation (feeding new boron carbide granules at the top and removing the spent ones from the bottom of the column) is practicable according to product demand. BCl_3 is a colourless mobile liquid, rapidly (and even violently) hydrolysed by water.

Sodium Borohydride. Metal borohydrides (sometimes called, pedantically and misleadingly, "hydroborates") are the only B-H bonded compounds in industrial tonnage production. By far the most common is $NaBH_4$, a white, crystalline and water-soluble solid. It is usually made from sodium hydride and trimethyl borate:

$$4NaH + B(OMe)_3 \longrightarrow NaBH_4 + 3NaOMe$$

The process is carried out in two stages: hydrogenation of a very fine dispersion of sodium in mineral oil, followed by reaction with the borate ester at $250-270^{\circ}C$. The product may be separated by extraction with liquid ammonia, or converted to potassium borohydride by precipitation with potassium hydroxide[10]. The extraction stage is expensive and, as ultimate use is often in aqueous solution, it can be cost-effective to ship and apply as a 12% solution. This is sold under the trade name of BOROL[R] by Ventron Corporation.

An alternative method was developed by Farbenfabriken Bayer A.G.[11]. This is a two-stage process in which anhydrous borax is fused with silica and the resulting sodium borosilicate is treated with hydrogen and molten sodium at $450-500^{\circ}C$ and 3 atmospheres pressure:

$$Na_2B_4O_7 + 7SiO_2 + 16Na + 8H_2 \longrightarrow 4NaBH_4 + 7Na_2SiO_3$$

The product is extracted with liquid ammonia.

Sodium borohydride is a powerful and selective reducing agent, employed in inorganic and organic systems. Its ability to reduce carbonyl groups ir-reversibly is of value both for organic syntheses and in pulp bleaching, but by far the most important use is for sodium hydrosulphite regeneration (or replace-ment of toxic zinc hydrosulphite) in connection with the latter. Borohydrides are also able to reduce heavy metal ions in solution to the elemental form, a quality useful in electrodeless plating, hydrometallurgical winning and effluent treatment.

Boranes. The boron hydrides, or boranes, are not produced commercially at present. They have been made by U.S. Government contractors on an industrial

scale for military purposes (high-energy fuels for missiles) and it is presumed
that stockpiles still exist.   The lower boranes are hazardous to handle, being
highly toxic as well as spontaneously inflammable and explosive.   For these
reasons together with their very high cost it is unlikely that they will find
serious industrial use;   it would be as illogical to consider boranes for
commercial fuels as it would be to make petrol from limestone.   The closest
approach to commercialisation has been with a carborane-siloxane high-temperature resistant polymer which has been sold under the trade name DEXSIL(R)[12].

Boron.   Elemental boron is not made on a large scale,   World-wide, probably
less than 10 tons are produced annually and most of this is impure material
required for flares and pyrotechnics.   Such material is made by magnesiothermic
reduction of boric oxide, in much the same way as Moissan prepared it originally.
The product is capable of being upgraded by vacuum melting after the acid-
soluble impurities have been removed, and a small amount of material is so
treated for electronics and research purposes.

The greatest advances in boron technology in recent years have been in connection with the production of reinforcement for plastics and metal composites such
as helicopter blades and aircraft wing sections.   This use is based on the
high specific stiffness of the boron filament;   it resembles carbon fibre in
its attributes for these special applications, and has an advantage in being a
less reactive reinforcement for aluminium, especially under moist conditions.
Boron filament is made continuously by chemical vapour deposition onto a fine
tungsten wire (similar to the filament of electric light bulbs) which is heated
resistively while being passed through a boron trichloride-hydrogen atmosphere.
The tungsten core remains in the product, adding marginally to its density.
This process represents by far the largest use for boron trichloride.

Boron Carbide.   Boron carbide, B_4C, is produced at a rate of several hundred
tons per annum, mainly for abrasive purposes and similar uses.   The bulk is
made by the electrothermal reduction of boric oxide with petroleum coke, the
cooled product being a clinker which is mechanically fractured into granules
and size-classified according to its final destiny.   Gravel-size material is
chosen for boron trichloride production, less coarse product being used as an
abrasive grit or pulverised even finer into polishing powder and hot-pressing
grades.

Like the metal borides (see below) boron carbide is a hard, refractory solid.
Its uses are mainly based on these qualities, but the properties at the same
time demand special techniques in order to fabricate usable objects.   Melting-
point is around 2,250°C, a temperature which no conventional mould material can
withstand.   Resort is therefore made to hot-pressing, a high-temperature
powder metallurgical technique which in this case is practised in graphite dies

at around 2000°C and pressures of about 2 tons/inch^2. Boron carbide is also so hard (over $9\frac{1}{2}$ on Moh's scale, or approximately 4000 kg/mm^2 V.P.N.) that it is imperative to press at as near as possible to the shape and size required in order to minimise the amount of diamond machining necessary. Currently the major use for pressings is shot-blast nozzles, which out-last tungsten carbide and are only about one-sixth of the weight. This low specific gravity (2.5 g/cm^3), coupled with high acoustic impedance, led to the use of boron carbide as lightweight ceramic armour. Scores of thousands of suits were used to protect battle personnel during the Vietnam war and the need greatly accelerated the development of hot pressing technology[13].

One other important application of boron carbide makes use of neither hardness nor refractoriness, but of the neutron-absorbing properties of the boron present. One fifth of this is of the ^{10}B isotope and, with 78% total boron, B_4C is much more cost-effective than the free element. The cost difference, appreciable even in the powder form, is greatly magnified by the relative ease of fabrication of boron carbide. A variety of hot-pressed boron carbide objects is shown in Plate 2.

<u>Metal Borides.</u> These compounds, like boron carbide, are interstitial solids which collectively fall into the class of substances known as "refractory hard metals". The inter-relationship and methods of manufacture are summarised in Figure 2. Chromium monoboride, CrB, and the three diborides CrB_2, TiB_2 and ZrB_2 are commercially the most important. Possessing melting points of just under 2000°C in the cases of the chromium compounds, and over 3000°C for the titanium and zirconium diborides, fabrication is by hot-pressing in a similar manner to boron carbide. The products are only marginally less hard (in the region of 3000 kg/mm^2 V.P.N.) and hence diamond machining is also necessary for the final shaping of components. Most metals form at least one boride within the series MB, MB_2, MB_4, MB_6 and MB_{12} and combination in other ratios is common. Small amounts of hexaborides of rare earth elements, LnB_6, are used in the electronics industry. The chemistry, fabrication and applications of metal borides, boron carbide and boron nitride have been reviewed fully elsewhere[14]. For this brief account of the contemporary industrial significance of borides it is appropriate simply to mention that the three important diborides are produced by the carbothermic reduction of the metal oxide in the presence of boric oxide, e.g.

$$TiO_2 + B_2O_3 + 5C \longrightarrow TiB_2 + 5CO$$

The reaction is endothermic and effected by heating in vented electric furnaces at 2000°C briquettes made from an intimate mixture of the finely powdered reactants. Chromium monoboride is made by the exothermic reduction of the parent oxides with aluminium powder. Titanium diboride, which is a good

electrical conductor (the conductivity is similar to that of iron) and is
chemically resistant to attack by most non-ferrous metals, including even
boiling aluminium, finds its main outlet as the standard crucible material
used in the vacuum metallising industry.   For this application the overall
conductivity of the boat-shaped crucible is reduced (to facilitate resistive
heating during use) by mixing the titanium diboride powder with boron nitride
(hexagonal) powder prior to hot-pressing.   Finished products are shown in
Plate 2.   The principal use for the chromium borides is as ingredients for
welding alloys, especially those employed for hard facing.   A crude calcium
boride has been used for de-oxygenating copper.

Boron Alloys.   These are stoichiometrically indefinite but invariably
metal-rich compositions of commercial importance formed between boron and
metals such as copper, manganese and the iron group.   Production of ferroboron
(10-25% B), manganese-boron (15-20% B) and nickel-boron (15-18% B) alloys is
by the reduction, usually aluminothermic, of boric oxide in the presence of the
alloying metal or its oxide.   Ferroboron is used to introduce small amounts
(0.002%) of boron into steel in order to improve hardenability, or larger
amounts (up to about 7%) for nuclear shielding.   The other boron alloys are
used in hard facing and the production of special alloy steels.

Boron Nitride.   Isoelectronic with carbon, boron nitride is its total
analogue in existing as a soft hexagonal form and an intensely hard cubic form.
Hexagonal BN is a white solid with a graphitic crystal structure, but unlike
graphite it is a non-conductor of electricity.   This property and its refractoriness (it begins to decompose at 1700°C) account   for its industrial applications.   One such outlet (as a constituent of vacuum metallising crucibles)
has already been mentioned;   its others are largely in the electrical and
related industries, where its high thermal conductivity as well as its high
resistivity is of especial value.   Boron nitride is usually made as a fine
powder by nitriding boric oxide (on a suitable acid-soluble carrier such as
calcium phosphate) with nitrogen or ammonia at around 800°C.   The leached
powder is fabricated by hot-pressing, as in the cases of the other binary
refractory compounds of boron.   Being much softer than either the metal
borides or boron carbide, it is customary to press large blocks in graphite
moulds and cut to shape using normal engineering tools.   A selection of shaped
products is shown in Plate 2.   Cubic BN is practically as hard as diamond but
is slightly more oxidation resistant.   It is made from the hexagonal form
in the same type of high-temperature, high-pressure equipment as is used for
synthetic diamond production.   It is sold as boart under the trade name
BORAZON[R].

REFERENCES

1. Thompson, R. "Boron Chemistry in Industrial Perspective", Chemistry in Britain, 7, 140, 1971.

2. Teeple, J.E. "The Industrial Development of Searles Lake Brines", American Chemical Society Monograph Series No. 49, 1929.

3. British Patent 910,541; U.S. Patents 2,969,275 and 3,111,383.

4. Bixler, G.H. and Sawyer, D.L. Ind. Eng. Chem., 49, 330, 1957.

5. British Patents 1,133,434 and 1,157,675.

6. U.S. Patent 3,060,049.

7. Woods, W.G. and Bower, J.G. "Firebrake ZB, A New Fire Retardant Additive", SPI Reinforced Plastics/Composites Division; 25th Technical & Management Conference, Washington, U.S., Feb. 1970.

8. U.S. Patents 3,549,316 and 3,718,615.

9. Bashkirov, A.N. et al., "The Oxidation of Hydrocarbons in the Liquid Phase" (Ed. N.M. Emanuel), Pergamon Press, p.183,1965.

10. Fedor, W.S., Banus, M.D. and Ingalls, D.P., Ind.Eng.Chem., 49, 1664, 1957.

11. Anon, Chemical Week, 63, 10th June 1961.

12. Mangold, D.J., Appl. Polym. Symp., 11, 157, 1969.

13. Hansen, J.V.E., Research/Development, Technical Publishing Company, Barrington, Illinois. 26, 1968.

14. Thompson, R., "Borides: Their Chemistry and Applications" R.I.C. Lecture Series, 5, 1965.

Soluble Silicates and Their Derivatives

By D. BARBY*, T. GRIFFITHS, A.R. JACQUES AND D. PAWSON*

J. Crosfield & Sons Ltd.
and (*) Unilever Ltd.

1. Introduction

Sodium silicate is one of the older heavy inorganic chemicals, with an industrial history which, because of its strong dependence on the alkali industry, parallels that industry.

Soluble silicate glasses were known to the Phoenicians. More recent studies can be traced to work early in the seventeenth century by Glauber and Agricola among others; the poet Goethe experimented with silicate solutions around 1768 [1]. The earliest industrial development is generally ascribed to Von Fuchs,[2,3] a Professor of Mineralogy in Munich who proposed the use of silicates for adhesives, cements and fire-proof paints. Industrial manufacture in Germany was begun by Walcker who dissolved sand in caustic soda in about 1828, and Kuhlmann[4-6] established manufacture in France in 1841.

In Britain, Gossage pioneered the use of sodium silicate for improving the properties of soap, and he started production in Widnes in 1854[7]. He produced sodium silicate by fusing soda ash and white sand in an open-hearth furnace. The glass was tapped into moulds and, after cooling and solidifying, it was broken up, ground to powder, and dissolved in boiling water in an iron pan into which steam was injected. The weak liquor was then concentrated in an evaporator. Manufacture of silicate in the USA was established by Elkington in 1863, and again an open-hearth furnace was used. Gossage's technique is still the most widely used production method throughout the world, but the glass is no longer powdered before dissolution.

From this, it can be seen that sodium silicate was one of the first chemicals to be made on a large scale, and that the industry grew up more or less alongside caustic soda and soda ash development. This early development was clearly related to the soapy feel of the product, which, combined with a lower alkalinity than that of **caustic soda**, made it a natural component as a builder for soap, that is, mixed with soap, "it had a greater detergent power than genuine soap"[7].

It is possible to make silicate glasses, and hence derived solutions and amorphous solids, over a whole range of compositions (not necessarily stoichiometric), and these materials are therefore usually defined in terms of the molar or weight ratio, silica:alkali metal oxide ($SiO_2:M_2O$), that they contain. The molar and weight ratios $SiO_2:M_2O$ are hereafter designated R_M and R_W respectively. For sodium silicates, the conversion factor is $R_M = 1.03\ R_W$, and here the terms are often used indiscriminately since the definitions are fortuitously similar. For potassium silicates, $R_M = 1.57\ R_W$.

Traditional names, still frequently used in industry, are associated with certain molar ratios. Table 1 gives the names used for sodium silicate solids (whether amorphous, crystalline, glassy or simply physical admixtures) and solutions of the specified ratio.

TABLE 1 - COMMON NAMES FOR SODIUM SILICATES

MOLAR RATIO	R_M	COLLOQUIAL NAME
$SiO_2:2Na_2O$	0.50	Orthosilicate
$2SiO_2:3Na_2O$	0.67	Sesquisilicate
$SiO_2: Na_2O$	1.00	Metasilicate
$\sim2SiO_2: Na_2O$	~2.00	Alkaline silicate or disilicate
$\sim3.4SiO_2: Na_2O$	~3.40	Neutral silicate

The structural chemistry of silicates cannot be understood simply in terms of R_M and therefore use of these names other than as labels may be misleading. IUPAC recommended names are given in Section 3.2, Table 3.

The soluble silicate industry today is one of the larger chemical enterprises. It is not possible to estimate precisely the amount of silicate glass manufactured throughout the world because a great deal is used 'in-house' as a feed-stock for further products. The authors' best estimate is in excess of 3 million tonnes per annum (expressed as tonnes of glass ($R_W = 3.3$)) with 1976 UK production estimated at 170,000 tonnes. Industrial output in both Europe and the USA has been rising consistently at about 4% per annum since 1949.

Selling prices vary widely, but as an example, a recently quoted
bulk ex-works price for a sodium silicate glass with R_W = 3.3 was $178/
long ton and for glass with R_W = 2.0 was $234/long ton[8]. There is little
export from or import into industrialised nations because the product is
cheap and transport costs are significant.

Only the soluble silicates of sodium and potassium are produced on a
large scale and production of sodium glass is a hundred times greater than
that of potassium glass. Lithium and quaternary ammonium silicates are
produced in limited commercial quantities, but up to now no other soluble
silicates have been manufactured. For this reason, most of this chapter will
be restricted to the sodium and potassium derivatives.

Many different soluble sodium and potassium silicates are sold
throughout the world, and their composition varies over a whole range of
$SiO_2:M_2O$ ratios from R_M = 0.50 to 3.98 for sodium silicates and from
R_M = 2.24 to 3.93 for potassium silicates. The products include powdered
and lump glass, aqueous solutions (technically called liquors) covering a
wide concentration range, and the so-called soluble powders which are made
by part-drying solutions to 85% solids. Figure 1 shows the range of
available sodium silicates and their derivatives in relation to the
Na_2O - SiO_2 - H_2O ternary phase diagram and Table 2 lists the more common
commercial soluble silicates.

2.   Raw Materials and Manufacture

The fusion process to make sodium silicate glass commonly uses sand
and soda ash and it is this process we will discuss in detail. However, in
parts of Europe it has sometimes been economically advantageous to use
sodium sulphate from the Stassfurt deposits together with sand and carbon.
The reaction is complex but may be represented:

$$2Na_2SO_4 + 2C + xSiO_2 \rightarrow 2Na_2O. xSiO_2 + S\uparrow + SO_2\uparrow + 2CO_2\uparrow$$

Sodium silicate can also be made by direct dissolution of sand in
caustic soda, but this route has limited commercial application. The
reaction between common salt and sand has also been studied as a potential
route to silicate, but it is not suitable for commercial exploitation[9]. One
more recent process is based on the reactions[10]:

FIGURE 1

MANUFACTURED PRODUCTS AND THE $Na_2O-SiO_2-H_2O$ PHASE DIAGRAM

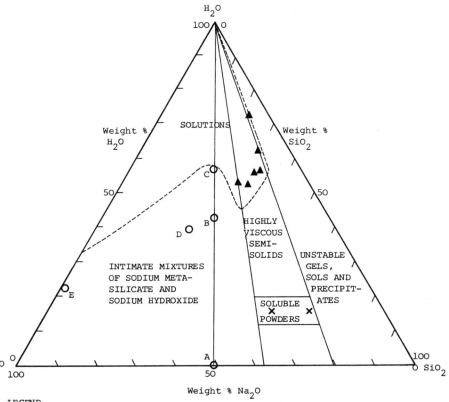

LEGEND

X   Typical commercial soluble powders.

▲   Typical commercial solutions.

O   Crystalline materials

    A. Sodium metasilicate
    B. Sodium metasilicate 'pentahydrate'
    C. Sodium metasilicate 'nonahydrate'
    D. Sodium sesquisilicate pentahydrate
    E. Sodium hydroxide

---' Solution boundary.

TABLE 2 - TYPICAL COMMERCIAL SODIUM AND POTASSIUM SILICATES

Commercial silicates	Wt ratio SiO_2/M_2O, R_W	Mole ratio, SiO_2/M_2O, R_M	M_2O, wt.%	SiO_2, wt.%	Soften- ing point, °C	Flow point, °C	H_2O, wt.%	Sp gr, d^{20}_{20}	Typical viscosity 20°C cP.or mPa.s
SODIUM SILICATES									
'Alkaline†' glass	2.00	2.06	33.0	66.0	590	874			
'Neutral' glass	3.35	3.46	22.90	76.72	655	960			
Solution	1.60	1.65	17.98	28.75			53.27	1.60	900
"	2.00	2.07	16.07	32.15			51.78	1.60	2000
"	2.50	2.58	12.45	31.10			56.45	1.50	400
"	2.85	2.94	11.20	31.95			56.85	1.48	500
"	3.30	3.41	8.85	29.25			61.90	1.39	300
"	3.85	3.98	5.75	22.20			72.05	1.26	20
'Alkaline†' soluble powder	2.00	2.07	28.0	56.0			16.0		
'Neutral' soluble powder	3.30	3.41	19.53	64.47			16.0		
Metasilicate beads	0.95	0.98	50.7	48.0	-	1089*			
'Metasilicate pentahydrate' crystals	0.97	1.00	29.5	28.5	-	72.2*	42.00		
'Metasilicate nonahydrate' crystals	0.97	1.00	21.8	21.1	-	47.85*	57.10		
'Sesquisilicate' powder	0.65	0.67	60.8	39.2	-	1122*			
'Orthosilicate' powder	0.48	0.50	67.4	32.6	-	1118*			
POTASSIUM SILICATES									
Glass	2.50	3.92	28.3	70.7	700				
Solutions	1.43	2.24	21.6	30.8			47.6	1.60	300
"	1.80	2.83	16.4	29.5			54.1	1.49	1300
"	2.10	3.30	12.5	26.3			61.2	1.38	1050
"	2.20	3.45	9.05	19.9			71.1	1.26	7
"	2.50	3.93	8.3	20.8			70.9	1.26	40

† Sometimes called disilicate.
* Definite melting points.

$$4NO_2 + 2NaCl \rightarrow 2NaNO_3 + 2NOCl$$
$$2NOCl + O_2 \rightarrow Cl_2 + 2NO_2$$
$$2NaNO_3 + SiO_2 \rightarrow Na_2SiO_3 + 2NO_2 + \tfrac{1}{2}O_2$$

2.1 Raw Materials. Sodium silicate production is very dependent on the alkali industry, in particular for its soda ash supply. Furthermore, caustic soda is often used to produce the more alkaline silicates ($R_W < 2$) from liquors derived directly from glasses.

The quality of the finished product is very sensitive to raw material purity. The sodium alkalies are much purer than even the best sands so little or no impurity enters the product from the alkali, but the process does demand high-purity silica sand. Potassium silicates are often made to a higher purity than the sodium analogues, in which case the especially pure sands and potash are used to ensure the quality of a premium product. There is a big deposit of almost pure sand requiring no processing in the Maastricht-Aachen area of Holland-Germany. Since the end-product is cheap, however, it is vital that indigenous sand be used. In Britain, the purer sands are at Oakamoor, Kings Lynn, Loch Aline and Redhill[11]. These contain 99% silica as granular quartz and a variety of impurities, of which iron and aluminium are the commonest, and thus the sands usually require some form of chemical processing. For British glass-making (of both soluble and insoluble glasses) to remain competitive, sands have therefore to be carefully selected for purity to keep processing costs to a minimum.

The purest sand reserves in Britain are becoming depleted, and in an industry so thoroughly worked as glass-making, there is little hope that further pure sand deposits will be found. However, in Europe and the rest of the world, there is no foreseeable shortage of sand of adequate purity. These remarks on raw material purity underline the fact that there is no cheap, simple way to remove trace impurities from the soluble silicates.

2.2 <u>Manufacture of Sodium Silicates</u>. Figure 2 shows the general flow scheme.

FIGURE 2. <u>FLOW DIAGRAM FOR THE SODIUM SILICATE PROCESS</u>

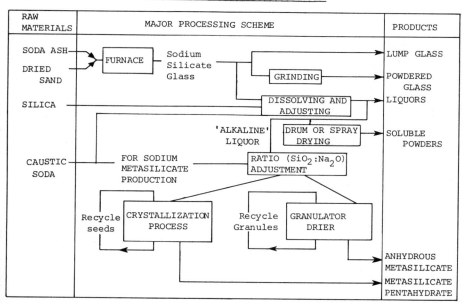

2.2.1 <u>Sodium Silicate Glass</u>. Sand and soda-ash are fused at a temperature
of approximately 1400°C.

$$Na_2CO_3 + xSiO_2 \rightarrow Na_2O. xSiO_2 + CO_2\uparrow$$

Figure 3a shows the liquidus for sodium silicate melts (i.e. the temperature
required to obtain a homogeneous melt at equilibrium) as a function of
silicate ratio R_W, and also includes the range of common manufacturing
conditions.

<u>FIGURE 3</u>

<u>LIQUIDUS CURVES FOR SODIUM AND POTASSIUM SILICATE MELTS</u>

a) SiO_2 - Na_2O System   b) SiO_2 - K_2O System

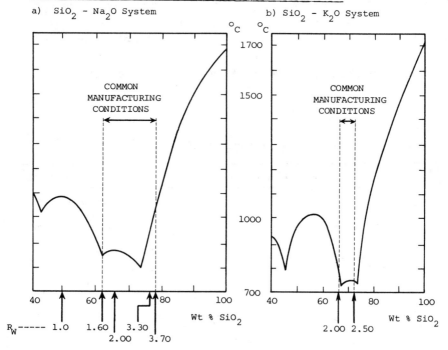

The melting points of sodium and potassium carbonates are 851 and
891 K respectively.

The high temperatures give a melt of suitably low viscosity - the viscosity
of sodium silicate melts falls by approximately two orders of magnitude on
increasing the temperature from 900 to 1400°C.

A ROTARY FURNACE

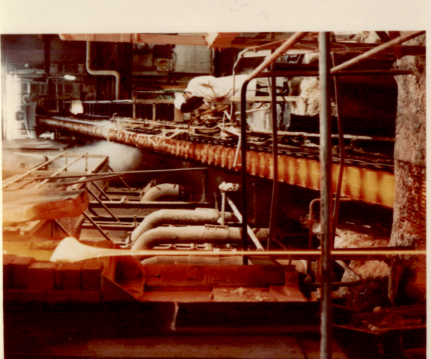

CONTINUOUS TAPPING OF A REGENERATIVE FURNACE

The relative proportion of sand to soda ash is varied to give a range of glasses of differing $SiO_2 : Na_2O$ ratios. For 'in-house' use, manufacturers may produce glass with ratio $R_W = 0.97$ and in the range $R_W = 1.60$ to 3.70, but most glass available commercially is restricted to the two ratios of $R_W = 2.00$ ('Alkaline' glass) and $R_W = 3.30$ ('Neutral' glass).

The fusion process may be carried out in electric arc furnaces if cheap electricity is available; there is also some production in oil or gas-fired rotary furnaces. However, large-scale production is usually in oil or gas-fired open-hearth regenerative furnaces of conventional design. Rotary and gas-fired regenerative furnaces are shown in Plate 1.

A typical regenerative furnace has a tank melting area of 40-60 m^2 and can be expected to produce approximately 150 kg of glass per square metre of tank area per hour at a specific fuel consumption of about 0.12 tonnes oil per tonne of glass ($\sim$ 5MJ kg^{-1}).

A typical rotary furnace for the continuous production of silicate glass is installed at a slight slope to the horizontal and rotates at 0.5 to 1.0 r.p.m. At the lower end of the furnace there is an outlet into which hollow plugs of varying apertures can be fitted. At the reaction temperature the product becomes more or less viscous depending upon the $SiO_2 : Na_2O$ ratio and the appropriate aperture can be used for exit of glasses of different ratios[12].

With both furnace types, charging and tapping are continuous. The molten glass stream is collected, cooled and solidified on a tray elevator/conveyor and the resultant lump glass is either fed to dissolving equipment or stored in silos. Alternatively, there is a continuous process whereby the molten glass is dissolved directly in water in a rotary dissolver operating at atmospheric pressure.

Furnace life depends on the quality of refractories used, but 3 years between rebuilds would be regarded as normal, with an intermediate repair after 18 months' operation. The selection and disposition of refractories are vital to refractory life because of the corrosive nature of the melt. Soluble silicate glasses are much more corrosive than the common bottle (soda-lime) glasses.

2.2.2 Sodium Silicate Liquor. The production of sodium silicate liquor by
dissolving molten glass directly in water at atmospheric pressure has already
been mentioned. However, it is more usual to dissolve lump glass using
either rotary or stationary pressure dissolvers.

A stationary dissolver may have a glass capacity of up to 40 tonnes and
operate at a steam pressure of up to 700 kN m^{-2} (100 p.s.i.). The vessel is
charged with glass and sealed. Water is added and steam injected until
the desired working pressure is obtained. The heat of solution assists in
maintaining working pressure. When the liquor reaches the required
concentration it is blown over to a collecting tank, whereupon the dissolver
may then either be topped up with fresh glass and the process repeated, or a
second boil may be taken off before recharging.

It is necessary to maintain a high glass-to-water ratio in the
dissolver to achieve a satisfactory dissolution rate and high throughput.
The dissolution rate increases as the operating pressure increases and as R_W
for the glass decreases, but there is no saturation point in the accepted
sense. Thus, if the dissolving operation is allowed to proceed indefinitely,
the contents of the dissolver set into an elastic mass which is difficult to
remove. The onset of this phenomenon depends on the ratio of the glass,
there being a minimum in the viscosity - ratio curve at any particular
concentration (Figure 4). As a result 'alkaline' glass (R_W = 2.00) may be
dissolved to give a handleable solution containing 54% solids, whereas the
upper practical solids content of a solution of 'neutral' glass (R_W = 3.30)
is only 39%; further small increases in concentration result in so rapid an
increase in viscosity that virtual solidification takes place. Thus the
viscosity of silicate solutions plays an important part in their manufacture,
and this property is also crucial to many of their industrial applications.

Glass concentration in dissolver liquor may be varied according to
quality requirements in the finished products. Generally, solid impurities
are removed either by sedimentation or filtration, and the liquor is then
concentrated in evaporators.

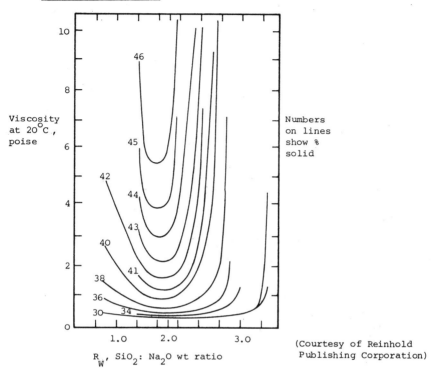

FIGURE 4

VISCOSITY OF SODIUM SILICATE SOLUTIONS AT DIFFERENT
TOTAL SOLIDS CONTENTS

(Courtesy of Reinhold
Publishing Corporation)

The standard liquors have the ratios of the common glasses, i.e.
R_W = 2.00 or 3.30; liquors of intermediate ratios may be prepared by
blending these standard liquors. Caustic soda may be added to obtain
liquors with R_W < 2.00 and silica to obtain liquors with R_W > 3.30. If the
demand for liquor of a particular ratio is relatively high, it may prove
economical to manufacture glass at the ratio required.

2.2.3 Sodium Silicate Soluble Powders. These are produced to satisfy a
demand for solid silicates which are much more readily soluble in water at
ambient temperatures than the equivalent anhydrous powdered glasses. The
powders normally contain 14% to 18% water (86% to 82% solids as glass). They
are prepared by removing part of the water content of sodium silicate liquors
either by drum-drying or spray-drying.

2.2.4 Sodium Metasilicates. Anhydrous sodium metasilicate is produced either
by the direct fusion route, followed by cooling, milling and grading, or by

drying a solution of metasilicate composition (prepared from 'alkaline' silicate liquor and caustic soda) in a special drum granulator/drier[13] or a fluid bed drier[14]. Both methods are commonly used in industry. In the drum granulator process small anhydrous metasilicate granules are fed into the end of a revolving drum in a current of hot air. The hot liquor sprayed into the drum is collected by the granular feed and dried out in the hot air stream. Thus, the granules grow and dry uniformly. At the end of the drum, they are screened, and the under-size material returned to feed for re-use as nuclei. To ensure uniform granule size with this sort of granulation, over 70% of the product is recycled.

The design and operation of the furnace for the direct fusion process require special care to minimise refractory attack by the highly alkaline melt. In fact, the melt is usually contained within a barrier of unfused feed. Because of the aggressive nature of the melt and the problems of handling it, metasilicate produced by direct fusion may contain an appreciable proportion of insoluble material such as unreacted sand, natural impurities in sand, and refractory particles.

In contrast, the spray-coated product, which is derived from a purified silicate liquor, has a much lower level of impurities and, by virtue of the granulating/layering process, has a superior particle size distribution, is relatively dust-free and more readily soluble in water.

There are four known so-called hydrates of sodium metasilicate, formally $Na_2SiO_3.nH_2O$ (n = 5,6,8,9), commonly called penta-, hexa-, octa-, and nona-hydrates respectively. Their correct formulae are $Na_2[SiO_2(OH)_2],(n-1)H_2O$ (see Sec. 3). Although the nona-hydrate is available in some countries, only the pentahydrate is produced in large-scale commercial quantities.

In a typical pentahydrate manufacturing process, 'alkaline' silicate liquor (R_W = 2.00) and caustic soda liquor are mixed together to give a solution of 'metasilicate' ratio. This is concentrated by evaporation to a composition equivalent to that of the pentahydrate. The mixture is cooled, crystallised as a mass, milled, graded and packed. Crystallisation is accelerated by seeding the liquor. The evaporation stage is often avoided by preparing metasilicate liquor of the final pentahydrate composition from a mixture of solid caustic soda and a concentrated alkaline silicate liquor at a temperature in excess of the melting point of pentahydrate (72.5°C).

Irrespective of the method used to prepare the liquor, the relative proportions of silica, sodium oxide and water in the mass prior to crystallisation must be carefully controlled, and supercooling avoided during the subsequent cooling/crystallisation process to ensure that the solid crystalline mass may be readily milled and graded and that the granulated product may be stored without caking.

2.2.5 Other Sodium Silicate Hydrates. In general, the commercially available hydrates of 'sodium sesquisilicate' and 'sodium orthosilicate' are not definite compounds, but are prepared by dry-mixing sodium metasilicate (anhydrous or pentahydrate) with caustic soda to the appropriate composition. Hot liquid mixtures of sodium silicate liquor and caustic soda, with an overall $R_W = 0.17$, can be chilled to give flakes using a water-cooled flaking drum.

2.3 **Manufacture of Potassium and Lithium Silicates.** Potassium silicate glass is prepared by the direct fusion of potassium carbonate and sand at a temperature of approximately $1400^\circ C$. The liquidus for potassium silicate melts is shown in Figure 3b, together with the range of common manufacturing conditions. The demand for potassium silicate relative to sodium silicate is small and although it may be produced in a conventional sodium silicate furnace, a small furnace specifically for potassium silicate is preferred to avoid the risk of contamination.

Glass is normally produced within the range $R_W = 2.00-2.50$. Potassium glasses are more hygroscopic than the equivalent sodium glasses. Lump potassium glass of $R_W = 2.00$ tends to agglomerate on storage, but glasses of $R_W = 2.30$ may be stored satisfactorily in covered silos for long periods.

The methods used to prepare potassium silicate liquors and soluble powders follow the same general principles as are applied to sodium silicate. The dissolution rate of potassium silicate glass is much greater than that of the equivalent sodium silicate glass. Liquors of especially low impurity content are prepared by dissolving chemically treated silica gel in caustic potash solution.

Lithium silicate glasses do not dissolve in water, and lithium silicates are therefore prepared by dissolution of silica in lithium hydroxide solutions.

2.4 <u>Analysis</u>. British Standard 3984,1966 (Amended 1971) gives details of
methods of analysis of soluble silicates for alkali content, silica content,
insoluble matter, iron, chlorides, and sulphates. Methods are also given
based on silica:alkali oxide ratio, relative density and viscosity. The
relationship between alkali oxide content, silica content, relative density
and ratio can be represented in a chart, part of which is illustrated in
Figure 5. If any two of the four variables are known, then a given
silicate solution can be identified by referring to the chart.

FIGURE 5

SPECIFIC GRAVITY OF SODIUM SILICATE SOLUTIONS

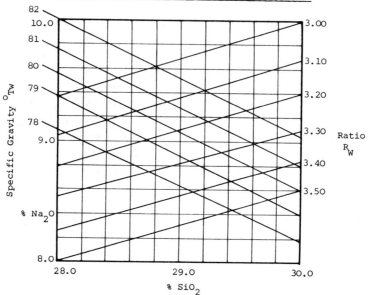

For example, a solution with 9.0% Na_2O and SG 81 ^{o}Twaddell
has R_W= 3.3 and contains 29.6% SiO_2.

<u>NB</u>: Specific Gravity on Twaddell scale,^{o}Tw
      ^{o}Tw = 200 (Specific Gravity - 1).

A more recent and rapid (< 5 minutes) determination[15] of the
constituents of sodium silicate uses the reaction:

$$Si(OH)_4 \; + \; 6NaF \; + \; 4HCl \; \rightarrow \; Na_2SiF_6 \; + \; 4NaCl \; + \; 4H_2O$$

A mixed indicator, 1:1 methyl red:bromothymol blue, is used.
Spectrophotometric techniques can be used to measure dilute concentrations
of silica and these depend upon the formation of molybdosilicic acids[16].

3. The Chemistry of Soluble Silicates

The early development of the soluble silicate industry was paralleled by many painstaking studies on soluble silicate chemistry. Results obtained up to 1951 are discussed in the monumental compilation of Vail[17], which is still a standard text on the subject, and summarised in reviews by Wills[18], Weldes and Lange[19], and Engler[20]. It is only recently that modern chemical and spectroscopic techniques have been used to elucidate the complex structural chemistry of crystalline silicates, silicate glasses, and their solutions, and here we aim to do no more than update parts of Vail's discussion using these new data. Thus for a concise treatment of the refractive indices, specific volumes, densities, viscosities, rates of solution and surface tension of silicate glasses as a function of R_M and temperature, the reader is referred to Wills[18].

3.1 Principles of Structural Silicate Chemistry.

Silicate ions in glasses, crystals and solutions are built from orthosilicate tetrahedra, $[SiO_4]^{4-}$, which may be discrete or joined to one, two, three or four other tetrahedra by sharing oxygen atoms to form Si-O-Si bridges. Two tetrahedra are never jointly bridged by more than one oxygen atom (thus $Si\overset{O}{\underset{O}{\diagdown\kern-1em\diagup}}Si$ units are unknown), and only rarely does the co-ordination number of silicon exceed four[21]. Anions may therefore have both Si-O-Si bridging and Si-O   non-bridging linkages, and any number of the latter may be protonated.

A variety of structural types can thus occur. All four sodium salts of mononuclear orthosilicic acid, $Si(OH)_4$, are known, although $[SiO_4]^{4-}$ ions are not found even in the most alkaline aqueous solutions. Orthosilicate tetrahedra may join to give linear chain ions (e.g. $[Si_2O_7]^{6-}$ [22] and $[Si_3O_5(OH)_5]^{3-}$ [23]), cyclic ions (e.g. $[Si_3O_9]^{6-}$ [23]), or cage ions (e.g. $[Si_6O_{15}]^{6-}$ [24]). Infinite chains of tetrahedra are found in Na_2SiO_3, and infinite sheets of tetrahedra in $Na_2Si_2O_5$. A convenient notation for these structures defines a silicate tetrahedron as Q^n, where n is the number of other tetrahedra to which it is joined (n = 0-4). Thus a linear trisilicate ion $[Si_3O_{10}]^{8-}$ or any of its protonated forms would be designated $Q^1Q^2Q^1$.

Structure is not directly related to ratio R_M. A compound whose ratio is greater than that of $[SiO_4]^{4-}$ (R_M = 0.50) is either protonated or has Si-O-Si bridges or both. Thus a crystalline material at 'sesquisilicate' ratio (R_M = 0.67), for example, may contain either dinuclear $[Si_2O_7]^{6-}$ ions (as in $Li_6[Si_2O_7]$) or $[SiO_3(OH)]^{3-}$ ions (as in $Na_3[HSiO_4]\cdot5H_2O$). The

average number of non-bridging oxygen atoms per silicon atom in a
completely deprotonated solid is given by $2/R_M$.

3.2 Soluble Silicate Solids. Vail has described the properties of many
crystalline alkali metal silicates[25] and most of the materials mentioned have
now been studied using single crystal X-ray diffraction techniques. The
results of these and other chemical studies (paper chromatography, reaction
with molybdic acid and trimethylsilylation) and spectroscopic studies (infra-
red, Raman, broad-band proton n.m.r.) are summarised in Table 3. Of
particular note is the different structural nature of sodium metasilicate,
Na_2SiO_3, and its 'hydrates' $Na_2\left[SiO_2(OH)_2\right].nH_2O$ (n = 4,5,7,8).

Table 3 also includes some crystalline materials with polynuclear cage
anions obtained from aqueous solutions using tetraalkylammonium and
transition metal ethylenediamine complexes as cations. (These cations
interact little with the anions.) There is at present no definite evidence
for the presence of these particular polynuclear species in silicate
solutions, although n.m.r. studies (see below) show that species of their
complexity are present.

In addition to the tabulated materials, there are several high ratio
sodium silicates reported by Wills[26] (R_M ~ 4, designated K in Vail[25]) and
McCulloch[27] (R_M = 4, 13.1 and 9.4 (this material was later shown to have
R_M = $8^{28,29}$)). Several high ratio sodium silicate minerals have recently
been discovered[30-36], but the structures of these materials and their
relationship to the synthetic ones are not generally clear. McCulloch's
R_M = 13.1 silicate has been identified with one of the minerals, magadiite[33].

The structures of silicate glasses have been investigated using a
variety of techniques (see, for example, reviews by Eitel[72] and Brawer[73]).
Of particular interest are recent Raman spectroscopic studies on alkali
metal silicate glasses having compositions with R_M between 1 and 4[63]. The
spectra are consistent with glasses with R_M = 1 and R_M = 2 having similar
structural units to the crystalline phases of the same composition. Thus at
R_M = 2, $\left[Si_2O_5\right]_\infty^{2-}$ sheets are present, but these are disordered in several
ways compared to the crystalline material: (i) there are random
fluctuations in Si-O-Si angles and bond lengths, tetrahedral angles, and the
relative orientation of the tetrahedra, (ii) the rings of tetrahedra from
which the sheets are formed vary in size, (iii) the number of non-bridging
oxygen atoms on each silicon atom varies, whilst in the crystal it is always
one. As R_M is decreased towards 1, the silicate sheets become increasingly

TABLE 3 - STRUCTURAL DATA FOR CRYSTALLINE ALKALI METAL SILICATES

DESIGNATION IN VAIL[25]	FORMULA	R_M	SYSTEMATIC NAME[37]	DESCRIPTION OF STRUCTURE
1. MONONUCLEAR IONS				
I.	$Li_4[SiO_4]$	0.5	Tetralithium orthosilicate	X-ray studies show discrete $[SiO_4]^{4-}$ tetrahedra[38]. Unit cell dimensions found for the sodium salt[39]
C	$Na_3[SiO_3(OH)] \cdot 5H_2O$	0.67	Trisodium hydrogenorthosilicate pentahydrate	X-ray studies show discrete $[SiO_3(OH)]^{3-}$ tetrahedra linked in pairs by strong hydrogen bonds[40]. The mono- and di-hydrates (A and B in Vail) are thought to contain similar anions.
	$Li_3[SiO_3(OH)]$	0.67	Trilithium hydrogenorthosilicate	Chemical and spectroscopic studies suggest $[SiO_3(OH)]^{3-}$ ions are present[41].
F,G,H,I	$Na_2[SiO_2(OH)_2] \cdot nH_2O$, n = 4,5,7,8	1	Disodium dihydrogen-orthosilicate hydrates	X-ray[42,43] and neutron diffraction studies[44] show discrete $[SiO_2(OH)_2]^{2-}$ ions in all hydrates. Broad band ^1H n.m.r. studies have been made for n=4 and n=8[45,46], I.r. studies for n=4[47,48] and 5[49].
	$Na[SiO(OH)_3]$	2	Sodium trihydrogen-orthosilicate	Broad band ^1H n.m.r. studies consistent with discrete $[SiO(OH)_3]^-$ ions[46,50].
2. DINUCLEAR IONS				
	$Li_6[Si_2O_7]$	0.67	Hexalithium disilicate	X-ray studies show discrete $[Si_2O_7]^{6-}$ ions are present[22]. Unit cell dimensions obtained for sodium, potassium, rubidium and caesium salt[51].
	$Li_4[Si_2O_5(OH)_2]$	1	Tetralithium dihydrogen disilicate	X-ray studies show discrete $[Si_2O_5(OH)_2]^{2-}$ ions[52].
3. POLYNUCLEAR IONS (anion names only given, with the appropriate structural prefix, when known)				
	$K_4[Si_4O_8(OH)_4]$	2	Tetrahydrogen cyclo-tetrasilicate (4-)	X-ray studies show cyclo-tetrasilicate rings $[Si_4O_8(OH)_4]^{4-}$ with all 4 Si-O and Si-OH groups involved in inter-anion hydrogen bonding[53]. Raman + i.r. studies made[54].
	$[Ni(dae)_3]_3[Si_6O_{15}] \cdot 26H_2O$	2	Triprismo-hexa silicate (6-)	X-ray studies show a cage anion with two six membered Si_3O_6 rings, linked by 3 Si-O-Si bridges[24]. A tetraethyl-ammonium salt may also contain hexa-nuclear ions[55].
	$[Me_4N]_{10}[Si_7O_{19}] \cdot nH_2O$	1.4	Heptasilicate (10-)	Chemical[56] and spectroscopic studies[55] show the anion is derived from the $[Si_8O_{20}]$ cage, with an SiO group removed (see below).
	$[Cu(dae)_2]_4[Si_8O_{20}] \cdot 30H_2O$	2	Hexahedro-octa-silicate (8-)	X-ray studies show a cage anion with two eight membered Si_4O_8 rings linked by 4 Si-O-Si bridges[57]. An Me_4N^+ salt contains a similar anion[55,58].
	$[Co(dae)_3]_2[Si_8O_{18}(OH)_2] \cdot nH_2O$	2.67	Hexahedro-di-hydrogenoctasilicate (6-)	X-ray studies[59] show the anion is a diprotonated derivative of $[Si_8O_{20}]^{8-}$ (see above).
	$[Bu_4N]_{10}[Si_{10}O_{25}] nH_2O$	2	Pentaprismo decasilicate (10-)	Chemical[60] and spectroscopic[55] studies show the ion has a cage structure with two 10 membered Si_5O_{10} rings linked by five Si-O-Si bridges.

Cont'd ......

TABLE 3 (cont'd)

DESIGNATION IN VAIL[25]	FORMULA	R_M	SYSTEMATIC NAME[37]	DESCRIPTION OF STRUCTURE
4. INFINITE CHAINS				
II.	$Li_2[SiO_3]$	1	Dilithium ino-polymetasilicate	X-ray studies show infinite $[Si_2O_6]_\infty^{4-}$ chains[61].
VI, D	$Na_2[SiO_3]$	1	Disodium ino-polymetasilicate	X-ray studies show the structure is similar to lithium salt[62]
VIII, L	$K_2[SiO_3]$	1	Dipotassium ino-polymetasilicate	Raman spectra show the structure is similar to the lithium and sodium salts[63].
5. INFINITE SHEETS				
III.	$Li_2[Si_2O_5]$	2	Dilithium phyllo-polydisilicate	X-ray studies show infinite $[Si_2O_5]^{2-}$ sheets built from rings of six tetrahedra[64].
VII, J	$Na_2[Si_2O_5]$	2	Disodium phyllo-polydisilicate	X-ray studies show the α form is isostructural with Li_2O_5[65]. $\beta Na_2Si_2O_5$ has a similar structure[66].
IX, 10	$K_2[Si_2O_5]$	2	Dipotassium phyllo-polydisilicate	Raman spectra show the structure is similar to the lithium and sodium salts[63].
Q	$KH[Si_2O_5]$	4	Potassium hydrogen phyllo-polydisilicate	X-ray studies show infinite $[Si_2O_5]^{2-}$ sheets[67,68].
	$Na_2[Si_3O_7]$	3	Disodium phyllo-polytrisilicate	X-ray studies show complex sheets[69].
	$Na_6[Si_8O_{19}]$	2.67	Hexasodium phyllo-polyoctasilicate	Unit cell dimensions obtained[70].
X	$K_4[Si_8O_{18}]$	4	Tetrapotassium phyllo-polyocta-silicate	X-ray studies show complex sheets[71].

Footnote: dae = diaminoethane.

ripped, until when R_M has reached 1, disordered, sometimes broken $[Si_2O_6]_\infty^{2-}$ chains are present.

There appear to be no structural studies on the species present in the hydrated amorphous powders prepared commercially by spray- or drum-drying silicate solutions.

3.3 **Silicate Solutions.** Careful structural studies on silicate solutions require that the solutions be specially purified and filtered, and kept in inert atmospheres. Since industrial chemicals are not prepared under such stringent conditions, they will not necessarily have the same properties as the nominally identical, laboratory-prepared solutions.

At room temperature, clear, stable, sodium and potassium silicate solutions with R_M up to 4 can be obtained, and all are moderately-to-strongly alkaline. The relationship between the pH of sodium silicate solutions and their concentrations (based on alkali content for comparison) at several ratios is shown in Figure 6.

FIGURE 6

pH OF SODIUM SILICATE SOLUTIONS AT 20°C

Numbers on lines are $SiO_2:Na_2O$ wt ratios

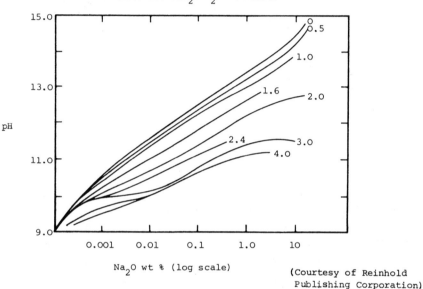

(Courtesy of Reinhold Publishing Corporation)

Potassium silicate solutions behave similarly but lithium silicate solutions do not[19].

A complex series of pH (effectively ratio R_M) and concentration dependent equilibria between many silicate anions are present in solution. Using published data on silicate equilibria, Stumm has discussed the types of species present in solution as a function of pH and concentration, as shown in Figure 7[74], and this diagram will be used as the basis for discussion.

FIGURE 7

SPECIES IN EQUILIBRIUM WITH AMORPHOUS SILICA AT 25°C

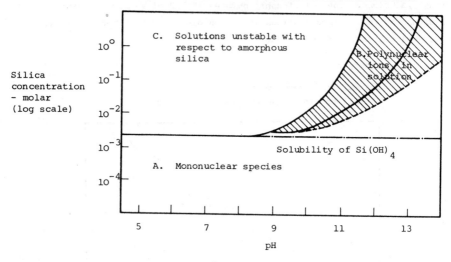

Silica concentration - molar (log scale)

There are three areas in the diagram labelled A (mononuclear species present), B (polynuclear anions in solution) and C (solutions unstable with respect to amorphous silica). More recent work has shown that the model for silicate polymerisation in solution used to construct the original diagram is not adequate (see below and Aveston[75]) and the boundary which separates areas A and B of the diagram has been amended to a position recent data suggest is more likely; this is sketched in the broken line.

(a) <u>Mononuclear species (Area A)</u>. The first two dissociation constants of orthosilicic acid have been measured using several techniques[76]. Values taken from potentiometric titration studies on dilute solutions in 0.5 molar sodium perchlorate at 25°C[77-79] are:

$$Si(OH)_4 \rightleftharpoons \left[SiO(OH)_3\right]^- + H^+; \; pK_1 = 9.46$$
$$\left[SiO(OH)_3\right]^- \rightleftharpoons \left[SiO_2(OH)_2\right]^{2-} + H^+; \; pK_2 = 12.56$$

Recent Raman spectroscopic studies[80] have provided evidence for $\left[SiO_2(OH)_2\right]^{2-}$ ions in solution. In concentrated highly alkaline solutions, spectral lines thought to be due to $\left[SiO_3(OH)\right]^{3-}$ ions were also seen. If this interpretation of the spectra is correct it would indicate that pK_3 for orthosilicic acid is approximately 15. Salts containing $\left[SiO_3(OH)\right]^{3-}$ ions can be crystallised from highly alkaline aqueous solutions[17,40].

(b) Polynuclear species (Area B). The complexity of the equilibria in silicate solutions has not permitted direct evaluation of the dissociation constants of polysilicic acids. If their behaviour follows that of poly- phosphoric acids, then polysilicic acids should have lower initial and final pK's as their size increases. The limit of this lowering can be judged from the acidity of silanol groups on silica gels, which have pK's of ~7.

Early studies on the degree of polymerisation of the species present in silicate solutions[17] (using, for example, light-scattering techniques[81]), showed that polymerisation increased with increasing R_M and concentration, and that no 'large particles', (i.e. aggregates, micelles, or colloidal material) were present even in concentrated high ratio solutions[81] (Wills[18] and Weldes[19] refer to all polynuclear ions as 'colloidal silicate'). There was some evidence for different degrees of polymerisation in corresponding sodium and potassium silicate solutions[19,82].

More modern techniques have shown the structure and distribution of silicate ions in solutions as ratios and concentrations are varied. [29]Si n.m.r. spectroscopy and trimethylsilylation have been especially useful and rapidly acidified silicate solutions have also been studied using paper chromatography[83], and by measuring their rates of reaction with molybdic acid[84]. Although several polynuclear anions have been precipitated from silicate solutions using large cations (Table 3), none of these has yet been identified in solution.

[29]Si n.m.r. spectroscopy has been used by four groups of workers to study sodium[85-87] and potassium[23] silicate solutions.

At the necessary working concentrations (> 1M) spectra are quite complex, showing that many different anions are present even when R_M is as low as 1. It is clear that as the ratio R_M of a silicate solution is increased at constant concentration, so the relative number of silicon atoms in ever more branched environments (Q^2, Q^3, Q^4) increases, indicating that larger, more complex ions are present[87]. In view of the complexity of the spectra, it has not yet been possible to assign resonances to particular anions for solutions with $R_M \gtrsim 1$, but careful high resolution studies on a 3M potassium silicate solution of $R_M = 0.75$ show that here as few as six anions, variously protonated, are present[23] (see Table 4).

TABLE 4 — SPECIES PRESENT IN 3M POTASSIUM SILICATE (R_M = 0.75)

Types of Species	Notation	Concentration, mol dm^{-3}	
		Silicon	Species
Orthosilicates	Q^0	1.45	1.45
Disilicates	Q^1Q^1	0.44	0.22
Cyclotrisilicates	$(Q^2)_3$	0.48	0.16
Silicatocyclotrisilicates	$Q^1Q^3Q^2Q^2$	0.32	0.08
Trisilicates	$Q^1Q^2Q^1$	0.21	0.07
Tetrasilicates	$Q^1Q^2Q^2Q^1$	0.12	0.03

The degree of protonation of these species cannot of course be obtained from the spectra since equilibria such as

$$\left[\text{SiO(OH)}_3\right]^{3-} + \text{H}^+ \;\rightleftharpoons\; \left[\text{SiO}_2\text{(OH)}_2\right]^{2-}$$

are rapid on the ^{29}Si n.m.r. time-scale. Thus only a weighted mean signal would be seen for these two anions, and accordingly the resonance for orthosilicates shifts significantly as the pH of the solution is changed[85], indicating that one or more such equilibria are present.

The different anions present in solution as the ratio is held constant and the concentration varied have been investigated using the method of trimethylsilylation, first described by Lentz[88]. Silicate solutions are rapidly acidified and the Si-OH groups capped by Me$_3$SiOH to give stable trimethylsilyl esters which can be separated by conventional chromatographic techniques[55,89,90]. It is thought that the rapid acidification does not affect the initial distribution of polysilicates and that the polysilicic acids thus formed are esterified before they decompose. Thus the distribution of esters reflects the distribution of anions originally present.

The relative proportions[91,92] and structures[55,90] of the esters derived from Q^0, Q^1Q^1, $Q^1Q^2Q^1$, cyclic $(Q^2)_4$ and hexa- and octa-nuclear anions recovered from a series of silicate solutions have been obtained. Many more complex esters have also been recovered, but not yet separated. These studies show that for solutions of both R_M = 1 and R_M = 3.4, low molecular weight ions condense to give ever higher polymers as the concentration increases; they are consistent with other studies in showing that anions are more polymerised in solutions of high ratios than in solutions of the same silica concentration but lower ratio. Thus, in 2.0M sodium silicate solutions (R_M = 3.4), almost all the silicon is present in species with more than eight silicon atoms, but in 0.1M solutions with $R_M = 1$ the principal species present are orthosilicates (Q^0), with some disilicates (Q^1Q^1) and small amounts of other species.

The time required for the polymerisation equilibrium to be established after the ratio or concentration of a solution is changed, or when a solid is dissolved, is not generally clear. In dilute solutions or those of low ratio (i.e. those with relatively low degrees of polymerisation), equilibration is thought to be quite rapid unless, for example by exposing the solution to atmospheric CO_2, or when titrating with acid[77], the solution is brought through, or locally taken into the insolubility region (see below). However, there is evidence that the properties of highly concentrated, high-ratio, commercial solutions depend on their method of manufacture, even after they have been stored for months[93].

(c) S̲o̲l̲u̲t̲i̲o̲n̲s̲ ̲a̲n̲d̲ ̲t̲h̲e̲ ̲i̲n̲s̲o̲l̲u̲b̲i̲l̲i̲t̲y̲ ̲r̲e̲g̲i̲o̲n̲ ̲(̲A̲r̲e̲a̲ ̲C̲)̲. Solutions can be brought into the insolubility region by acidification or simply by diluting a high ratio silicate solution, which will then change slowly with time[77,82]. Solutions in this region are thermodynamically unstable with respect to amorphous silica[74]. Acidification leads to sols and gels, which may be indefinitely stable. Their formation involves protonation of the silicate anions in solution, polymerisation, ordering and aggregation of the species so produced, and structuring of the water. These processes are further discussed in Section 5.1.

4. Uses and Applications of the Soluble Silicates

Soluble silicates have properties not shared by other alkaline salts and these, together with their relatively low cost, have resulted in their widespread use in many industries. One of the earliest uses for sodium silicate was to improve the properties of soap, and even today household detergent and soap powders comprise one of the biggest single applications. Fabric-washing powders usually contain 5-15% sodium silicate of R_W 1.6-2.4. The role of silicates in current powders has never been clearly defined, but they are said to assist in the deflocculation of soil and prevention of its redeposition, to inhibit the corrosion of a range of metals, and to help structure the spray-dried powder; they also act as a source of alkali. From the discussion on properties in solution, it follows that under the conditions used in fabric detergency, i.e. relatively dilute solutions (5 x 10^{-3}M) at pH 10.2, then the silicate tetrahedra exist in a low state of polymerisation.

In contrast to household detergents, industrial detergents contain
much greater amounts of silicate and dissolve to give solutions of much
higher pH than domestic formulations. They are usually based on anhydrous
or crystalline sodium metasilicate; sometimes caustic soda is added to
increase the alkalinity. The silicate is compounded with yet other materials
such as water-softening agents (e.g. phosphates), and detergent actives.
Both solid, liquid, and flaked crystalline products are sold to a wide
variety of industries (e.g. brewing, dairying, food-processing, metal-
processing) for uses such as keg- and bottle-washing, plant-cleaning,
conveyor lubrication and in-process metal cleaning. Furthermore, detergents
for automatic dishwashing machines are based largely on silicate, and there
is a large trade in silicated detergents for general hygiene cleaning in
industry. This wide spectrum of need is met by an equally wide range of
purpose-made materials, structured to suit the application. The technology
is strongly based on applicational know-how, i.e. the ability to choose sets
of mutually compatible materials specifically for a given application.

A relatively new and growing use for silicates is as deflocculants in
the cement industry, where silicates at very low concentrations allow more
concentrated slurries to be produced with lower viscosities than would
otherwise be the case. Thus, with less water to remove, there are
substantial savings in the energy costs of calcination[94].

A great deal of silicate is sold to the foundry industry, where it is
neutralised in situ using either CO_2 gas[95] or glycerol esters under carefully
controlled conditions to give highly aggregated silica networks, and thus
permit the rapid production of sand cores and moulds from which castings may
be produced. Partial neutralisation of sodium silicate is used to produce
siliceous hydrosols which are very effective flocculants in potable water and
effluent processes[96]. It appears that the hydrosol behaves as an anionic
polyelectrolyte. This is the opposite effect to that shown in deflocculation
applications.

Silicate liquors are widely used as adhesives. The largest single
off-take for this purpose is the paper and board industry, although recently
use of silicate has been declining against competitive materials such as
starch. Silicates are increasingly used to bond insulating materials such as
vermiculite and perlite; for this purpose potassium silicate is sometimes
preferred because, in general, it has higher softening temperatures than the
equivalent sodium silicate.

Silicate liquors can also be used to form surface coatings for a variety of purposes, and examples include the sealing of porous surfaces (such as asbestos and fibrous materials) and the formulation of paints. Although sodium silicate can be used for these applications it has a tendency to form an unsightly white bloom by interacting with atmospheric carbon dioxide. Potassium silicate, which does not show this effect, is therefore preferred. Potassium silicate is also widely used for the fabrication of welding rods, and for deposition of phosphors on 'black and white' television tubes.

A considerable proportion of the sodium silicate manufactured is used directly by the industry to produce derivatives whose manufacture, properties and uses are discussed in the following section.

5. Soluble Silicate Derivatives

There are two major groups of silicate derivatives, amorphous silicas and silica-alumina compounds.

5.1 Amorphous Silicas - Precipitates and Gels.

The precipitates include insoluble metal silicates such as calcium, aluminium and magnesium silicates; these aluminium silicates are distinguished from the silica-alumina compounds which are discussed in Section 5.2. It will be convenient to discuss the precipitates first.

Precipitates are made from concentrated, high ratio silicate solutions (R_W = 3.3), which contain many complex polysilicate anions (see Section 3.3). Negative charges are distributed in the polymeric anions which are surrounded by a solvation sheath, and any additive which reduces this sheath causes precipitation. Thus drying, or adding a water-soluble organic compound (e.g. acetone), or a soluble salt (e.g. $CaCl_2$, NaCl), will cause a non-gelatinous precipitate to separate. The charge on the polymer prevents the precipitate from compacting. In normal industrial practice, a combination of mineral acid and salt solution is used to precipitate almost all the silicate from solution.

Various techniques are used to give precipitates of different, albeit undefined, morphologies. The precipitate formed by adding a sodium salt and some mineral acid will contain trapped soluble salts, as well as silicate anions and sodium cations. Washing with a weak acid (in this context water behaves as such) - removes the salts and neutralises the sodium silicate

precipitate to give a polysilicic acid. Drying converts the acid to one
or other of the precipitated silicas.

Had a calcium salt been used instead of a sodium salt, then water
washing and drying would leave calcium bound into the silicate structure.
Although such compounds are sold as calcium silicates, they are by no means
clearly defined chemical compounds since the amount of calcium needed to bring
about precipitation is quite small. They are in fact mixtures of calcium
silicate and silica.

Aluminium silicates are prepared similarly except that aluminium
sulphate (which is conventionally used to precipitate the silicate) requires
no added mineral acid because of its own intrinsically acidic nature. The
aluminium silicates have structures similar to the calcium derivatives.
Mixed silicates (e.g. calcium aluminium silicate) may be prepared by using a
mixture of precipitating salts.

World-wide, there are about 10 manufacturers of these precipitates and
annual production is in excess of 300,000 tonnes. The precipitates are widely
used as fillers for rubber, as extenders for titania in emulsion paints, and
as carriers for liquids (high quality materials can carry up to 4.0 cc/g and
2.0 cc/g is common). Significant amounts are used in tabletting machines as
powder lubricant and also as aids to the free flow of 'sticky' powders -
notably hygroscopic materials. They form a base for many pastes and
medicaments and are also used in heat-insulating applications.

Whereas precipitates are formed under alkaline conditions, the mixing
of silicate and acid produces a liquid called a sol which rapidly gels to
give a clear glassy material (hydrogel) which is a mixture of porous silica,
water and alkali salt (commonly sodium sulphate). Okkerse[97] shows how gel
time (i.e. time for a mechanical structure to form in the sol) is related
to pH for dilute systems. Industrial production does not use such dilute
systems, since they have long gel times, and lead to weak, jelly-
like materials, but rather uses highly concentrated reagents. These are
thoroughly mixed in a time period short by comparison with gel time. There
is high heat output from the reaction and a correspondingly high temperature
rise. The sol rapidly condenses to a gel, which then contracts and forces
out some salt solution (syneresis). The mechanical strength of the gel
increases and it begins to fracture. When the gel has reached an
appropriate mechanical strength the massive lumps are reduced to roughly

centimetre-sized pieces, which are washed free of electrolyte, but left
mildly acidic (pH 4-5).

This hydrogel can then be treated in two ways. Firstly, it can be
either milled to a fine free-flowing powder containing 66% water, or dried
and roasted to form a desiccant gel, which is graded by particle size to suit
the use. Particle size ranges from 6 to 0.25 mm and specially comminuted
micron-sized particles can be made. This gel has a surface area of ~800 m^2/g,
a pore volume ~0.4 cc/g and a water vapour capacity of ~28%. Secondly, the
hydrogel can be washed to neutrality and steeped in boiling water. Drying
then gives a xerogel of much lower surface area (~ 350 m^2/g) and higher
pore volume (~1.2 cc/g). By carefully controlling drying conditions, higher
pore volume materials known as aerogels can be made. These differ from
xerogels in that their porosity diminishes upon wetting and redrying.
Supercritical drying can give aerogels with porosities of up to 4.0 cc/g;
the industrial process was pioneered by Monsanto[98]. The more common
aerogels have porosities of about 1.4-2.0 cc/g, the variation being a function
of drying procedure. There are a number of patents on making gels of higher
than normal porosity[99].

The different properties of the various gels have been discussed by
Barby[100]. It is suggested that during gelation polysilicic acids with
molecular weights of ~2000 separate from solution, and are compressed in the
high salt concentration to approximately spherical particles 15Å in diameter.
In a mildly acidic hydrogel these spheres, which are packed with six co-
ordination, are covered with a sheath of two monolayers of water, and further
water occupies the space between the spheres. It can be shown that
dehydration of such a hydrogel would lead to a material with the properties
of a desiccant xerogel. Further washing of the hydrogel to neutrality
disrupts the stabilising sheath which surrounds the spheres, and allows
particles to rearrange to ten co-ordination, giving secondary particles
approximately 80Å in diameter. After drying, these have the lower surface
area of 350 m^2/g and are non-porous to nitrogen. Varying the drying
conditions changes the co-ordination of the large secondary particles, and it
is in this way that the porosities of the materials are altered. Any
procedure which minimises pore contraction (caused by the compressive force
of the surface tension of the water in the capilliaries) permits higher pore
volume materials - aerogels - to be made.

The porosity of silica xerogels and aerogels allows them to be milled and used as small (micron) sized particles. Milling is often accomplished in a microniser, which is a fluid energy mill in the form of a flat cylinder. Air, or more often high pressure steam, enters at the circumference via a number of tangential jets and the feed is forced into an approximately circular orbit. As particles hit each other, they break and fall into a smaller orbit. The steam plus milled powder emerges from the centre of the cylinder, where the product is classified to remove oversize particles, collected and bagged. Particles can be coated in the microniser simply by adding the coating material (e.g. wax) to the feed. It is also possible to make aerogels in the machine by feeding hydrogel and using super-heated steam as the microniser fluid. This dries as it mills and gives higher porosity products.

In addition, the porosity of silica gels reduces their effective density making them among the lightest of minerals. A gel of pore volume 1.2 cc/g has a particle density of 0.6 g/cc and pores of 120Å mean diameter. To summarise, the pores can be structured to make the silica gels suitable for a variety of uses, which are exemplified below:

(i)   Desiccants.

(ii)   Abrasives and liquid-structuring agents in translucent toothpastes. Their use here stems from their similar refractive index to conventional toothpaste humectants, and their ability to give humectants an appropriate rheology.

(iii) Beer clarifiers - an appropriate porosity is chosen to adsorb the haze-forming precursors from beer and the beer is then filtered free from the gel[101]. There is a large trade in these silicas, particularly hydrogels, in countries where food legislation bans the use of soluble clarifiers such as proteolytic enzymes.

(iv)   To minimise inter-sheet adhesion in plastic sheets without affecting film clarity; and as light-scattering agents in lacquers - here they give a matt finish to the lacquered article. In both these uses, accurate sizing is essential because large particles are visible as blemishes and small particles are ineffective. The preferred size range lies between 1 and 10 microns.

Other uses for gels include acting as carriers for catalysts, particularly nickel in fat hydrogenation, and as chromatographic adsorbents. Further details can be found in two authoritative reviews[102,103]. Note that the pyrogenic silicas, widely used as thixotropes, etc., are not derived from sodium silicate but are made by the steam hydrolysis of silicon tetrachloride.

5.2 Silica-alumina Compounds. This section excludes the so-called
aluminium silicates referred to in the preceding section, and the natural
clays, which, of course, are not derived from sodium silicate. There are
two broad types of synthetic silica-alumina compounds, silica-alumina cracking
catalysts and sodium aluminosilicates. Silica-alumina cracking catalysts are
used in the petroleum industry for fluid bed cracking of residual oils. There
is an international trade in these materials, and about 10 major manufacturers
world-wide; Grace, Crosfield, Houdry, Filtrol, and Akzo in the Western
hemisphere, and Japanese, Australian and Chinese manufacturers in the East.

The manufacturing process differs from that used to prepare aluminium
silicates (5.1) in that much more aluminium salt and also acid is used to
give a gel rather than a precipitate. Sodium silicate liquor is acidified
with CO_2 and reacted with either sodium aluminate or aluminium sulphate to
form a mixed gel, with concentration and choice of reagents varied according
to the grade being made. The gel is neutralised, washed, filtered and spray-
dried to fix the structure and form microspheres for the fluidised catalytic
process. It is then rewashed to remove the pH-controlling soluble ions which
were needed to avoid degradation of the structure during the primary drying,
and redried. Catalyst quality is controlled in terms of surface area, pore
volume, particle size and also the level of molecular sieves and rare
earths - additives included in more recent times to improve gasoline yield.
The structure of the catalysts is that of a gel with order intermediate
between the amorphous precipitates and crystallographically pure chemicals.

Until the recent dramatic increase in crude oil prices, the business
had been stable, since the major outlet for cracking catalyst was simply to
replenish loss. The economic and environmental pressures now faced by the
oil industry require a higher yield of gasoline from crude oil, and if the
use of lead compounds as octane number improvers is banned, gasoline of
different quality will be required. Refiners meet these pressures by
changing their pattern of operations, including their use of catalytic
crackers. Currently more crackers are being built and there is therefore an
expanding trade.

The second type of silica alumina compounds are the sodium aluminosilicates, i.e. the zeolite molecular sieves, and their amorphous precursors
(permutites). The structures of zeolites are now well classified, and their
properties, such as cation exchange capacities, the subject of many recent
studies[104]. The structures of the amorphous precursor materials have not yet

been clearly defined. Knowledge of the ability of these materials to soften
water goes back to 1852[105].

Thus, when pressures were recently brought to remove phosphorus-based
water-softeners from household detergents for environmental reasons, attention
returned to the use of aluminosilicates for this purpose, and over the last
few years at least 50 patent applications covering this area have been
filed. Both Procter and Gamble[106] and Henkel[107] have announced their
intention to use these materials as alternatives to phosphate.

6.   Health and Safety Factors

The ordinary liquid soluble silicates are inorganic, non-flammable,
non-explosive and relatively non-toxic, and consequently are not regarded
as hazardous industrial chemicals. A commercial liquid silicate of
$R_W \simeq 3$ has an LD_{50} value for mice of about 4.5 g/kg. Since they are
alkaline, none of the commercial silicates should be allowed to come into
contact with the eyes and liquids more alkaline than $R_W = 2.4$ should
be treated with caution. Silicates with $R_W < 1$ are strong alkalies and
release about 50% of the heat of solution of caustic soda. Unlike the
crystalline silicas, amorphous silica powders do not cause silicosis. Dusts
of the soluble silicates do not accumulate in the lungs.

7.   The Future

Even though soluble silicate manufacture ranks among the oldest of
industrial chemical enterprises, a healthy future can be foreseen. The
current technology is well established and reliable; likely change is in the
area of improved furnace design. Raw materials are in plentiful local supply.
Since silicates and their derivatives have so many diverse applications,
natural decline of any one market has always been offset by new and
developing markets. There are minimal environmental or safety problems
bearing directly on the industry. The environmental and economic pressures
on the petroleum industry in turn affect the cracking catalyst business;
changes to date have however favoured silicate manufacturers.

The long-term growth trend of the industry has been consistent at about
a 4% increase in tonnage per annum, and although the business, like most
others, is subject to variations in world trading, steady expansion can be
foreseen, accelerated by low environmental risks, ready availability of raw
materials, and ever-widening applications for the derivatives.

References

1   W. Goethe, <u>Dichtung und Wahreit</u>, 1768, <u>8</u>.

2   J. von Fuchs, <u>Dinglers Polytech. J.</u>, 1825, <u>17</u>, 465.

3   J. von Fuchs, <u>Dinglers Polytech. J.</u>, 1856, <u>142</u>, 365.

4   C.F. Kuhlmann, <u>Ann. Chim.</u>, 1842, <u>41</u>, 220.

5   C.F. Kuhlmann, <u>Ann. Chim.</u>, 1848, <u>64</u>, 289.

6   C.F. Kuhlmann, <u>Compt. rend.</u>, 1855, <u>41</u>, 980.

7   W. Gossage, Brit. Pat. 762, 1854.

8   <u>Chemical Marketing Reporter</u>, 29th November 1976.

9   R.K. Iler and E.J. Tauch, <u>Trans. Am. Inst. chem. Engrs</u>, 1971, <u>37</u>, 853.

10   Compagnie de Saint-Gobain, French Pat. 1,371,968, 1964.

11   H.D. Segrove, <u>Glass Technol.</u>, 1976, <u>17</u>, 41.

12   J.F. Sugranes, Ger. Offenlegensschrift 2501850, 1975.

13   C.L. Baker and P.W. Holloway, US Pat. 3,208,822, 1965;
     Brit. Pat. 1,149,859, 1969.

14   S.L. Bean and A.W. Mouton, US Pat. 3.748,103, 1973.

15   V.E. Sokolvich, <u>Steklo Keram.</u>, 1975, <u>10</u>, 36.

16   V.W. Truesdale and C.J. Smith, <u>Analyst</u>, 1975, <u>100</u>, 797.

17   J.G. Vail, <u>Soluble Silicates</u>, Reinhold, New York 1952.

18   J.H. Wills, <u>Kirk Othmer Encyl. Chem. Technol.</u>, 1969, <u>18</u>, 134.

19   H.H. Weldes and K.R. Lange, <u>Ind. and Eng. Chem.</u>, 1969, <u>61</u>, 29.

20   Engler, <u>Seifen-Öle-Fette-Wachse</u>, 1974, <u>100</u>, 165, 207, 269, 298.

21   R.A. Edge and H.F.W. Taylor, <u>Nature</u>, 1969, <u>224</u>, 364.

22   H. Voellenkle, A. Wittmann and H. Nowotny, <u>Monatsh.</u>, 1969, <u>100</u>, 295.

23   R.K. Harris and R.H. Newman, <u>J. Chem. Soc. Faraday I</u>, submitted.

24   Yu. I. Smolin, <u>Chem. Comm.</u>, 1969, 345.

25   J.G. Vail, <u>Soluble Silicates</u>, Reinhold, New York 1952, 134.

26   C.L. Baker, L.R. Jue and J.H. Wills, <u>J. Amer. Chem. Soc.</u>, 1950, <u>72</u>, 5369.

27   L. McCulloch, <u>J. Amer. Chem. Soc.</u>, 1952, <u>74</u>, 2453.

28   R.K. Iler, <u>J. Colloid Sci.</u>, 1964, <u>19</u>, 648.

29   V.G. Il'in, N.V. Turutina, K.P. Kazakov and V.M. Bzhezovskii,
     <u>Doklady Chem.</u>, 1973, <u>209</u>, 296.

30   H.P. Eugster, <u>Science</u>, 1967, <u>157</u>, 1177.

31   G.W. Brindley, <u>Am. Miner.</u>, 1969, <u>54</u>, 1583.

32   E.K.H. Wittich, J. Voitländer and G. Lagaly, <u>Z. Naturforsch.</u>, 1975
     <u>30A</u>, 1330.

33   G. Lagaly, K. Beneke and A. Weiss, <u>Am. Miner.</u>, 1975, <u>60</u>, 642.

34   R.A. Sheppard and A.J. Gude III, <u>Am. Miner.</u>, 1970, <u>55</u>, 358.

35   G.F. Maglione and M.H. Maglione, <u>Bull. Inst. Geol., Univ. Louis Pasteur
     Strasbourg</u>, 1973, <u>25</u>, 231.

[36] I.M. Timoshenkov, Yu. P. Menshikov, L.F. Gannibal and I.V. Bussen, Zap. Vses. Mineral. O-va, 1975, 104, 317.

[37] Nomenclature in Inorganic Chemistry, Butterworths, London 1971.

[38] H. Völlenkle, A. Wittmann and H. Nowotny, Monatsh, 1968, 99, 1360.

[39] K. Kautz, G. Muller and W. Schneider, Glastech. Ber., 1970, 377.

[40] Yu. I. Smolin, Yu. F. Shepelev and I.K. Butikova, Soviet Phys. Cryst., 1973, 18, 173.

[41] H. Scheler, Z. anorg. alleg. Chem., 1968, 363, 51.

[42] K. Jost and W. Hilmer, Acta Cryst., 1966, 21, 583.

[43] P.B. Jamieson and L.S. Dent Glasser, Acta Cryst., 1967, 22B, 507; 1966, 20, 688; 1976, 32B, 705.

[44] P.P. Williams and L.S. Dent Glasser, Acta Cryst., 1971, 27B, 2269.

[45] A.G. Brekhunets, I.M. Kiselev and V.V. Mank, Russ. J. Phys. Chem., 1964, 17, 335.

[46] C. Dorémieux-Morin and E. Freund, Bull. Soc. chim. France, 1973, 418.

[47] F. Miller and C. Wilkins, Analyt. Chem., 1952, 24, 1253.

[48] F.A. Miller, G.L. Carter, F.F. Bentley and W.H. Jones, Spectrochim. Acta, 1960, 16, 135.

[49] Ya. I. Ryskin, G.P. Stavitskaya and N.A. Mitropolskii, Izv. Akad. nauk., 1964, 416. See also "Vibrational Spectra and Structure of Silicates", A.N. Lazarev, Consultants Bureau, New York, 1972.

[50] E. Thilo and W. Miedreich, Z. anorg. alleg. Chem., 1951, 267, 76.

[51] W. Schartau and R. Hoppe, Naturwiss, 1973, 60, 256.

[52] H. Voellenkle, A. Wittmann and H. Nowotny, Monatsh, 1970, 101, 684.

[53] W. Hilmer, Acta Cryst., 1964, 17, 1063; 1965, 18, 574.

[54] W.P. Griffith, J. Chem. Soc. (A), 1969, 1372.

[55] D. Hoebbel, G. Garzo, G. Engelhardt, H. Jancke, P. Franke and W. Wieker, Z. anorg. alleg. Chem., 1976, 424, 115

[56] D. Hoebbel and W. Wieker, Z. anorg. alleg. Chem., 1974, 405, 267.

[57] Yu. I. Smolin, Yu. F. Shepelev and J.K. Butikova, Kristallografiya, 1972, 17, 15.

[58] D. Hoebbel and W. Wieker, Z. anorg. alleg. Chem., 1971, 384, 43.

[59] Yu. I. Smolin, Yu. F. Shepelev, R. Pomes, D. Hoebbel and W. Wieker, Kristallografiya, 1975, 20, 917.

[60] D. Hoebbel, W. Wieker, P. Franke and A. Otto, Z. anorg. alleg. Chem., 1975, 418, 35.

[61] W. Hilmer, Dokl. Akad. nauk. SSSR, 1968, 178, 1309.

[62] W.S. McDonald and D.W.J. Cruickshank, Acta Cryst., 1967, 22, 37.

[63] S.A. Brawer and W.B. White, J. Chem. Phys., 1976, 63, 2421.

[64] F. Liebau, Acta Cryst., 1961, 14, 389.

[65] A.K. Pant and D.W.J. Cruickshank, Acta Cryst., 1968, 24B, 13.

66 A.K. Pant, Acta Cryst., 1968, 24B, 1077.

67 H. Schweinsberg and F. Liebau, Z. anorg. alleg. Chem., 1972, 387, 241.

68 M.T. LeBihan, A. Kalt and R. Wey, Bull. Soc. fr. Miner. Cristallogr., 1971, 94, 15.

69 P.B. Jamieson, Nature, 1967, 214, 794.

70 J. Williamson and F.P. Glasser, Science, 1965, 148, 1559; Phys. and Chem. Glasses, 1966, 7, 127.

71 H. Schweinsberg and F. Liebau, Acta Cryst., 1974, 30B, 2206.

72 W. Eitel, Silicate Science Vol. VII, Academic Press, New York, 1976.

73 S. Brawer, Phys. Rev. B, 1975, 11, 3173.

74 W. Stumm, H. Hüper and R.I. Champlin, Environmental Science and Technology, 1967, 1, 221.

75 J. Aveston, J. Chem. Soc., 1965, 4444.

76 Stability Constants, Chemical Society Special Publication No. 17, 1964.

77 G. Lagerström, Acta Chem. Scand., 1959, 13, 722.

78 N. Ingri, Acta Chem. Scand., 1959, 13, 753.

79 N. Ingri and H. Bilinski, Acta Chem. Scand., 1967, 21, 2503.

80 E. Freund, Bull. Soc. chim. France, 1973, 2238, 2244.

81 R.V. Nauman and P. Debye, J. Phys. and Colloid Chem., 1951, 55, 1; 1961, 65, 5.

82 A.P. Brady, A.G. Brown and H. Huff, J. Colloid Sci., 1953, 8, 252.

83 W. Wieker and D. Hoebbel, Z. anorg. alleg. Chem., 1969, 366, 134.

84 E. Thilo, W. Wieker and H. Strade, Z. anorg. alleg. Chem., 1965, 340, 261.

85 H.C. Marsmann, Z. Naturforsch., 1974, 29B, 495.

86 R.O. Gould, B.M. Lowe and N.A. MacGilp, Chem. Comm., 1974, 720.

87 G. Engelhardt, D. Zeigan, H. Jancke, D. Hoebbel and W. Wieker, Z. anorg. alleg. Chem., 1975, 418, 17.

88 C.W. Lentz, Inorg. Chem., 1964, 3, 574.

89 D. Hoebbel and W. Wieker, Z. anorg. alleg. Chem., 1974, 405, 163.

90 F.F.H. Wu, J. Götz, W.D. Jamieson and C.R. Masson, J. Chromatog., 1970, 48, 515.

91 L.S. Dent Glasser, E.E. Lachowski and G.G. Cameron, J. Appl. Chem. Biotech., 1977, 27, 39.

92 L.S. Dent Glasser and S.K. Sharma, Br. Polymer J., 1974, 6, 283.

93 A.J. Walker and N. Whitehead, J. Appl. Chem., 1966, 16, 230.

94 T. Griffiths, The deflocculation of cement slurries, In press.

95 K.E.L. Nicholas, The CO_2 silicate process in foundries. (BICRA 1972).

96 F. Smith, Brit. Pat. 1,399,598, 1975.

97 C.K. Okkerse, Physical and Chemical Aspects of Adsorbents and Catalysts, B.G. Linsen ed., Academic Press London, 1970, 213.

[98]   Monsanto, Brit. Pat. 649 896, 1939.

[99]   W.R. Grace and Co., US Pat. 3,243,262, 1966 and 3,526,603, 1970;
       Brit. Pat. 1,077,908, 1967, 1,219,877, 1971 and 1,284,085, 1972.

[100] D. Barby, Characterisation of Powder Surfaces, eds. G.D. Parfitt and
       K.S.W. Sing, Academic Press London, 1976, 353.

[101] D. Barby and J.P. Quinn, Brit. Pat. 1,215,928, 1967.

[102] N. Burak, Chem. Processing, Supplement August 1967 "Powder
       Technology", p. 510.

[103] R.K. Iler, Surface and Colloid Science, ed. E. Matijevic,
       Wiley New York, 1973, 6, 1.

[104] D.W. Breck, Zeolite Molecular Sieves, Wiley, New York, 1974.

[105] T.J. Way, Jl. R. agric. Soc., 1852, 13, 123.

[106] Chemical Week, Dec 1st, 1976.

[107] Chemical Week, Dec 8th, 1976.

Titanium Dioxide Pigments

By R.S. DARBY and J. LEIGHTON

Laporte Industries Ltd.

Introduction

Titanium dioxide is the most widely used white pigment, principally in paint, paper, plastic and rubber applications. Its predominance is due to the high refractive index, lack of colour and chemical inertness. High refractive index in the visual portion of the spectrum results in strong scattering of visible radiation of particles in the correct size range. This provides the opportunity for producing films with high opacity. Among those substances that are available in quantities sufficient to meet the requirements of industry, no other material has the high refractive index and other desirable pigmentary properties possessed by titanium dioxide[1].

The following table shows the approximate average refractive index values for a group of representative pigments and substances of interest to the user industries:

Air	1.00
Water	1.33
Linseed Oil	1.48
Barium Sulphate	1.64 - 1.65
Calcium Carbonate	1.53 - 1.68
Silica	1.54 - 1.56
Zinc Oxide	2.00
Zinc Sulphide	2.36 - 2.38
Diamond	2.42
Titanium Dioxide	
Anatase	2.49 - 2.55
Rutile	2.61 - 2.90

The manufacture of titanium dioxide pigments results in the production of either the anatase or rutile crystal structure, depending on the processing. Rutile pigments have the greater hiding power and also an improved durability - or less chalking - in paint medium. Compared with anatase pigments, they have the disadvantages of costing rather more to produce and a slight inferiority in colour.

Historical

The element titanium was discovered by the Reverend William Gregor in Cornwall, England in the year 1790. Although the fundamental chemical reactions on which most of the present day titanium pigment is based were known by the pioneer investigators before 1800, it was not until 1918 that these pigments became commercially available. In fact commercial production began almost simultaneously in Norway and the U.S.A.[2]

Initially composite pigments were produced, based on either barium or calcium sulphates. These have almost disappeared and have been replaced by titanium dioxide alone. The original basic process for the manufacture of titanium dioxide itself was and is known as the Sulphate Process. This is still employed by the majority of manufacturing facilities. However by 1959, an alternative known as the Chloride Process had been developed by Dupont in the U.S.A. The latter now represents a significant proportion of the two processes employed, particularly in the U.S.A.

The rapid growth of titanium dioxide production is shown by the following:

Year	World Production (Tons per annum)
1925	5,000
1937	100,000
1975	2,000,000

Whilst the original process technology was initially U.S.A. based, the growth in production has been matched by a proliferation of manufacturing Companies brought about by licensing and in particular the applications of the Anti-Trust Laws in the U.S.A. in 1945.

The following major Companies exist today:

Country	Name of Company	Countries in which Companies have subsidiaries
U.S.A.	American Cyanamid	Mexico
	Du Pont	
	Kerr McGee	
	Glidden Co.	
	National Lead	Belgium, Canada, W. Germany, Norway
	New Jersey Zinc	
Great Britain	British Titan Products	Australia, Canada, France, South Africa, Spain
	Laporte Industries	Australia

Country	Name of Company	Countries in which Companies have subsidiaries
France	Thann and Mulhouse	
Finland	Vuorikemia	
Holland	Titaandoxydefabriek	
Italy	Montedison	
Germany	Bayer	Brazil, Belgium
	Pigment Chemie	
Japan	Fuji Titanium Ind.	
	Furuhawa Mining Co.	
	Ishihara Sangyo	
	Sakai Chemical Ind. Co.	
	Teikoku Kako	
	Titanium Industry Co.	

Up to 1939, titanium dioxide pigments were commercially manufactured in the anatase crystal form. Rutile pigments were then developed, which have improved properties in most applications. The majority of pigments produced today are rutile. Since 1946 another development occurred in that the base pigment was processed further to enhance its pigmentary properties.

Production Statistics

The likely production of titanium dioxide in 1977 and a forecast for 1980, is shown by process route, as follows[3] :-

	1977 ('000 ton)		1980 ('000 ton)	
	Sulphate	Chloride	Sulphate	Chloride
N. America	370	540	380	750
Europe	900	85	1050	135
Japan & Asia	210	15	250	50
Oceania	50	–	60	–
Africa	25	–	30	–
S. America	22	–	40	–
	1577	640	1810	935
Grand Total	2217		2745	

The selling price of premium quality titanium dioxide pigments in 1976 was £400/500 per ton. The manufacturing cost and hence the selling price is sensitive to inflationary trends. The increase in selling prices over recent years is shown in Figure 1.

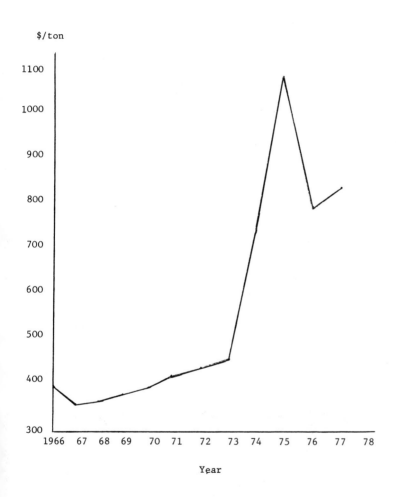

Figure 1 – Average Selling Price for
European Surface Treated Rutile Pigment

Raw Material Sources

Titanium is ninth in abundance of the elements making up the lithosphere. Further there are ample supplies of mineral deposits that can economically yield titanium bearing ores. There are two main types of mineral exploited, ilmenite and rutile.

It is most important to distinguish between the raw material requirements of the Sulphate and the Chloride process. The Sulphate process relies on digesting the ore with sulphuric acid as completely as possible. However, sulphuric acid does not attack rutile or high grade ores.

This restriction does no apply to the Chloride process. Traditionally natural rutile has been used. The lack of adequate rutile reserves has however, led to the use of raw materials containing smaller percentages of titanium dioxide such as mixtures of ilmenite with rutile or beneficiated ilmenite.

Certain other restrictions should also be noted. Ilmenites with a high chromium content cannot be used in the Sulphate process, because of the deterioration caused in the colour of the pigment produced. This is due to residual chromium in the pigment. Certain compounds, such as those containing magnesium, manganese and the alkaline earths, can create operating difficulties in the chlorination stage of the Chloride process.

For either process, less chlorine or sulphuric acid is required and less by-products or waste products are produced, as the percentage titanium dioxide of the feed material increases. There is therefore, an economic optimum depending on the relative costs of these three factors.

The following table indicates the main ilmenite deposits which are being worked today:-

Country	Reserves of Titanium Bearing Ore (Million Tonnes)	TiO_2 Content of deposit (%)	TiO_2 Content of Mineral after treatment (%)
Canada	200 - 220	35	37
U.S.A.			
New York	100	17	43
Virginia	5	8.5	44
Florida	180	0.5 to 3.0	63
Norway	250	17	43
Finland	50	13.5	44
India	100	20 - 40	59
Sth. Africa	20	5	49
Australia	15	30	50 - 60

The most significant development for further up-grading ilmenite as a feed material for the Sulphate process is the slag process. The use of slag reduces both the quantities of sulphuric acid required and the quantity of waste produced. It is an electrothermal process originally based on Canadian ilmenite, which produces a slag containing about 70% titanium dioxide. A proportion of the iron oxide content is smelted out and sold as pig iron. Slag is an exception of a high percentage titanium dioxide material, which is attacked by sulphuric acid, because the titanium dioxide is in the form of a mixed magnesium iron dititanate and not rutile.

Unlike ilmenite the known reserves of rutile (titanium dioxide content about 94 to 96%) are very small. There has therefore been an incentive to develop synthetic rutiles or beneficiated ilmenites for the Chloride process, as a possible alternative to the total chlorination of ilmenite, in order to minimise chlorine usage. These processes can be divided into leaching and chlorination routes.

The leaching processes involve the oxidation and reduction of ilmenite followed by leaching out either iron oxides with hydrochloric acid or sulphuric acid, or metallic iron with aerated water. The reactions can be represented as follows:-

$$2 \ FeTiO_3 + \tfrac{1}{2} \ O_2 = Fe_2TiO_5 + TiO_2$$
$$Fe_2TiO_5 + TiO_2 + \tfrac{1}{2}C = 2 \ FeO + 2 \ TiO_2 + \tfrac{1}{2} \ CO_2$$
$$FeO + 2 \ HCl = H_2O + FeCl_2$$
$$2 \ FeCl_2 + 2H_2O + \tfrac{1}{2} \ O_2 = Fe_2O_3 + 4 \ HCl$$

OR

$$Fe_2TiO_5 + TiO_2 + 2C = 2 \ Fe + 2 \ TiO_2 + CO_2 + CO$$
$$2 \ Fe + 3 \ H_2O + \tfrac{1}{2} \ O_2 = 2 \ Fe(OH)_3$$

The chlorination process consists of selective chlorination of the iron content, followed by regeneration of chlorine from the iron chloride produced.

$$FeTiO_3 + Cl_2 + \tfrac{1}{2} \ C = FeCl_2 + TiO_2 + \tfrac{1}{2} \ CO_2$$
$$4 \ FeCl_2 + CO_2 = 4 \ Cl_2 + 2Fe_2O_3$$

In all cases, products of 92% titanium dioxide or above are produced.

The present supply of the different raw material sources to the titanium dioxide industry is shown in the following table:-

Mineral	%
Rutile (Australian)	9.5
Other rutiles	1.5
Synthetic rutiles (beneficiated ilmenite)	1.5
Slags	25
Australian ilmenite	16.5
U.S.A. ilmenite	21.5
Norwegian ilmenite	15.0
Other ilmenites	9.5
	100.0

Process Technology

The Chloride and Sulphate processes represent alternative means of extracting essentially pure titanium dioxide from the titanium bearing ores, slags or beneficiates and at the same time providing a base pigment within a very closely defined particle size range. This base pigment is normally further processed to provide products tailored to their end-use.

Titanium dioxide pigments are not marketed according to a chemical specification but have to reach certain physical characteristics such as hiding power, colour and stability, which confirms durability on the paint film. These characteristics can be affected by the presence of trace quantities of metallic oxides, as well as by the particle size.

The Sulphate Process

The typical analyses of three sources of raw materials commonly used for the Sulphate process are as follows:

	Norwegian Ilmenite (%)	Australian Ilmenite (%)	Canadian Slag (%)
TiO_2	44.6	54.2	60.1
Ti_2O_3	–	–	8.3
Metallic Fe	–	–	0.2
FeO	34.6	23.3	12.6
Fe_2O_3	11.8	17.3	Nil
Cr_2O_3	0.08	0.04	0.2
V_2O_5	0.2	0.1	0.5
SiO_2	5.0	1.0	4.6
Al_2O_3	0.7	0.7	4.0
P_2O_5	0.04	0.08	0.03
MgO	3.7	0.2	5.0
MnO	0.3	2.00	0.2

Apart from the impurity levels, the ratio of ferrous to ferric in an ilmenite is important. The efficiency of extraction by sulphuric acid tends to improve with an increase in this ratio and the quantity of reductant needed is less.

In the Sulphate process, ground ilmenite or slag is digested with concentrated sulphuric acid at an elevated temperature. The resultant sulphates of titanium and iron are then leached from the reactant mass with water or dilute acid, and any ferric salts present are reduced to ferrous by treatment with iron scrap.

The solutions are clarified and cooled. Ferrous sulphate crystallises out and is separated from the mother liquor by centrifuging.

The mother liquor is further clarified by filtration after addition of filter aid and then concentrated by vacuum evaporation. Seed crystals or other nuclei are added and the concentrated liquor is treated with steam to hydrolyse the titanyl sulphate present. This precipitates the acidic hydrated titanium. The precipitate is collected by filtration, washed several times and calcined to yield titanium dioxide. The calcined product is ground. Until 1946, the process finished here, the product being packed off as either rutile or anatase pigment. Now the majority of pigment produced undergoes further treatment, which will be discussed later.

A standard process flow diagram for the Sulphate process to base pigment is shown in Figure 2.

Each stage of the base process is specified so as to provide a liquor or solid product within specified chemical or physical limits. Certain of the stages are of special interest.

The digestion of dried and ground ilmenite or slag can be carried out with 96%, 98% sulphuric acid or oleum. The quantity of sulphuric acid is calculated by determining the equivalents of each of the elements in the ore together with an approximately 60% excess based on the titanium dioxide content alone. The excess of free acid is a critical factor in the eventual liquor supplied to the hydrolysis stage for the production of optimum size pigment particles. The presence of this excess acid also permits a digestion efficiency, based on the soluble titanium produced, of 92 to 96%.

The reaction itself can be initiated by either steam or water and is strongly exothermic - the resultant mixture can reach temperatures up to 200°C.

Figure 2

Sulphate Process Flow Diagram

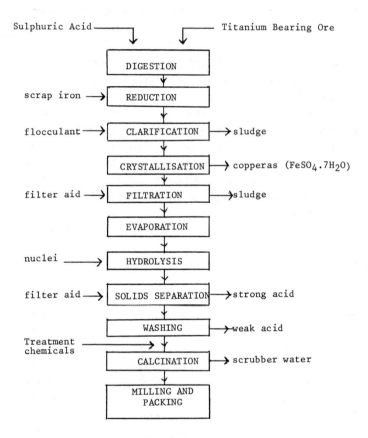

TiO$_2$ Pigment - Anatase or Rutile

Under certain conditions and if not controlled carefully, it can become violent. During the reaction, the slurry becomes thick and viscous, then pasty and finally sets up to a comparatively dry, porous mass, called the sulphate cake.

The reaction proceeds stoichiometrically according to the equation.

$$TiO_2 + H_2SO_4 = TiOSO_4 + H_2O$$

However, the ionic and physical properties of the resultant liquor are not those which would be predicted from a solution of iron and titanium sulphates. In particular, the conductivity and viscosity levels are much higher. It has been inferred that titanium and iron sulphates, interact in a sulphuric acid medium to form complexes. Because they are extremely dissociated, these complexes do not affect the subsequent hydrolysis stage and hence do not affect pigmentary properties.

The process stages immediately following digestion have three purposes. The fraction of the undigested ilmenite, which is present as insolubles in the liquor, is removed from the system. A proportion of the soluble ferrous sulphate is removed after crystallisation - any ferric sulphate present having been previously reduced to the ferrous state with scrap iron.

Finally the concentrations of titanium, iron and free acid are raised to those required for the hydrolysis reaction. These concentrations affect the particle size eventually produced and also the ease with which the remaining iron and other impurities can be removed in the later washing stages.

Rigid control of the thermal hydrolysis of this liquor with the formation of colloidal hydrous oxide and of the conditions during its subsequent coagulation, is of primary importance in the manufacture of titanium dioxide of highest pigmentary properties.

There are variations of the basic hydrolysis process. Nuclei have to be included. They may be formed in situ, or can be prepared separately. In the latter case the seed may be designed to orientate the crystals towards either potential anatase or rutile formation. Either live steam or indirect heating can be employed to bring the liquor to the boil, at which it is held for some hours under specified conditions. At the end of this period over 90% of the titanium dioxide in the liquor has been hydrolysed.

The acidic hydrated titania is precipitated in a liquor containing large quantities of free sulphuric acid, ferrous sulphate and trace-metal sulphates.

These are essentially completely removed in a series of washing stages, such that a typical analysis of the resultant titanium dioxide would contain

$$Fe_2O_3 \quad : \quad < \quad 30 \ ppm$$
$$Cr_2O_3 \quad : \quad < \quad 3 \ ppm$$

The sulphate process leads to the production of base pigment, based on either the anatase or rutile crystal lattice. Originally only anatase was produced, but this has been largely superseded by rutile, which has very significant advantages in durability (or chalking) and hiding power.

For anatase pigments it is necessary to add rutilisation inhibitors such as phosphates. For rutile pigments it may be necessary to add a further seed and small quantities of zinc or aluminium salts to aid rutile formation and other desirable pigmentary quantities. In both cases, alkali salts can be added to make the resultant pigment from the calcination stage more readily grindable.

The calcination stage itself is carried out in long (usually over 50 m) slowly rotating kilns. The fuels used, normally natural gas or distillate oil, must be essentially free from vanadium, to avoid possible discolouration through pick-up of the vanadium. Three zones can be identified on the kiln - removal of water - removal of residual sulphuric acid - and growth of the crystals to the correct particle size. The titanium dioxide can take up to 18 hours to pass through the kilns (Figure 3). The hot end temperature, which is in the region of 800-900°C, has to be controlled to ± 5°C. The principle of calcination control is that a given temperature profile and maximum temperature is required to develop all the required pigmentary properties. Exceeding this temperature will lead to discolouration and sintering.

The coarse product from the calciner is ground in heavy roller mills so that no more than 0.01% is retained on a 200 mesh.

The overall efficiency of the process, comparing the quantity of titanium dioxide produced with the quantity present in the ilmenite or slag is usually 80 to 85%. Further, for ilmenites (but less with slag) between 4.0 and 5.0 tons of sulphuric acid are required per ton of pigment. This acid leaves the system as crystalline ferrous sulphate (copperas), soluble ferrous sulphate and other metallic sulphates and free sulphuric acid.

The Chloride Process

The Chloride process for the manufacture of titanium dioxide consists of production of titanium tetrachloride from the raw material, purification of this titanium tetrachloride, and subsequent oxidation in oxygen or oxygen enriched air to titanium dioxide.

The chlorine released is recovered and re-issued to chlorination.

Preparation of titanium tetrachloride is carried out by chlorination of the raw material in a reducing atmosphere at a temperature of approx. 950°C.

$$2 \ TiO_2 + 3C + 4Cl_2 = 2 \ TiCl_4 + CO_2 + 2CO$$

Combustion of titanium tetrachloride in oxygen or oxygen enriched air requires the mixture to be heated to a temperature of 900-1000°C before any substantial reaction will occur, but once reaction has been initiated the temperature will rapidly rise to as high as 1400-1500°C.

$$TiCl_4 + O_2 + N_2 = TiO_2 + 2Cl_2 + N_2$$

There are potentially three naturally occurring raw materials suitable for the chloride process:-

> Mineral rutile
> Ilmenite
> Luncoxene

Mineral rutile contains 94-96% titanium dioxide. Ilmenite contains 45-60% titanium dioxide, while luncoxene, which is naturally weathered ilmenite, contains 65-80% titanium dioxide. Typical analyses of Australian rutile and ilmenite are given below:-

	Ilmenite (%)	Rutile (%)
TiO_2	54.0	95.2
Fe_2O_3	17.7	0.65
FeO	23.3	-
Cr_2O_3	0.1	0.25
V_2O_5	0.42	0.80
Al_2O_3	0.35	0.55
Nb_2O_5	0.24	0.36
ZrO_2	0.42	0.76
SiO_2	1.0	1.0
MnO	1.4	-
MgO	0.35	-

Mineral rutile is the most attractive raw material as it requires only 0.77 ton chlorine per ton titanium tetrachloride produced, of which 0.73 ton is recovered in oxidation, ilmenite requires 1.1 ton chlorine per ton titanium tetrachloride and only 0.73 ton is recovered. A plant for producing titanium tetrachloride from ilmenite must therefore be larger than one using rutile and must handle and dispose of larger quantities of impurities, mainly iron chlorides.

There are also more technical problems involved in the chlorination of ilmenite but it can be, and is, used by some producers.

As stated previously, various processes are known, and are in commercial operation, for up-grading ilmenite to synthetic rutile and this can be used as a raw material.

The choice of raw material is ultimately an economic one involving capital cost of plant, cost of raw material, rutile, ilmenite or beneficiate, and the cost of other raw materials, coke and chlorine. If the chlorine could readily be recovered from the iron chlorides, economics could considerably enhance the use of ilmenite. However, although various processes have been claimed for recovering chlorine from iron chloride there does not appear to be any economic process in operation at the present time.

A standard process flow diagram for the Chloride process to base pigment is shown in Figure 4, and a view of a typical plant in Figure 5.

Reaction of the raw material with chlorine in the presence of coke takes place at approximately 950°C, the temperature varying slightly with the reactivity of the raw material. In earlier processes reaction was carried out in a fixed bed in vertical retorts but these have been superseded by fluid bed reactors which consist of vertical cylindrical carbon steel vessels lined with blast furnace type refractory.

The gas leaving the chlorinator contains titanium tetrachloride vapour, carbon monoxide and dioxide and other impurity chlorides. Only under abnormal conditions is there any unchanged chlorine. In some processes these gases are partially cooled and a large portion of the impurities removed as dry solids. The vapours are then condensed in water-cooled and refrigerated condensers.

The crude titanium tetrachloride produced contains impurity chlorides, either in solution or in suspension, which must be removed before oxidation or the colour of the titanium dioxide produced will be seriously affected.

Most of the impurities can be removed directly by distillation and fractionation, the serious exception being vanadium chloride, which has a boiling point too near that of titanium tetrachloride. It is therefore necessary to reduce it to a lower valency, and this is carried out either by treating with gaseous hydrogen sulphide or boiling with oil, preferably unsaturated prior to distillation and fractionation.

In order to treat titanium tetrachloride vapour with oxygen, or oxygen enriched air, to a satisfactory level of efficiency it is necessary to preheat the reacting gases to approximately 1000°C.

Photograph of Calciner Discharge in the
Sulphate Process.

FIGURE 3.

Figure 4

Chloride Process Flow Diagram

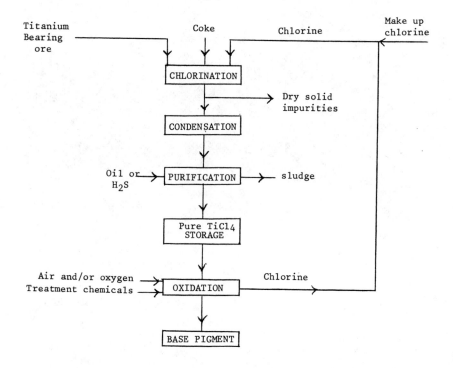

Various methods of direct heating have been used, all with their own problems of materials of construction. Other processes have been developed in which the reactants are heated by passing through a supplementary flame such as carbon monoxide-oxygen.

To produce good pigment, particle size and size distribution must be carefully controlled. Study of the reaction shows it to be two stage, consisting of nucleation followed by growth. To promote nucleation it is usual to add various agents to the reactants, and other agents are added to promote production of the rutile crystal structure. Rapid mixing of the two reactants is important so that nucleation occurs under homogeneous conditions, and various devices are employed to bring this about. These include impinging jets, concentric burners and fluid beds, each posing their own particular problems. Hot gases leaving the chlorinator are cooled and recycled to the chlorinator after removal of the solid titanium dioxide. In some cases a chlorine recovery unit is interposed between oxidation and chlorination.

The base pigment freed of chlorine is passed forward for further treatment.

Base Pigment Improvement

The titanium dioxide process used to end with the production of base pigment. Over the last twenty-five years, further processes have been developed to enhance the properties of the base pigment by further treatment. These are commonly known as Surface Treatment or Coating processes.

The complexity of these processes may be judged by the fact that approximately one third of the total capital cost of a manufacturing plant for titanium dioxide pigments is required for the Surface Treatment equipment.

A number of important properties are given to the pigment by this additional processing, e.g. the dispersion of the pigment into paint medium is vastly improved, thereby saving the customer time and equipment. The opacity or hiding power is improved. The durability of the paint in which the pigment is used is improved. The actual process can be tailored according to the user's application, but broadly follows similar principles.

The base pigment, from either the Chloride or Sulphate process, is slurried in water, dispersed and intensively ground. The product is classified by hydrocyclones or centrifuges and the oversize recycled. The product should conform to a particle size distribution which provides optimum optical properties, with a complete absence of oversize material, normally known as nibs.

Photograph of Chlorination in the Chloride Process.

FIGURE 5.

The individual particles are then coated by selective precipitation of small quantities of colourless, inert oxides e.g. silicon dioxide, aluminium oxide and titanium dioxide. The quantities and mode of precipitation are varied according to the final properties required.

The resultant slurry is washed over vacuum filters, to remove the soluble salts produced during the coating procedure (Figure 6). The pulp is dewatered and dried either through a tunnel or spray drier.

The dried product is fed to fluid energy mills, which generally use high pressure steam. The intensity of milling is varied according to the properties required.

Process control of the surface treatment plant requires accurate determination of particles in the 0-10 µm range. Over the years sophisticated direct and indirect means of measurement have been developed for this purpose.

Within the process itself, the wet fineness of the pigment used to be measured directly by a Hegmann gauge. The necessity to improve pigmentary properties further had led to the use of more critical techniques such as the Coulter Counter and Disc Centrifuge. The final pigment is checked to establish the state of aggregation, rate of dispersibility in the user's medium, the particle size distribution (either directly or through performance tests) the hiding power and the colour.

A process flow diagram for the Surface Treatment process is shown in Figure 7.

Waste Disposal and Environmental Issues

The very large quantities of waste products discharged from the processes for manufacturing titanium dioxide pigments have led to the industry being singled out for special attention by Government Environmental Agencies over the last ten years.

The main products from the Sulphate process are copperas and weak solutions of waste sulphuric acid containing ferrous sulphate and other sulphates and some solubles. The quantities involved can be as high as follows, based on a plant producing 50,000 tons per annum of pigment.

Waste Product	Tons per annum
Copperas (as $FeSO_4,7H_2O$)	140,000
Strong acid (as 100% H_2SO_4)	75,000
Weak acid (as 100% H_2SO_4)	40,000
Fe $SO_4,7H_2O$ (soluble)	83,000

Photograph of Washing Filter in Surface
Treatment Process.

FIGURE 6.

Figure 7

Surface Treatment Process

Flow Diagram

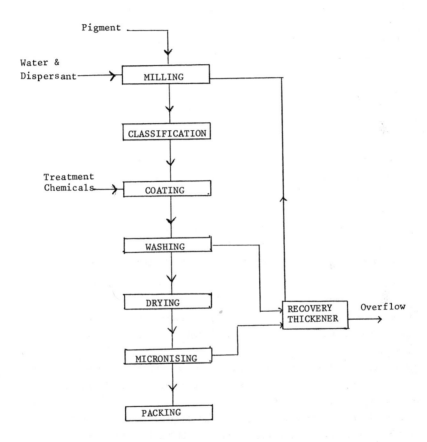

The quantities of waste product are strictly related to the type of raw material used. For example, slag produces less wastes than ilmenite. Theoretically, it is possible to recover the sulphur values from copperas or the waste acid or alternatively convert them into saleable products. Unfortunately these processes are not economic.

The degree of difficulty and cost involved in recovering sulphur values are due to the following:-

a)   The strong acid rejected from the process is between 10-20% strength. For re-use it must be concentrated to 96% strength. Since vast quantities of water have to be evaporated, these processes are highly energy intensive.

b)   During the evaporation step, metallic salts precipitate in the form of slurries which are difficult to separate from the liquid. Certain salts remain in solution even at high acid strength, and recycling of such salts leads to deterioration of colour of the pigment.

c)   High temperature roasting of the various salts is also energy intensive, and both this step and the acid evaporation lead to high quantity gaseous emissions of water vapour containing residual sulphur dioxide.

d)   Process economies have been plagued with significant variations in fresh sulphur prices. Since 1973 energy costs have quadrupled. The ratio $\dfrac{\text{energy cost}}{\text{sulphur value}}$ has fluctuated widely and recently in a worsening direction. The situation is further complicated because in the production of sulphuric acid from sulphur, a very significant quantity of energy is released which can be easily utilised in the associated titanium dioxide plant.

Many paper studies have been produced on alternative processes. Mention will be made here only of the limited number which have achieved some commercial application by 1976.

a)   Neutralisation with limestone/lime and dumping of neutralised products - this requires large scale availability of limestone and disposal areas for the neutralised solids. It is operated by one small Company.

b)   Neutralisation with limestone to produce saleable gypsum - this requires availability of limestone and suitable market. It is operated on a small scale in the U.S.A. and in Japan.

c)   Neutralisation with metallic iron in the production of
      beneficiated ilmenite. This is operated by one Japanese
      Company, but leads to significantly higher discharge of
      iron salts.

d)   Limited quantities of copperas are used for water treatment,
      or can be converted into soil conditioners.

e)   Production of iron oxide pigments from copperas - this is
      operated in Germany.

f)   Neutralisation with other alkalies to produce saleable
      by-products (ammonium sulphate, sodium sulphate) - this
      requires close proximity to markets, because of relatively
      low value of these chemicals. It is operated by one Japanese
      Company.

The Chloride process, employing rutile as the feed-stock, produces
a trivial quantity of waste products. However, the lack of availability of
rutile means that alternatives have to be found without creating waste
disposal problems. Two approaches can be broadly identified. The first
approach is to use synthetic rutile (beneficiated ilmenite) which has been
described in the section dealing with Raw Materials.[*]

A number of commercial plants - using this type of process - are
now in operation in Taiwan, Australia, Japan, U.S.A., India and Malaysia.
The other approach is uo use leaner ores such as mixtures containing
ilmenite, with the consequent discharge of increasing quantities of iron
chlorides. Ferric chloride has some application for water treatment or
alternatively it is technically possible to convert these chlorides into
hydrochloric acid or chlorine (for re-cycle) and iron oxides.

Most pigment Companies have sited their plants so that they can
discharge their waste products by pipe-line into fast-flowing water or
barge the wastes out to sea for disposal. Monitoring of these areas, over
many years, has shown little impact on the ecology, providing that the
disposal site has been selected wisely.

The Future of the Industry

It is a truism to say that the industry has never stopped developing,
and that is particularly true of today. There are three important areas:

1)   New pigments are being continually produced to meet market
      requirements. An example is the trend from oil-based to water-
      based paints. Primarily, these are produced through changes in
      the Surface Treatment process procedures.

[*] See also N.C. Cabeldu, J.H. Moss and A. Wright, in 'High Temperature Chemistry
of Inorganic and Ceramic Materials', The Chemical Society, (Special Publication
No. 30), London, 1977.

2)   Both the Chloride and Sulphate processes are energy intensive.
     The energy requirement for the latter is about 45 million BTU's
     per ton. Clearly there is now a great incentive to find means
     of improving the energy efficiency of the various stages of the
     process.

3)   However, the most important area of development arises directly
     from the fact that more stringent waste disposal regulations
     are likely to be imposed in the future. The options open to
     the industry are capital intensive. The capital cost of an
     economically sized pigment plant of say 50,000 tons per annum
     is now approaching £50 million. The capital costs of ilmenite
     beneficiation plants or effluent treatment facilities must be
     measured in terms of tens of millions of pounds.
     Further, the traditional cost of rutile compared to ilmenite,
     with respect to its titanium dioxide content is extremely high.
     Certain tentative conclusions may be reached:

a)   The Sulphate process will tend to use richer titanium dioxide feedstock
     to minimise ferrous sulphate wastes - such as slag - unless a
     technical breakthrough is found for economically dealing with
     the wastes.

b)   The Chloride process will tend to use less rich titanium dioxide
     feedstocks, particularly if ways are found to treat and recycle
     ferrous/ferric chloride effluent. The future for further
     ilmenite beneficiation plants is less certain and will depend
     on the eventual price for natural rutile.

Principal Uses for Products

The major outlets for titanium dioxide pigments is in the paint
industry. The following two tables indicate the principal users in 1973
for both Europe and the USA.

Europe		USA	
Paint	62%	Paint	51%
Plastics	13%	Paper	23%
Paper	8%	Plastics	10%
Textiles	5%	Ceramics	3%
Rubber	3%	Rubber	3%
Various	9%	Various	10%

Titanium dioxide used presently is threatened by substitute products
in only one market segment - paper. There titanium dioxide enjoys the
advantage of being an efficient opacifier, but it is at a cost disadvantage

to alumina and silica clays, some of which offer adequate brightness in
particular paper applications. In the paint industry, titanium dioxide is
by far the most effective white pigment in terms of hiding power. While
pigment research is extensive, no equally effective substitute has been
found. In plastics and rubbers, titanium dioxide offers the best
combination of white pigment coat, dispersion and resistance to discolouration.
In other product application areas, no substitute products represent serious
threats to titanium dioxide's established position.

References

1   American Cyanamid Company (Unitani Titania Dioxide Pigments,
        U.S.A. 1956), Chapter 2, p. 21.

2   J. Barksdale,Titanium (The Ronald Press Company New York,
        1949), Chapter 1, p. 3.

3   ECN Large Plants Supplements, October 23rd, 1976 p. 58.

Bibliography

Dr. Robert Powell, Titanium Dioxide and Titanium Tetrachloride (Noyes
Development Corporation, 1968).

Kirk-Othmer, Encyclopedia of Chemical Technology 2nd Edtn. (Interscience
Publishers, a Division of J. Wiley and Sons, Inc., New York, 1969) Vol. 20,
pp. 342-502.

Barksdale, Titanium 2nd Edtn. (The Ronald Press Company, New York, 1966).

Phosphorus, Phosphoric Acid and Inorganic Phosphates

By A.F. CHILDS

Albright & Wilson Ltd.

1. Occurrence of Phosphorus

Phosphorus is the eleventh most common element in the earth's crust, and phosphorus compounds of many kinds are essential for all plant and animal life. Because of its reactivity, phosphorus only occurs in the combined form in nature, almost entirely as phosphates of various metals.

Of the several hundred known phosphate minerals, the apatites, $Ca_{10}(PO_4,CO_3)_6 (F,OH,Cl)_2$ collectively known as phosphate rock, are by far the most important commercial source of the element, and more than 100 million tonnes are mined annually (Table 1). The United States is the principal producer, the mines of Florida alone having produced about one third of total world output in 1975.

TABLE 1.   Annual World Production of Phosphate Rock[1,2]

(Megatonnes per annum)

	Year		
	1973	1974	1975 (est)
Europe	0.1	0.1	0.1
Africa	27.4	32.9	25.3
America (N)	38.2	41.4	43.9
America (Central/S)	0.5	0.6	0.6
Asia/USSR	28.4	32.5	33.4
Australia/ Oceania	3.1	2.8	2.0
TOTAL	97.7	110.3	105.3 Mte p.a.
Equivalent to	9.2	10.3	9.9 Mte Phosphorus p.a.

Phosphate minerals were laid down in most geological periods, the earliest being hard alkaline igneous rocks, but by far the commonest are relatively soft sedimentary rocks. Deposits of sea-bird guano have also been worked.

The quality of phosphate rock is usually expressed in terms of its percentage content of phosphorus pentoxide (% P_2O_5) or more frequently, of tricalcium phosphate, $Ca_3(PO_4)_2$, known commercially as "bone phosphate of lime" or BPL. Commercial phosphate rock generally varies in grade from about 60 to 83 percent BPL (28-38% P_2O_5), marketable material generally containing more than 30% P_2O_5 or 13.1% P.

Roughly 80 per cent of world phosphate rock production comes from open-cast and quarrying operations using draglines or bucket-wheel excavators[4]. Rock containing as little as 5 per cent P_2O_5 can be mined, and usually the crude ore needs to be "beneficiated" before use to improve its grade. Operations used in beneficiation may include:

- grinding, washing and screening the crude ore
- desliming with cyclones or hydro-separators
- magnetic separation to reduce the iron oxide content
- flotation, to separate apatite from quartz
- calcination to reduce the content of calcium carbonate and organic impurities
- drying
- blending

The price of the resulting concentrate depends on its phosphate content on a dry basis. After remaining roughly constant for many years, the price of Moroccan rock rose sharply in October 1973 (70/72% BPL from US \$12.10 to \$40.00/tonne), and by 1975 had reached \$65.00. Fortunately, the price has subsequently fallen considerably, thus improving the economics of using phosphorus compounds.

Although apatite was first mined in England early in the 19th century, there are no longer any commercial deposits in the United Kingdom, and few in Western Europe. The largest reserves are found in North Africa (particularly Morocco), North America and the USSR (Table 2), while big, almost unexploited, deposits occur in Australia and South America. Production of phosphate rock increased at about 7% per year for the period

1960-1972, and there have been suggestions that the world's phosphate deposits may be exhausted within a centry. However, most estimates of reserves (e.g. Table 2) are based on present technology and the use of high grade rock at relatively low prices. It has, for example, been calculated that world reserves could increase nearly tenfold by raising the price of rock from $8 to $20 per recoverable ton. Reserves in new and unexploited deposits, together with low grade material which might be exploited by new technology, amount to many hundreds of gigatonnes, and should be sufficient for centuries to come[5].

TABLE 2.   Reserves of Phosphate Rock[3]

(Gigatonnes phosphorus content)

Continent	Rock Type (a)	Area	Identified Resources (b)
Africa	s	Morocco; Senegal; Tunisia; Algeria; Sahara; Egypt; Togo; Angola.	3.5
	i	South Africa	0.02
America (N)	s	U.S. (Florida, Georgia, Carolina, Tennessee, Idaho, Montana, Utah, Wyoming); Mexico	1.4
America (S)	s	Peru; Brazil; Chile; Colombia	0.08
	i		0.1
	g		0.01
Asia/Near East	s	U.S.S.R. (Kazakhstan, Siberia); Jordan; Israel; Saudi Arabia; India; Turkey; Christmas Island	0.2
	i	U.S.S.R. (Kola Peninsula)	0.2
Australia/ Oceania	i	Queensland	0.2
	s,g	Nauru; Makatea	0.01
		Rounded Total	5.7

(a)   s = sedimentary,   i = igneous,   g = guano

(b)   Specific identified mineral deposits that may or may not be evaluated as to extent, grade or possibility of profitable recovery with existing technology and economic conditions.

2. Elemental Phosphorus

2.1. Manufacture

Phosphorus is made by smelting phosphate rock, coke and
silica, in a closed electric furnace. The main reaction taking place is:

$$2Ca_3(PO_4)_2 + 6SiO_2 + 10C \longrightarrow 6CaSiO_3 + 4P + 10CO - 3.06MJ$$

A mixture of phosphorus vapour, carbon monoxide and hydrogen is driven
off from the top of the furnace, while the remaining products of the
reaction, mainly calcium silicate and ferrophosphorus, are tapped off as
a slag at intervals at the bottom.

Since its invention by Readman in 1888, the phosphorus furnace has
grown steadily in size. The first furnace erected by Albright & Wilson
at Oldbury near Birmingham in 1893 was of only 80kW capacity. By contrast,
their largest furnaces today, at Long Harbour in Newfoundland, are each
rated at 60MW. They are 12m in diameter, and can produce phosphorus at
the rate of over 30,000 tonnes per year. 500V/70 MW furnaces, operated
by Hoechst in Europe, and Monsanto in America, are the largest in current
use.

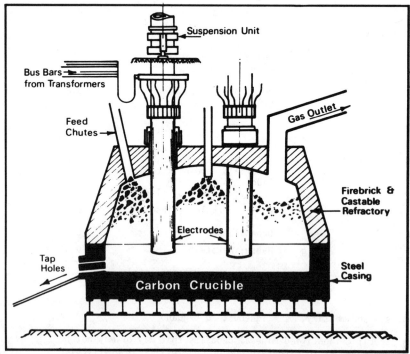

FIGURE 1. Cross-section of a Typical Phosphorus Furnace

The cross-section of a typical phosphorus furnace is shown in Figure 1. It consists of a cylindrical steel shell, often lined with firebrick, and having a carbon hearth, and an arched refractory roof through which pass holes for the three electrodes, feed shutes, and a gas off-take. The electrodes are provided with seals, to enable them to move up and down while keeping the furnace gas-tight. In the Albright & Wilson furnaces at Long Harbour in Newfoundland, the graphitised carbon electrodes are built up of sections 1.4m in diameter and 2.75m long which are screwed together to form a column with an average height of 16.75m and weighing 60 tonnes. Each column is supported by hydraulic rams, which raise or lower it automatically in response to the electrical requirements of the furnace. As the electrode is used up in the furnace, further sections are screwed on to the column top, to compensate for the loss. Farbwerke Hoechst in Germany, however, use continuous self-baking Söderberg electrodes.

The furnace mix must be controlled, not only in chemical composition, but also in physical form, which must be open and porous, to allow the furnace gases to cool to 300-400°C by heat transfer with the down-coming feed. If it contains too much fine material or shatters on heating in the furnace, the furnace burden sinters above the reaction zone, and the hot gases may be channeled at a high temperature and velocity into the gas off-take system.

The silica and coke are easily prepared for use by drying, crushing and screening to 4-40mm, but most phosphate rocks need some form of agglomeration. Typically they are ground, moistened with water, and formed into small pellets, briquettes or nodules. The lumps thus made are fired at a high temperature in a kiln, cooled, and discharged to storage.

The three constituents are weighed out on continuous band weighers, mixed, and stored in large bins, for feeding to the furnace. The quantities of raw materials needed to produce 1 tonne of white phosphorus depend on their purity and the efficiency of the furnace operation (usually about 90% on input phosphorus), but lie in the ranges:

Phosphate rock	6	- 10	tonnes
Silica	0.5	- 3.5	tonnes
Coke	1.2	- 1.8	tonnes
Electrodes	0.02	- 0.06	tonnes
Electricity	42	- 57	GJ (11.6-16MWh)

Phosphorus manufacture is therefore a very energy-intensive process, and is usually carried out where hydroelectric power or other relatively cheap energy sources are available. For this reason, phosphorus is no longer made in Britain. It has been estimated[6] that just over 50 per cent of the energy supplied to a phosphorus furnace is used for chemical reactions, 35 per cent is lost as sensible heat of the slag and 9 per cent by radiation, conduction and convection, leaving 5 per cent unaccounted for. Because of the high cost of electrical energy, many efforts have been made to operate a blast furnace to make either phosphorus or phosphoric acid, but without any commercial success.

The phosphorus vapour which distils in the stream of carbon monoxide at the top of the furnace is passed through hot electrostatic precipitators to remove dust, and then condensed by water sprays (Figure 2). The first condensers are run at about $70^{o}C$, and collect the bulk of the phosphorus as a liquid (m.p. $44.1^{o}C$) while the last traces are removed by cold water. The residual carbon monoxide passes on and is flared or used after purification for process heating.

The crude molten phosphorus is stored under warm water in large tanks. Here it settles and a viscous phase called "phosphorus mud" floats to the top. This is a complex colloidal dispersion of phosphorus and water. The phosphorus content of the mud is recovered by distillation, either by pumping the mud back to the phosphorus furnace, or in special "mud furnaces".

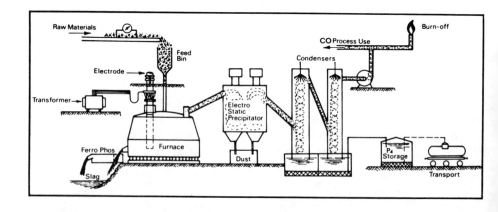

FIGURE 2.   Manufacture of Elemental Phosphorus

Plate 1

A cut-away drawing of the 10,000 ton
Albright Pioneer, the world's first
bulk-phosphorus carrier launched in 1969.
This ship's four steel tanks
each carry 1,250 tons of phosphorus
which is maintained in liquid form
by circulating warm water round them.
The phosphorus, which is spontaneously
inflammable in air at ordinary
temperatures, is blanketed under water.

By courtesy of Albright & Wilson Ltd.

Phosphorus is moved about in the molten state by water displacement through heated pipes. Over longer distances, it is carried in railway or road tankers, and more recently in specially designed ships which carry 5,000 tons of molten phosphorus in four insulated and heated tanks (Plate 1). Commercial phosphorus usually has a purity of 99.8 per cent or better. A specification for good quality material is:[7]

Insoluble in benzene	0.006 - 0.025%
Oil	0.001 - 0.020%
Arsenic as As	38 - 136 ppm
Iron as Fe	1 - 8 ppm
Fluorine as F	1 - 4 ppm

Up to 135 ppm of antimony and larger amounts of silica may also be present.

2.2. By-products of Phosphorus Manufacture

The major by-products of phosphorus manufacture are shown in Table 3. The slag is tapped intermittently at about 1550°C. It is mainly monocalcium silicate, but also contains small amounts of unchanged phosphate, and much of the fluorine from the original fluorapatite. The remainder of the fluorine is evolved as silicon tetrafluoride, which hydrolyses in the condensers

$$2CaF_2 + 3SiO_2 \longrightarrow SiF_4 + 2CaSiO_3$$
$$3SiF_4 + 2H_2O \longrightarrow 2H_2SiF_6 + SiO_2$$

This by-product fluorine could be recovered, but so far this has not proved economic. The slag is normally sold as hard core, or as a concrete aggregate. Other applications, as a raw material in cement or slag wool manufacture or as a crude agricultural lime, have not been widely adopted. Ferrophosphorus is used in the manufacture of special steels or cast irons, especially for non-sparking railway brake shoes[8]. It has also been proposed as a heavy aggregate in concrete used for shielding nuclear reactors[9].

Phosphorus manufacture presents environmental problems, the most serious of which is the "phossy water" arising from the condensers. Besides dissolved and suspended elemental phosphorus this contains phosphorus oxy-acids, hexafluosilicic acid, ammonia, and silica. After settling, the water is neutralised and precipitated with lime or ammonia, settled once more, and finally treated with chlorine to remove the last traces of phosphorus. Precautions are also taken against the emission of gaseous hydrogen fluoride from furnaces or ore pre-treatment plant.

TABLE 3.   By-Products of Phosphorus Manufacture

(tonnes per tonne phosphorus output)

Slag	7.1-8.9
Ferrophosphorus	0.09-0.38
Phosphorus mud	0.01-0.25
Precipitator dust	0.06
Condenser gas	2.8-3.4

2.3. World Production of Phosphorus

World capacity for elemental phosphorus production is about 1.2 megatonnes p.a., of which about 50% is in the US (Table 4). Production in the US rose steadily until 1969 when it began to decline because of decreasing demand for its use in phosphates, and has since fallen by about 25% to the 1975 value of 413 kilotonnes. Environmental problems of manufacture aided this decline.

TABLE 4. World Elemental Phosphorus Production Capacity (1976)

(Kilotonnes p.a.)

NORTH AMERICA

U.S.	Monsanto Company	(Idaho, Tennessee)	215
	F.M.C. Corporation	(Idaho)	128
	Stauffer Chemicals	(Florida, Montana, Tennessee)	102
	Hooker Chemicals	(Tennessee)	45
	Electrophos	(Florida)	18
	Mobil Chemicals	(Florida, S. Carolina)	20
Canada	Albright & Wilson Ltd.	(Newfoundland, Quebec)	85
			613

EUROPE

Germany (DBR)	Hoechst	(Knapsack-bei-Köln)	80
(DDR)	State	(Bitterfeld, Piesteritz)	15
Holland	Hoechst	(Vlissingen)	90
France	Pierrefitte	(Nestalas)	12
	Kuhlmann	(Epierre)	14
	Montedison	(Crotone)	13
			224

TABLE 4. (Continued)

ASIA

U.S.S.R.	State	(Chimkent, Dzhambul, Kuybyshev, Perm, Togliatti)	250
China			35
Japan	Nippon Kagaku Kogyo K.K.	(Koniyama)	12
	Rin Kagaku Kogyo K.K.	(Toyama)	8
India	Star	(Gujerat)	3
	Excel	(Gujerat)	3

311

AFRICA

| South Africa | Amcor | (Kookfontein) | 6 |

6

TOTAL   1154

3. Phosphoric Acid

3.1. Introduction

   Phosphoric acid is the most commercially important derivative of
phosphorus, and about 90% of all phosphate rock mined is converted to the
acid, either by treatment with a stronger acid, usually sulphuric (the
"wet" or "gypsum" process) or via phosphorus, which by oxidation and
hydration gives "thermal" or "furnace" phosphoric acid. Wet process acid
is more dilute (25-50% P_2O_5 = 34.5-69% H_3PO_4) and contains many of the
impurities present in the original phosphate rock, while thermal acid
can be made very pure, and suitable for all applications, including food.
However, because of the complexity and high energy usage of phosphorus
manufacture, thermal acid is much more expensive, 1976 UK prices being about
£85/tonne for wet process acid, and £265/tonne for thermal acid of
equivalent concentration. Hence thermal is being increasingly substituted
by wet process acid in all but the most demanding applications (Figure 3).

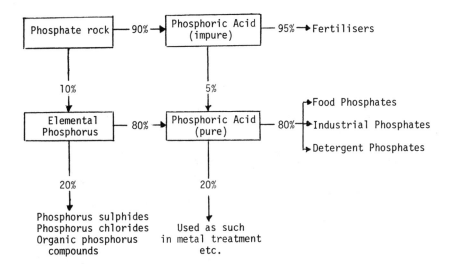

FIGURE 3.   Phosphate Rock Use Pattern

3.2. The Thermal Process

The two reactions involved in this process are:

$$P_4 + 5O_2 \longrightarrow P_4O_{10} - 3.015 \text{ MJ}$$
$$P_4O_{10} + 6H_2O \longrightarrow 4H_3PO_4 - 0.293 \text{ MJ}$$

Many attempts have been made to burn the gases coming from an electric or blast furnace, but they contain impurities which give an impure acid, and in most modern plants the phosphorus is condensed separately before being burnt.

In older processes for thermal acid, the combustion and hydration steps were kept separate, which had the advantage that much of the heat of combustion could be used to raise steam. However, it is simpler and cheaper to do both steps in one vessel, and this is now the usual practice. In a typical plant (Figure 4), a spray of molten phosphorus is burnt in air and steam in a combustion chamber made of stainless steel externally cooled in water. Acid-proof brick or impervious carbon may be used in place of steel. The exit gases are passed immediately into a hydrator and thence through a venturi scrubber or electrostatic precipitator to remove mist.

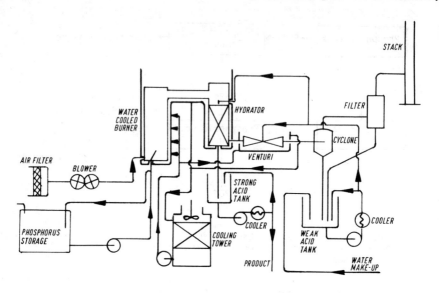

FIGURE 4.   Thermal Phosphoric Acid Manufacture

The product acid normally contains up to 85% H_3PO_4 (62% P_2O_5), but concentrations as high as 84% P_2O_5 may be obtained, and such acids contain pyrophosphoric and higher polyphosphoric acids in addition to orthophosphoric acid[7].

Roughly 90% of the phosphorus charged to the electric furnace is recovered in the phosphoric acid[10]. The main impurity is arsenic, which may be precipitated by hydrogen sulphide, or volatilised by treatment with hydrogen chloride. The analysis of typical thermal acid is shown in Table 5.

3.3.   The Wet Process

The manufacture of fertilizer by treating bones with sulphuric acid was patented by Lawes in England in 1842, and this basic reaction, applied to phosphate rock is still the major method of making phosphoric acid:

$$3H_2SO_4 + Ca_3(PO_4)_2 \longrightarrow 3CaSO_4 + 2H_3PO_4$$

Many processes have been operated, but nearly all fall into one of two main classes, according to whether the calcium sulphate is precipitated as its dihydrate (gypsum) $CaSO_4.2H_2O$, or hemihydrate $CaSO_4.\frac{1}{2}H_2O$. The wet process has undergone continuous development to improve the yield, concentration and purity of its product. A high yield depends on completely reacting phosphate rock with sulphuric acid, and efficiently separating the phosphoric acid and calcium sulphate formed. A high concentration requires a minimum of wash water to be used in the separation. On both counts, it is essential to obtain a good filterable calcium sulphate.

TABLE 5.   Typical Analyses of Phosphoric Acids

	Thermal (Technical)	A(Moroccan)	Gypsum B(Florida)	Purified*
P_2O_5 %	62	56.5	54	60
SO_3 %	0.002	0.7	3.0	0.75
F p.p.m.	<1	1400	4000	100
Ca p.p.m.	60	800	1000	<50
Mg p.p.m.	<10	2800	2000	<50
Al p.p.m.	<10	1900	700	<10
Fe p.p.m.	10	2300	12000	50
Cr p.p.m.	<0.5	400	140	<10
Cu p.p.m.	<0.5	<70	<30	<1
As p.p.m.	<0.3	19	10	<10

*   Sample A purified by MO solvent extraction process.

(i)   Gypsum Processes

The earliest wet processes operated at relatively low temperatures and sulphuric acid concentrations. Gypsum was precipitated and the concentration of the product acid was only 8-10% P_2O_5. By the 1930's an improved process, developed largely by the Dorr-Oliver Company, gave a 30-32% P_2O_5 acid. In such a process, powdered phosphate rock was added to a stirred mixture of sulphuric acid, with dilute phosphoric acid and recycled gypsum slurry, which acted as a seed for formation of good

crystals. The slurry then passed through a train of reactor tanks with
conditions of temperature and concentration controlled to grow gypsum
crystals of optimum size and shape for filtration. Finally, part of the
fully reacted slurry was filtered on a continuous belt or tilting-pan
vacuum filter (Figure 5), while the bulk was recycled to the first stage.
Modern processes of this type use a single reaction tank fitted with
baffles. Numerous other variations of this process have been introduced
and patented by such companies as Fisons, Prayon, Pechiney-Saint Gobain
- U.C.B. and Kellogg Lopker, and single-train plants with capacities of
nearly 1,150 tonnes/day P_2O_5 have been built[11]. However, although the
process is widely used, the highest concentration of acid produced by it
is about 43% (30-32% P_2O_5) and further concentration of the corrosive
product is expensive in both energy and materials. For this reason,
attempts have been made for many years to work at higher initial
concentrations of sulphuric acid.

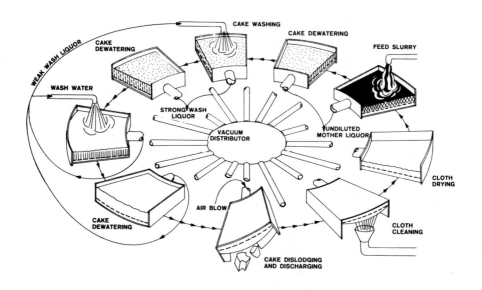

FIGURE 5.   Operation of the Bird-Prayon Tilting-pan
Vacuum Filter

(ii) Hemihydrate and Anhydrite Processes

Nordengren and his co-workers, in a series of basic studies established
the conditions under which calcium sulphate (dihydrate, hemihydrate and

anhydrous) is precipitated from phosphoric acid slurries (Figure 6). The
work, which led to a patent[12], showed that while dihydrate and anhydrite
were the thermodynamically stable forms, there was a considerable region
of temperature and P_2O_5 concentration in which the metastable hemihydrate
was precipitated and could be filtered off.

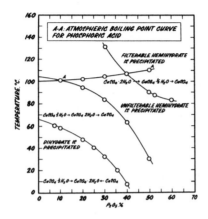

FIGURE 6.   The System H_3PO_4 - $CaSO_4$ - H_2O

Early attempts to commercialise this process failed, because of the
poor filterability of the hemihydrate, and its reversion to dihydrate on
the filter. However, in recent years, interest has revived in processes
in which the hemihydrate is recrystallised to dihydrate before or after
filtration. Such processes, developed in Japan by Nissan, Nihon Kokan
Kaisha and Mitsubishi, in the UK by Fisons[13], and in the US by Singmaster-
Breyer, are claimed to give, not only product acid of 40% or more P_2O_5,
but also a lower loss of P_2O_5 in the dihydrate, which is therefore more
suitable for such uses as plaster-board manufacture.

The Davison Chemical Company in the US in 1950-51 studied a process[14]
in which anhydrous calcium sulphate was formed by treating phosphate rock
with concentrated sulphuric acid. The work did not get beyond the pilot.
stage, because of separation problems.

Although sulphuric acid is the most widely used for attacking phosphate
rock, hydrochloric acid and nitric acids have also been used. Because of
the high solubility of the calcium salts of these acids, their separation
presents a problem, although a process involving the solvent extraction of
a hydrochloric/phosphoric acid mixture has been commercialised by Israel
Mining Industries Limited[15].

A typical gypsum process acid is dark green or brown in colour.
As well as hydrofluoric acid and sulphuric acids, it contains most of the
metals present in the original rock, especially calcium, iron, aluminium
and magnesium. It is usually concentrated to about 54% P_2O_5 in carbon-
lined vacuum evaporators and a typical analysis of such "merchant grade"
acid is given in Table 5. The product may be transported in road or rail
tankers, or in specially built ships which discharge at a world-wide chain
of terminals. It is stored in tanks made of stainless steel or rubber-
lined mild steel.

By-product gypsum and hydrogen fluoride give rise to environmental
problems during wet process phosphoric acid manufacture. The gypsum is
usually pumped as a slurry to large ponds, where it settles and solidifies.
It is sometimes used for plaster-board manufacture. Hydrogen fluoride in
waste water is neutralised with calcium carbonate. It may be recovered
from phosphoric acid by adding sodium carbonate or chloride, to precipitate
sodium silicofluoride, used in the fluoridation of tap water. About 100 ppm
of uranium is present in most wet process acid, equivalent to an annual
quantity of about 3,000 tonnes in the US alone. Methods for its recovery
are known, when the price of uranium rises high enough to make this an
economic proposition.

3.4. Purification of Phosphoric Acid

The extent to which it is necessary to purify wet process phosphoric
acid depends on the use to which it is put. For solid fertilizers, it can
be used without further treatment. In detergent phosphate maufacture,
most of the metallic impurities are precipitated when the acid is neutralised
by alkali.

Recently, several companies have proposed solvent extraction. In one
type of process, crude acid is mixed with a water-miscible solvent like
methanol, ethanol or acetone. The solvent dissolves phosphoric acid, but
not the metallic impurities, which are precipitated as an acid phosphate,
having some fertilizer value. The purification is assisted by adding
ammonia[16].

Unfortunately, this class of process needs a great deal of energy to distil the solvent from the phosphoric acid and water, and the use of such highly volatile solvents involves excessive solvent loss. Use of a water-immiscible solvent avoids these problems, since the phosphoric acid can be separated from the solvent by extraction with clean water. The MO process developed by Albright & Wilson is typical of this second class. Concentrated wet process acid is reported[17] to be extracted by methyl isobutyl ketone in mixer settlers. The solvent, containing roughly two-thirds of the phosphoric acid,passes forward to a purification train where it is washed counter current with clean acid to remove still more impurities, and the purified acid is re-extracted into water in two final stages. After concentration, the acid has the analysis shown in Table 5. The residue, which contains most of the impurities,can be used to make fertilizers. Processes using water-insoluble solvents are being operated by Toyo Soda in Japan, I.M.I. in Israel,and Prayon, APC and Rhone-Poulenc in France. In some cases, alkali is used in the re-extraction stage.

The product acid from such processes is pure enough for most purposes, but if necessary can be further purified by ion exchange, or by crystallisation as the hemihydrate, $2H_3PO_4.H_2O$ (m.p.29.2^{o}C).

TABLE 6.   World Phosphoric Acid Capacity (1975)

(Kilotonnes P_2O_5 p.a.)

Western Europe	8,350
North America	5,275
Central/South America	855
Africa	1,315
Middle East	535
Asia/Oceania	2,340
Eastern Bloc Countries	5,000
TOTAL	23,670

Phosphoric acid is made worldwide by many processes and in very many plants. Figures of annual production capacities can therefore only be estimates. The most probable values for 1975 are given in Table 6, the world total being about 23.7 megatonnes, expressed as P_2O_5. Capacity by the thermal route is 10-15% of the total[18].

4. Phosphates

Phosphoric acid is tribasic and forms three series of salts, for example, NaH_2PO_4, Na_2HPO_4 and Na_3PO_4. With a first dissociation constant of 1.1×10^{-2}, it is relatively weak for an inorganic acid, and its alkali metal salts can act as buffers.

Mono- and di-sodium phosphates are made by adding anhydrous sodium carbonate (soda ash) to stirred phosphoric acid. The operation can be done batchwise or continuously. In some large plants the Na:P ratio is analysed automatically and used to control the operation of the plant. The resulting orthophosphate solution is filtered and crystallised, or evaporated to dryness in spray or drum driers, if the anhydrous salt is required. Besides the anhydrous mono-and di-sodium phosphates, the hydrates $NaH_2PO_4.2H_2O$, $Na_2HPO_4.2H_2O$ and $Na_2HPO_4.12H_2O$ are made commercially.

Trisodium phosphate has to be made by neutralising disodium phosphate solution with sodium hydroxide. The so-called dodecahydrate is really a double salt, of approximate composition $(Na_3PO_4.12H_2O)_4.NaOH$[19] and the isomorphous salt $(Na_3PO_4.11H_2O)_4.NaOCl$, "chlorinated trisodium phosphate", can be crystallised from solutions containing sodium hypochlorite. The crystals are centrifuged and dried in a current of warm air in rotary driers.

By far the most commercially important phosphate is the fertiliser grade monocalcium phosphate made by treating phosphate rock with sulphuric acid ("single superphosphate"):

$$Ca_{10}(PO_4)_6F_2 + 7H_2SO_4 \longrightarrow 3Ca(H_2PO_4)_2 + 7CaSO_4 + 2HF$$

After being crushed and ground, the rock is mixed with sulphuric acid (68.5%), and the wet mass is pushed slowly through a large rectangular "den". In this, the bulk of the chemical reactions occur, giving a solid product, which is then "cured" in large piles, to complete the reactions.

If phosphoric acid is used in place of sulphuric acid, "triple superphosphate" is obtained, so called because it contains 45-50% P_2O_5 available for plant growth instead of the 16-20% available P_2O_5 content of single superphosphate. The principal reaction occurring is:

$$Ca_{10}(PO_4)_6F_2 + 14H_3PO_4 \longrightarrow 10Ca(H_2PO_4)_2 + 2HF$$

but many complex side reactions take place, and the chemistry is still not completely understood. In the modern "granular" process, rock and acid are mixed in a continuously stirred tank reactor. The slurry overflows to a second reactor from which it is pumped to a granulator, usually a rotary drum containing dry recycled product. Here it forms granules which are dried and screened to 1.0 - 2.8 mm. Oversize material is crushed and returned to the granulator with the fines.

For food applications, a pure monocalcium phosphate monohydrate is made by treating lime with thermal phosphoric acid and drying the product in spray or rotary driers. An alternative is to crystallise the product from a concentrated solution.

Dicalcium phosphate dihydrate, $CaHPO_4.2H_2O$ for toothpaste use is made by treating a dilute slurry of lime with dilute phosphoric acid (30-40% H_3PO_4) at a temperature controlled at about 25°C. The product is filtered, dried in circulating warm air driers and ground in ring mills to a mean particle diameter of about 10 micrometres. Small quantities of pyrophosphate or magnesium salts are usually added, to prevent loss of water of hydration, which would shorten the shelf life of the finished toothpaste.

Anhydrous dicalcium phosphate is formed by the same reaction at higher temperatures (70°C). It needs no stabilisation, but is more abrasive than the dihydrate, and is normally used mixed with milder abrasives such as tricalcium phosphate. Calcium pyrophosphate $Ca_2P_2O_7$ is also used as an abrasive in fluoride toothpastes, where its very low solubility prevents the precipitation of calcium fluoride.

Mono- and di-ammonium phosphates are made by passing ammonia into stirred, cooled phosphoric acid. They crystallise from the solution, and are centrifuged or filtered off and dried in warm air in a rotary drier. By carrying out the reaction under a pressure of 7 atmospheres, a temperature appreciably above the normal boiling point of saturated monoammonium phosphate solution may be maintained. The melt is then sprayed into a tower where the water evaporates, leaving a solid product. This "Minifos" process, devised by Fisons Limited has the further advantage that the exothermic reaction between ammonia and phosphoric acid provides some of the process heat[20].

4.2. Polyphosphates

If inorganic acid phosphates are heated to a temperature between $250^{\circ}C$ and $500^{\circ}C$, they eliminate water to give polymeric compounds called condensed or polyphosphates. Of particular value are the sodium salts formed by the general reaction

$$2NaH_2PO_4 + nNaH_2PO_4 \longrightarrow (NaO)_2\overset{O}{\overset{\|}{P}}-O(\underset{ONa}{\overset{O}{\overset{\|}{P}}}-O)_n\overset{O}{\overset{\|}{P}}(ONa)_2 + (n+1)H_2O$$

Homologous series of both linear and cyclic polymers are known[21], though only the former find industrial use. Unlike most other polymeric inorganic anions, they are relatively stable to hydrolysis, except at extremes of pH, and may be separated by paper or ion exchange chromatography.

The simplest sodium polyphosphate is tetrasodium pyrophosphate (diphosphate), $Na_4P_2O_7$, which is made by calcining anhydrous disodium phosphate at a temperature between 300 and $900^{\circ}C$ usually in an oil- or gas-fired rotary calciner. The acid pyrophosphate, $Na_2H_2P_2O_7$, is made similarly from monosodium phosphate, but the temperature of conversion must be closely controlled at around $250^{\circ}C$, to avoid either incomplete reaction, or further condensation to insoluble polyphosphates of very high molecular weight.

Sodium triphosphate ("tripolyphosphate"), $Na_5P_3O_{10}$, is a major component of heavy duty laundry detergents, and more of it is produced annually than any other pure phosphorus compound with the exception of phosphoric acid. It is made by calcining a mixed acid orthophosphate equivalent to two moles of disodium hydrogen phosphate plus one mole of monosodium dihydrogen phosphate. The reaction is carried out in a gas- or oil-fired rotary calciner at a temperature of $450^{\circ}C$.

Sodium triphosphate occurs in two crystalline forms which hydrate at different rates to give the same hydrate, $Na_5P_3O_{10}\cdot6H_2O$. This difference is important in the manufacture of synthetic detergents and the process must be carefully controlled to give a product which contains the desired ratio of the two forms, as well as having a high content of triphosphate and appropriate density and flow characteristics.

Sodium triphosphate was first made from thermal phosphoric acid, and this is still largely true in the US. However, for reasons already mentioned, there is an economic advantage in using gypsum acid, and many European companies operate such a process.

Higher molecular weight glassy polyphosphates, $NaO(\overset{\text{ONa}}{\underset{O}{\overset{|}{\underset{\parallel}{P}}}}-O)_n Na$, can be made by calcining acid sodium phosphates with Na:P ratios of less than 1.667. A solution of slurry of orthophosphate of the appropriate ratio is fed to large shallow gas- or oil-fired furnaces. Water rapidly boils off, and the remaining solid then melts and undergoes the condensation reaction. The molten glass is chilled in water-cooled drums and broken into convenient lumps, or ground to a powder. The soluble water-softening polyphosphate "Calgon" is a glass of this kind, with a mean chain length n of 12 phosphate units (67% P_2O_5), while similar products with n=5 and 20-30 are also manufactured.

By tempering melts with a M:P ratio very near to 1, insoluble sodium and potassium polyphosphates with molecular weights of several millions can be made[20].

5. Applications of Phosphorus Compounds

Phosphorus and its compounds have very many uses, some of which have already been mentioned. Most of these uses, however, stem back to three characteristic properties of phosphoric acid and its inorganic and organic derivatives, shown in Figure 7.

5.1 Biological Activity

All animal and plant soft tissue contains between 0.1 and 0.4% phosphorus, while bone has up to 13%. The phosphorus is present as simple or condensed phosphates which are essential to both the structure and metabolism of the organism.

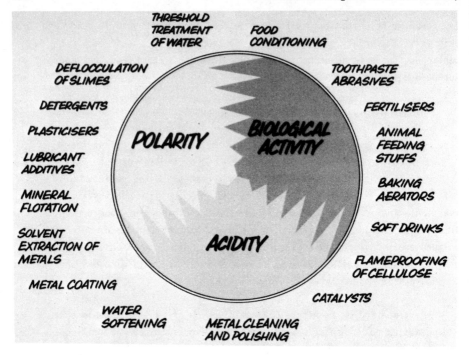

FIGURE 7.   Properties and Applications of Phosphoric
                    Acid and Derivatives

Plant Fertilisers and Animal Foodstuffs. Crude single superphosphate
was first used to stimulate the growth of plants in the early nineteenth
century. This was augmented by basic slag from steel manufacture and by
triple superphosphate. Together these still account for the bulk of
fertiliser used worldwide. However, the rising cost of transport has led
to an interest in fertilisers containing a higher proportion of plant
nutrient, such as the ammonium phosphates and, more recently, ammonium
polyphosphates. The world consumption of phosphate fertiliser fell slightly
in 1974/75 to 24.3 megatonnes P_2O_5 content, after having risen steadily
for many years, reflecting the increase in phosphate rock prices already
mentioned (Table 7).

     Scientific methods of animal husbandry have led to a demand for animal
feed supplements containing phosphate, especially calcium phosphates, which
are used to avoid dietary imbalance. The main problem is to make a cheap
product with a satisfactorily low fluorine content (less than 1% of the
phosphorus content). Much effort has gone into making suitable material
from gypsum acid or phosphate rock.

TABLE 7.   World Phosphate Fertiliser Consumption (Megatonnes P_2O_5)

	1971/72	1972/73	1973/74	1974/75
West Europe	6.1	6.2	6.3	5.5
East Europe	5.9	6.2	6.6	7.2
North America	4.7	5.1	5.1	4.5
Central/South America	1.1	1.5	1.5	1.7
Africa	0.6	0.7	0.7	0.7
Asia/Oceania	4.0	4.5	5.4	4.7
TOTAL	22.4	24.2	25.7	24.3

Food Phosphates. The biggest use in human food is in chemical aerators or leavening agents for dough and batter. The earliest of these comprised mixtures of sodium bicarbonate and an organic acid, usually tartaric or lactic, but in 1856, the use of monocalcium phosphate as an inorganic acid was suggested. The adoption of phosphates was hastened by the first World War, when a shortage of tartaric acid led to the use of acid sodium pyrophosphate. This was found to give a cheaper and more stable baking powder, and is still used in nearly all British baking powder, while monocalcium phosphate is the standard aerating material in self-raising flour. Very dilute phosphoric acid is also used as an acidulant in soft drinks such as colas. Polyphosphates are used for their effect on the mechanical properties of foodstuffs. Thus, in the presence of calcium ion, tetrasodium pyrophosphate causes milk casein to gel, forming an instant pudding similar to junket. Other condensed phosphates are used in meat and cheese processing, where they retain the meat structure during cooking, improve fat distribution and give a final product of improved colour and flavour. In frozen fish and poultry, phosphates keep the meat moist, prevent drip loss and retain nutrients, while again improving appearance and texture.

5.2 Acidity

Phosphoric acid is widely used in the de-rusting and pickling of steels.
Being less corrosive than sulphuric or hydrochloric acids, it does less
damage to the metal surface, and furthermore leaves a thin protective film
of phosphate bonded to the metal. A similar application of growing
importance is in the electrolytic or chemical polishing of metals, especially
aluminium and copper and their alloys, and stainless steel. Solutions used
for this purpose contain phosphoric and nitric acids, sometimes with organic
additives. In chemical polishing the article is simply immersed in the
solution, while in electrolytic polishing, it is made anodic. In both cases,
metal is removed selectively from high spots, without any etching, the
result being a brilliant specular finish. The treatment is especially
suited to the polishing of small articles of intricate shape, such as costume
jewellery, but is also used for larger items like domestic hollow-ware and
car trim.

Phosphoric acid is also used in a number of catalysts. Phosphoric
acid on kieselguhr will catalyse the telomerisation of ethylene and
propylene, whilst a mixture of condensed phosphoric acids of average
composition $H_6P_4O_{13}$ acts as a Lewis acid in promoting organic condensations
and cyclisations. An interesting property of phosphoric acid is its ability
to catalyse the dehydration of cellulose to carbon and water while inhibiting
the formation of inflammable tars and volatiles, and hence to render it
flameproof.

$$(C_6H_{10}O_5)_n \longrightarrow 6nC + 5nH_2O$$

Based on this fact, a completely wash-fast treatment has been developed for
cotton and rayon, based on the phosphine derivative tetrakis(hydroxymethyl)-
phosphonium chloride, $[(HOCH_2)_4 P]^+ Cl^-$. Fabric treated by this process
has unaltered physical properties, but if exposed to a flame smokes and
chars without burning. When the flame is removed, the residual char does
not continue glowing, and this dual effect of flame- and glow-proofing is a
unique property of phosphorus compounds.

5.3. Polarity

Because the phosphoryl group is highly polar, compounds such as
phosphates containing it show a powerful affinity for other polar materials,
including metals. We have already met an application of this, in the use
of mixtures of acid phosphates of iron, zinc, manganese and other metals

to form a corrosion-resistant adherent layer on steel. Most motor-car
bodies made in Britain are treated in this way. Alkyl thiophosphates show
similar properties, and if added to lubricating oils, are absorbed on to
bearing surfaces, giving anti-wear, corrosion inhibition and high load-
bearing properties. Organic thiophosphates can similarly be used in ore
flotation, since they are absorbed by many sulphide minerals, whilst other
phosphate esters have been used more recently in the solvent extraction
of metals, especially uranium, from aqueous solutions.

Detergents. After fertilisers, the biggest use of phosphates is in
water treatment and detergents. Alkali-metal polyphosphates are powerful
sequestering agents, especially for calcium and magnesium ions. They have
therefore been used for many years both to soften water and as "builders",
i.e. compositions which increase the detergency of soaps or synthetic
anionic surface active agents. The first soap builders to be used in
Britain were sodium orthophosphates and pyrophosphates, and these are still
used in hard surface cleaners. The sequestering power increases with
molecular weight, and polyphosphate glasses such as "Calgon" are used for
applications where their hygroscopic nature does not put them at a
disadvantage.

The biggest market, however, is in heavy duty laundry detergents, and
here sodium triphosphate has an almost ideal combination of properties for
use with surfactants of the alkylbenzene sulphonate type. Besides softening
water, and overcoming the inhibitory effect of calcium and magnesium ions
on the surfactant, the triphosphate acts by dissolving lime soaps already
on the cloth, and by assisting in dispersing, and preventing the
re-deposition of oily dirt.

In recent years, detergent phosphates have been said to cause
"eutrophication" or excessive growth of algae in lakes, especially in the
US and Canada. Arbitrary limits have been set in some areas on the use of
phosphates in detergents, in spite of the fact that possible alternative
systems are inferior and often themselves biologically or ecologically
suspect. Furthermore, over half the phosphate in the lakes comes from
other sources, including human and animal excreta. Hence, where (as is
not the case in Britain) the problem is serious, it must be dealt with by
complete removal of all phosphates from sewage by suitable treatment.

Another large-scale application of polyphosphates is in the "threshold"
treatment of water. Long-chain polyphosphates, in concentrations of only

a few parts per million are found to inhibit the deposition of calcium carbonate from hard water. The explanation of the effect appears to be that polyphosphate anions are absorbed at the growing surfaces of calcite crystals, inhibiting their further development. The deposition of iron oxide from ferruginous water is hindered similarly.

A final example of this kind is the effect of polyphosphates on the properties of clays. Between 0.1 and 1.0% of a condensed phosphate will reduce a firm putty-like clay to a free flowing slurry. Polyphosphate anions are absorbed on the surface of the clay particles, making them repel each other. This deflocculant action is of value in preparing clay coatings for paper manufacture, modifying slurry viscosities in cement manufacture and improving the flow properties of oil-well drilling muds.

In the short time available, I have only been able to discuss a few, albeit the most important, compounds of phosphorus and their applications. In particular, I have not mentioned the commercially important sulphides or chlorides, and only briefly touched on the organic compounds of phosphorus, which include one of the most useful classes of insecticides. However, I hope that I have given you some idea of the importance of this element to nearly every aspect of life, and shown how the British chemical industry by the size and efficiency of its operations can both meet competition at home and make a direct export contribution worth millions of pounds every year.

REFERENCES

1. The British Sulphur Corporation Limited, Phosphorus and Potassium, (Statistical Supplement No.12), 1975, 80 (6), 11.

2. The British Sulphur Corporation Limited, Phosphorus and Potassium, 1976, 81 (1), 11.

3. J. B. Cathcart and R. A. Gulbrandson, U.S. Geol. Survey Prof. Paper 820, Phosphate Deposits, (1973).

4. J. G. Notholt, Phosphate Rock World Production, Trade and Resources, (Proceedings of the first "Industrial Minerals" International Congress), (London 1974).

5. Anon., European Chemical News, 1977, 30 (775), 22.

6. H. A. Curtis, J. Electrochem. Society, 1953, 100, 81.

7. J. R. van Wazer, Phosphorus and its Compounds, (Interscience, New York, 1961), Vol. 2, passim.

8. Albright & Wilson Limited, Brit. Pat., 1,203,242 (1970).

9. H. S. Davis, Mining Engineering, 1957, 9, 544.

10. Ed. A. G. Slack, Fertiliser Science and Technology Series, Marcel Dekker, (New York, 1968), Phosphoric Acid, Vol. 1.

11. Anon., Chemical Marketing Reporter, 14th June 1975, 27.

12. S. G. Nordengren, U.S. Pat., 1,776,595, (1930).

13. J. D. Crerar, Chemical Engineering, 1973, 80 (10), 62.

14. C.C. Legal Jr., T. O. Tongue and E. H. Wight, U.S. Pat., 2,504,544, (1950).

15. A. Baniel, R. Blumberg, A. Alon, M. El-Roy and D. Goniadski, Chemical Engineering Progress, 1962, 58 (11), 100.

16. J. F. McCullough and L. L. Frederick, Agricultural and Food Chemistry, 1976, 24 (1), 180.

17. A. Barney, The Engineer, 1974, 23 (6153), 33.

18. The British Sulphur Corporation Limited, Phosphorus and Potassium, 1974, 71 (3), 35.

19. B. Wendrow and K. A. Kobe, Chem. Revs., 1954, 54, 891.

20. M. W. Ranney, Noyes Data Corporation Chemical Process Review No.35, Ammonium Phosphates, 1969, 30.

21. J. R. van Wazer, Phosphorus and its Compounds, (Interscience, New York 1958), Vol. 1, passim.

Inorganic Cyanogen Compounds
by J.B. FARMER
Borax Consolidated Ltd.

The chemistry of cyanogen compounds dates back to 1704 when Diesbach discovered
Prussian blue, but it was not until after the mid-18th century that the
properties of this interesting class of compounds were developed. In 1754,
Macquer prepared potassium ferrocyanide and in 1782-3, Scheele and Berthollet
isolated hydrogen cyanide; the relationship between these compounds was
demonstrated by Gay-Lussac in 1811-5. The name cyanogen is of Greek origin
meaning "dark blue", describing the colour of Prussian blue. Hydrogen cyanide
occurs naturally in combination with several glucosides, among which amygdalin
is probably the best known.

Cyanogen compounds possess the carbon-nitrogen triple bond group. The
inorganic compounds that are commercially important are hydrogen cyanide (HCN)
and its salts, cyano complexes such as ferro- and ferri-cyanides and the
"Iron Blues", cyanamide ($H_2N.CN$) and its calcium salt, and cyanogen chloride
(ClCN). Cyanogen itself, $(CN)_2$, is of minor importance other than as a
reactive intermediate. Growth in the synthetic fibres and plastics industries,
and the demand for nitrogenous materials for agricultural use have resulted in
a constant increase in production of these compounds.

The distinction between these compounds and organic nitriles, RCN, is marginal.
Thus hydrogen cyanide, being the nitrile of formic acid, gives many typical
reactions of organonitriles, cyanamide can be dimerised to cyanoguanidine
(dicyandiamide) or trimerised to melamine, an amino triazine, and cyanogen
chloride trimerises to give cyanuric chloride, also a triazine.

A comprehensive review of cyanogen compounds has been published[1]. Selected
physical properties of the commercial products are given in Table 1.

Table 1   Selected Physical Properties

	Molecular Weight	Melting Point °C	Boiling Point °C	Density	Solubility in g in 100 g. water
Hydrogen cyanide HCN	27.03	-14	26	0.699^{22}	∞
Sodium cyanide NaCN	49.01	565	-	-	48^{10}
Sodium ferrocyanide $Na_4Fe(CN)_6 \cdot 10H_2O$	484.04	-	-	1.45	32^{20}, 156^{98}
Potassium ferricyanide $K_3Fe(CN)_6$	329.26	decomp.	-	1.85	33^{4}, 77^{100}
Calcium cyanamide $CaCN_2$	80.1	1300 subl >1150	-	-	decomp.
Cyanogen chloride ClCN	61.47	-6	13	1.19	-

HYDROGEN CYANIDE

Hydrogen cyanide (prussic acid, hydrocyanic acid) is a colourless and volatile liquid possessing the characteristic odour of bitter almonds. In aqueous solution it is a weak acid (K_a = 7.2 x 10^{-10}) and is therefore liberated when soluble cyanides are treated with acids, this being the usual laboratory preparative method. In strongly acidic solutions hydrogen cyanide is hydrolysed to formic acid [2]. Liquid hydrogen cyanide polymerises to a brown-black sparingly soluble solid in the absence of trace quantities of acid. Diaminomaleonitrile $H_2NC(CN):C(CN)NH_2$, the hydrogen cyanide tetramer [3] is the most prominent low molecular weight product formed during the polymerisation, and it is believed that aminomalononitrile is a key intermediate [4].

Some of the principal reactions that demonstrate the versatility of hydrogen cyanide within the cyanogen industry are illustrated in Figure 1.

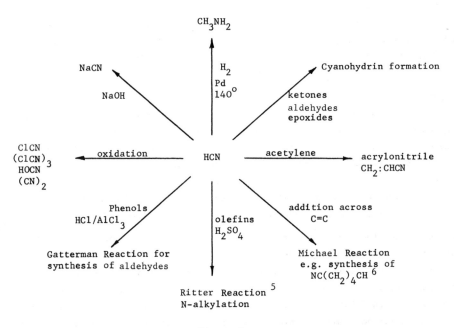

Figure 1

Acrylonitrile production was formerly the major consumer of hydrogen cyanide, according to the reaction shown, but now the chief outlet for the cyanide is the manufacture of methyl methacrylate. In this process, acetone cyanohydrin is produced by reacting hydrogen cyanide with acetone in the presence of an alkaline catalyst [7]. Treatment of the cyanohydrin with sulphuric acid gives methacrylamide sulphate which is subsequently reacted with methanol to yield methylmethacrylate.

<u>Manufacture of Hydrogen Cyanide</u>   Literature on the production of hydrogen
cyanide is widespread but only a few processes are economically viable.
Hydrogen cyanide can be recovered as a by-product from coal gas or coke oven
gas and can be synthesised by the dehydration of formamide in a three stage
process:

$$CH_3OH + CO \longrightarrow HCOOCH_3$$
$$HCOOCH_3 + NH_3 \longrightarrow HCONH_2$$
$$HCONH_2 \longrightarrow HCN + H_2O$$

Present day processes are based on the direct reaction between ammonia with
alkanes, or indirectly as a by-product from the acrylonitrile and related
nitrile industries.

In the Andrussow Process [7-10], hydrogen cyanide is manufactured by passing a
preheated mixture of air (74-78% by volume), ammonia (11-12%) and methane
(12-13%) over a platinum-rhodium [11] or -iridium catalyst at 20-30 psi at a rate
of 1-3 ft/sec and a temperature of 1000-1200°. In the presence of the catalyst,
the mixture rapidly burns and hydrogen cyanide is formed:

$$CH_4 + NH_3 + 1\tfrac{1}{2}O_2 \longrightarrow HCN + 3H_2O + 144 \text{ kcal} \text{ mol}^{-1}$$

Figure 2 shows a simplified flow diagram for the process. The converter exit
gases are rapidly cooled in a waste heat boiler to less than 400° to prevent
cracking or polymerisation of the hydrogen cyanide. In a single pass through
the converter a yield of hydrogen cyanide of up to 53% and 67% based on methane
and ammonia respectively can be achieved, and with recycling of ammonia a
conversion of 87% is possible. The off gases typically contain by volume
hydrogen cyanide 6%, ammonia 1-2%, nitrogen 55-57%, water 24%, hydrogen 7% and
small quantities of carbon monoxide, carbon dioxide, methane and oxygen.

Process conditions and subsequent purification of off gases are critical to
obtain high recovery of hydrogen cyanide. Feedstock air is scrubbed to remove
impurities, and methane from natural gas or coke oven gas is carefully treated
to remove sulphur compounds. Converter off gases can be freed of ammonia by
absorption with dilute sulphuric acid although the two favoured methods employ
a) a 35% solution of monoammonium phosphate that forms the diammonium salt with
ammonia and, after removal of hydrogen cyanide, can be regenerated by steam [12];
or b) a 2½% boric acid solution containing a polyhydroxy compound - the boron-
polyol complex acid absorbs ammonia and can be recovered through its
dissociation at elevated temperatures [13]. The ammonia free gases are then fed
to a water absorber and the dilute hydrogen cyanide solution concentrated by
distillation.

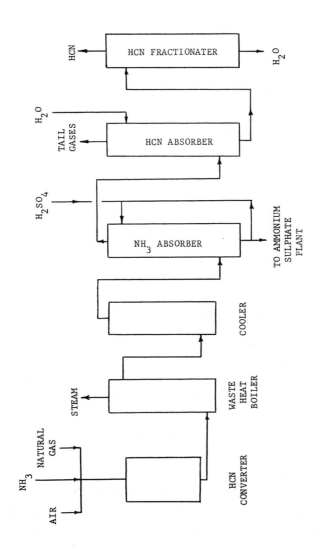

Figure 2 Hydrogen Cyanide Synthesis (Andrussow Process)

Formation of hydrogen cyanide from methane and ammonia in the absence of
oxygen is the basis of the Degussa process [14]. The reaction is endothermic,

$$CH_4 + NH_3 + 60 \text{ kcal/mol} \longrightarrow HCN + 3H_2O$$

and requires a temperature of 1200-1300°. Alumina tubes lined internally with
a platinum catalyst constitute the converters. A process efficiency of 90%
based on methane can be attained and the off gases contain over 20% hydrogen
cyanide, a distinct advantage over the Andrussow method. Further, the hydrogen
produced (70% of off gas) is a useful by-product.

High temperature synthesis of hydrogen cyanide through use of electrically
heated fluidised beds of carbon is a modification to the Degussa process and
has been successfully developed by Schawinigan Chemicals Ltd. [15,16]. Ammonia
and methane or propane are reacted at 1400-1650° to give an off gas consisting
of over 20% hydrogen cyanide and very little ammonia. Yields of 90% hydrogen
cyanide are reported.

The use of plasma arc reactors for the conversion of methane and nitrogen or
ammonia to hydrogen cyanide has been described [17,18] but the technique has not
yet proved economically feasible for large scale production.

Of increasing importance is the production of hydrogen cyanide as a by-product
in the manufacture of organonitriles. These materials were formerly made by
the reaction of hydrogen cyanide with unsaturated hydrocarbons, as, for example
acrylonitrile from hydrogen cyanide and acetylene. This method is now strongly
rivalled by the relatively new route to acrylonitrile using propylene, ammonia
and air [19]. Simultaneous production of hydrogen cyanide as a by-product in
this process probably represented up to 50% of the hydrogen cyanide plant
capacity in the USA in 1975-6 [20]. Examples of claims on the formation and
isolation of hydrogen cyanide as a secondary product are frequently reported in
the literature [21-24].

Commercial hydrogen cyanide is available with a purity of at least 99%. Small
quantities of acid (up to 0.1% as H_2SO_4) are added as a stabiliser. The liquid
is transported in steel containers capable of withstanding up to 100
atmospheres or in tanker shipments for large scale consumers.

Economic Aspects and Statistics   Figures for the production and consumption
of hydrogen cyanide are only readily available for the USA, but the known
statistics probably reflect world trends. In 1963, the USA production was
estimated at 140,000 short tons and the consumption was distributed as follows:

Acrylonitrile	51%	Sodium cyanide	7%
Methylmethacrylate	18%	Others	10%
Adiponitrile	14%		

By 1976, the USA demand for hydrogen cyanide reached 235,000 tons [20] with methylmethacrylate providing the chief outlet:

Methylmethacrylate	60%	Chelates	10%
Cyanuric chloride	15%	Others	5%
Sodium cyanide	10%		

Minor uses of hydrogen cyanide include the manufacture of ferrocyanides, sequestering agents and pharmaceuticals.

The main producers of hydrogen cyanide in the USA are DuPont, Monsanto, Rohm and Haas, Vistron, American Cyanamid and Dow, and Imperial Chemical Industries in the U.K.

SODIUM CYANIDE

The cyanide ion in crystalline alkali cyanides is freely rotating and can be considered as effectively spherical with an ionic radius of 1.92 Å. Thus, for example, sodium cyanide is isomorphous with sodium chloride. Dissolved in water, alkali metal cyanides give an alkaline solution as a result of partial hydrolysis:

$$NaCN + H_2O \rightleftharpoons NaOH + HCN$$

The cyanide ion is isoelectronic with carbon monoxide, and being negatively charged, it is a better Lewis σ- donor. Evidence of this donor character is prevalent, particularly in the ability of cyanide to form complex transition metal ions (see below) and in its role as a nucleophilic reagent in organic synthesis.

Manufacture of Sodium Cyanide   Since 1834 when the first process for preparing sodium cyanide from Prussian Blue was developed by Rodgers, many patents based on fusion techniques have been published. A summary of the former economical processes is given in the following equations.

Prussian blue + $Na_2CO_3 \longrightarrow KCN + NaCN$		Rodgers Process 1834
$K_4Fe(CN)_6 + K_2$ or $Na_2CO_3 \longrightarrow KCN + NaCN$		" " "
$K_4Fe(CN)_6 + 2\,Na \longrightarrow 4KCN + 2NaCN + Fe$		Erlenmeyer Process 1876
$K_2CO_3 + 4C + 2NH_3 \longrightarrow 2KCN + 3CO + 3H_2$		Beilby Process 1891

Until the mid-1960's when hydrogen cyanide became widely available, the Castner process was the preferred method for sodium cyanide manufacture: Metallic sodium, charcoal and ammonia are reacted at elevated temperatures.

$$2Na + C + 2NH_3 \longrightarrow 2NaCN + 3H_2$$

The reaction proceeds in three stages commencing with the formation of sodamide

(NaNH$_2$) which then reacts with carbon to form, initially, sodium cyanamide (Na$_2$NCN) and finally sodium cyanide. The charcoal and molten sodium are heated to 750° and ammonia is blown through the mixture until the reaction is complete. The molten reaction product is filtered at about 650° in an atmosphere of nitrogen to avoid oxidation (occurring above 370°). Yields of sodium cyanide of up to 94% based on sodium or ammonia can be achieved.

Most sodium cyanide is now produced by the neutralisation of hydrogen cyanide with caustic soda and is often a secondary process to hydrogen cyanide manufacture. Gaseous hydrogen cyanide is absorbed in a solution of about 50% sodium hydroxide to afford a solution of about 30% sodium cyanide. A slight excess of sodium hydroxide is desirable to minimise contamination by polymeric hydrogen cyanide. Crystalline sodium cyanide is recovered by evaporation under reduced pressure with careful control of temperature to avoid decomposition to sodium formate. Drying of the paste can be effected at low temperatures at reduced pressures or by passing a current of hot (400-500°) air over the slurry.

Absorption of hydrogen cyanide by sodium carbonate [25] and the preparation of sodium cyanide in a non-aqueous medium from alkali metal alcoholates and hydrogen cyanide [26] are typical variations of the neutralisation process.

Sodium cyanide is commercially available as the powder, flake or "egg" shaped pieces, or as a 30% aqueous solution. The solid products are obtained by melting and moulding sodium cyanide in an inert atmosphere. The current price is 36½ cents per lb. in the U.S.A. and £0.27 per kilo in the U.K.

Uses   One of the main uses of sodium cyanide is in the extraction of gold and silver from their ores. In the presence of oxygen, soluble complex salts of these metals are formed,

$$8 \ NaCN + 4Ag + 2H_2O + O_2 \longrightarrow 4 \ NaAg(CN)_2 + 4 \ NaOH$$

from which the previous metals can be regenerated by precipitation with zinc dust.

Large quantities of sodium cyanide are used in the flotation of ores, acting as a depressant for zinc sulphide (sphalerite) and iron sulphide (pyrites) in the separation of lead sulphide (galena), and for copper sulphide containing iron or zinc sulphides.

Other important uses of sodium cyanide include the production of Prussian blue pigments, the electrodeposition of gold, silver, copper, zinc or cadmium, case-hardening and heat-treatment of steel (nitriding or carbonitriding), and the synthesis of organic chemicals such as pharmaceuticals, dyes and complexing reagents.

Other Commercial Metal Cyanides   Potassium cyanide is prepared by the

neutralisation process. It is a white crystalline hygroscopic solid with a

melting point of 634°. Its chemical properties are similar to those of sodium

cyanide. Although potassium cyanide is more expensive than the sodium salt

(77 cents per lb. cf $36\frac{1}{2}$ cents for NaCN) it is sometimes preferred in electro-

plating.

Calcium cyanide has similar industrial applications to sodium cyanide and can

replace the sodium salt in the separation of precious metals and as a depressant

in flotation. It may be manufactured by heating crude calcium cyanamide (see

below) in an electric furnace at 1000°:

$$CaNCN + C \longrightarrow Ca(CN)_2$$

Excess carbon in the product imparts a dark grey colour, giving rise to the

common name "black cyanide". Neutralisation of lime with hydrogen cyanide is

also a preparative route.

Thiocyanates   The reaction of sulphur with boiling

solutions of alkali cyanides results in the formation of thiocyanates, MSCN.

Ammonium thiocyanate can be prepared by the reaction between ammonium and

carbon disulphide. Ammonium thiocyanate (m.pt. $149\text{-}150^{\circ}$) is used as a rust

inhibitor, a defoliant for cotton, in photography and as a starting material

for the preparation of other thiocyanates (e.g. potassium). Sodium thiocyanate

is a hygroscopic white crystalline solid (m.pt. 287°) and is used in the

dyeing and printing of textiles, as a weed killer and in processing of colour

films. The potassium salt (m.pt. 172°) has the same uses as its sodium

analogue but its main application is in the synthesis of antibiotics.

COMPLEX CYANIDE COMPOUNDS

Cyanide ions have a strong tendency to complex with metals, especially those of

the transition series [1,27]. The cyanide complexes of iron, ferrocyanide

$Fe(CN)_6^{4-}$ and ferricyanide $Fe(CN)_6^{3-}$ are the best known examples and their salts

are valuable industrial materials. These two species are also referred to as

hexacyanoferrate(II) and hexacyanoferrate(III) ions respectively.

a. Sodium and Potassium Ferrocyanides When excess cyanide solution is added

to ferrous salt solutions ferrocyanide is formed,

$$6\ CN^- + Fe^{2+} \longrightarrow Fe(CN)_6^{4-}$$

The stability of the complex is evidenced by the lack of chemical reactions for

either ferrous or cyanide ion. The two processes for the manufacture of the

sodium and potassium salts involve an indirect synthesis with calcium cyanide

and ferrous sulphate or direct with sodium (or potassium) cyanide and ferrous

salts:

Solutions of calcium cyanide and ferrous sulphate are heated with a current of steam to produce soluble calcium ferrocyanide and insoluble calcium sulphate. Potassium chloride is added to the filtered solution and the separated precipitated calcium potassium ferrocyanide treated with potassium carbonate to yield a solution of potassium ferrocyanide that can be evaporated to form the crystalline trihydrate product. A modification that can be used to prepare both sodium and potassium salts employs an alkali metal carbonate instead of chloride in the first stage.

The more direct route is to react alkaline metal cyanide solutions with ferrous salt solutions at 70-80° [28]. On cooling, yellow crystals of the sodium salt, "yellow prussiate of soda" $Na_4Fe(CN)_6.10H_2O$, or the potassium salt, "yellow prussiate of potash" $K_4Fe(CN)_6.3H_2O$ crystallise.

Ferrocyanides are used in the preparation of Prussian blue, in metal treating, electroplating, dyeing and printing processes, in tanning leather and in photography.

b.   Sodium and Potassium Ferricyanides   The ferricyanide ion is less stable than ferrocyanide and its solutions give reactions for cyanide. These salts are normally prepared by oxidation of the corresponding ferrocyanide with, for example, chlorine or by electrolysis. After oxidation the ferrocyanide solutions are evaporated to the point of crystallisation, made alkaline, filtered and recrystallised. The potassium salt "red prussiate of potash" $K_3Fe(CN)_6$ is used in colour photography.

c.   Iron Cyanide Blues   The term "Iron Blue" covers an extensive range of iron cyanide complex compounds. When a ferrous salt is added to an alkali ferricyanide, "Turnbull's" blue is formed; this was originally formulated as ferrous ferricyanide $Fe_3^{II}\left[Fe^{III}(CN)_6\right]_2$. Prussian blue can be prepared by adding a ferric salt to a ferrocyanide solution and for many years was assigned the structure $Fe_4^{III}\left[Fe^{II}(CN)_6\right]_3$. Literature on the structure of these two compounds and on the identity of the iron atoms has been controversial, but X-ray studies [29] have revealed that the two blues have nearly, if not the same, identical structures in which the iron atoms lie at the corners of a cubic lattice.

The Iron blues are more correctly formulated as $(MFe)Fe(CN)_6$ where M is potassium, sodium or ammonium. The presence of varying quantities of these univalent cations can subtly alter the properties of the blues.

In a typical process for the manufacture of Iron blue, [30,31] an aqueous solution of sodium ferrocyanide and ammonium sulphate is heated to 70°. Ferrous sulphate solution, also at 70° is slowly added to the ferrocyanide mixture to form a white precipitate named "Berlin White".

$$FeSO_4 + Na_4Fe(CN)_6 + (NH_4)_2SO_4 \longrightarrow Fe(NH_4)_2Fe(CN)_6 + 2Na_2SO_4$$

The precipitate is digested with sulphuric acid and the solution oxidised with sodium chlorate. The Iron blue product is washed by decantation to pH 3 to 3.5 and ammonia added to pH 4.5 before the solid is pressed, dried and pulverised.

The preparation of Prussian blue by the reaction of hydrogen cyanide with ferrous sulphate at room temperature, followed by addition of ammonia, acidification and oxidation is an alternative method[32]. A variation to increase the depth of colour and lower the viscosity of the Iron blue product by heating the Berlin white intermediate to 100-150° at 3-5 atmospheres pressure prior to oxidation is claimed [33].

The majority of Iron blues manufactured consists of types such as "Chinese", "Prussian", "Milori" and "Bronze" blues. In combination with lead chromate (chrome yellow), the Iron blues produce chrome green. The products are very stable in acid solution but in the absence of stabilisers such as nickel salts[34] are liable to decompose rapidly in strongly alkaline solutions.

The blues offer good colour and working properties for printing inks and are used in lithographic, letterpress, publication and flexographic solvent inks. Water dispersible ("soluble") blues can be used for paper making.

Production of Iron blue pigments in the U.S.A. has been at a level of 5000 tons per annum over the years 1950-1970 [31]. The current U.S.A. price is \$1.175 per lb and \$1.375 per lb for alkali resistant blue. In the U.K., sales of Iron blue in 1976 were in excess of 1900 tons, at a price of £1.128 per kilo.

CALCIUM CYANAMIDE

World consumption of calcium cyanamide fertilisers and the development of melamine and other organic materials has made calcium cyanamide a large tonnage inorganic cyanogen compound product; Europe's production, alone, in 1975 was 73,000 tons [35] (as nitrogen).

Manufacture of Calcium Cyanamide Calcium cyanamide is prepared by nitrogenation of calcium carbide [36]. The carbide, usually prepared on site, is produced in the endothermic reaction between lime and coke in an electric furnace:

$$CaO + 3C + 111 \text{ kcal/mol} \longrightarrow CaC_2 + CO$$

The molten carbide is tapped from the furnace, allowed to cool and crushed to a fine powder. The reaction between calcium carbide and nitrogen (from liquifaction of air) is exothermic but external heating is necessary to initiate the reaction and to increase yield of product. Small quantities (2%) of calcium chloride or fluoride are added to lower reaction temperatures and increase the reaction rate

$$CaC_2 + N_2 \xrightarrow{1000-1100^o} CaNCN + C + 72 \text{ kcal/mol}$$

The nitrogenation can be conducted batchwise in cylindrical steel shells lined with a refactory material and heated electrically. The process takes 4 to 6 days to go to completion. After cooling, the calcium cyanamide ingot is removed from the furnace and crushed.

Continuous processes using rotary kilns have been developed [37]. A current of nitrogen is passed over granulated calcium carbide containing 2% calcium chloride. The process is initiated by firing and the temperature of 1100° is maintained through the heat of reaction.

Tunnel ovens, in which a feed of calcium carbide in perforated steel plated boxes is moved along rail tracks against a flow of nitrogen, have also been used. Conversions of up to 90% based on calcium carbide can be reached.

Commercial calcium cyanamide is approximately 65% pure and its price in the U.S.A. is $300 per ton.

Uses and Statistics   The widely publicised use of calcium cyanamide is as a fertiliser. Data known for the production of fertiliser cyanamide in the year 1972/3 shows that four countries are the main producers[38].

	metric tons nitrogen
Germany	72,030
Japan	34,500
Italy	16,310
Belgium	11,380

In 1975, only Japan and Germany were producing calcium cyanamide in bulk - 36,000 tons [39] and 70,000 tons [35] (as nitrogen) respectively.

Calcium cyanamide is being challenged by other nitrogen fertilisers and in 1973, American Cyanamid decided to shut down its Niagara Falls Plant [40]. Other than its use as a fertiliser, calcium cyanamide is also a weed killer and as a defoliant for spreading on cotton, tomatoes etc. In industrial applications, it can be used for the production of dicyandiamide for melamine, and provides a source of crude calcium cyanide when fused with sodium chloride for the extraction of gold and silver from their ores.

Cyanamide and Dicyandiamide   Cyanamide crystallises as colourless orthorhombic crystals with a melting point of $45-46^o$. It is a linear molecule, isostructural and isoelectronic with carbon dioxide. It dimerises readily in alkaline solution to form dicyandiamide (cyanoguanidine)

$$2 \text{ } H_2N.CN \longrightarrow \begin{array}{c} H_2N \\ \diagdown \\ H_2N \diagup \end{array} C = N - CN$$

The commercial importance of cyanamide is found in the manufacture of dicyanadiamide as a raw material for the production of melamine, guanidine salts and cyanogen halides. It is prepared by hydrolysis of calcium cyanamide with steam and carbon dioxide [41].

$$CaNCN + H_2O + CO_2 \longrightarrow CaCO_3 + H_2N.CN$$

If cyanamide solution is the desired product, the final solution is kept at pH 6 to 6.5. By this method an 87% conversion to cyanamide as a 7% solution can be achieved. Further recycling of the filtered solution raises the concentration of cyanamide to 25%. Dicyandiamide is manufactured by allowing dimerisation of cyanamide solution (25%) at pH 8 to 9 and a temperature of 80°.

Melamine, the cyclic trimer of cyanamide

is formed in up to 90% yield by the pyrolysis of cyanamide or dicyandiamide in an atmosphere of ammonia [42].

CYANOGEN AND CYANOGEN CHLORIDE

Cyanogen   Cyanogen $(CN)_2$ and the cyanogen halides XCN can be prepared by the oxidation of hydrogen cyanide. Cyanogen is obtained by the direct oxidation of hydrogen cyanide in the gas phase by air over a silver catalyst, by chlorine over activated carbon or silica, and by nitrogen dioxide over a calcium oxide glass [43] or a cupric salt [44]:

$$2 HCN + \tfrac{1}{2}O_2 \longrightarrow (CN)_2 + H_2O$$
$$2 HCN + Cl_2 \longrightarrow (CN)_2 + 2HCl$$
$$2 HCN + NO_2 \longrightarrow (CN)_2 + NO + H_2O$$

Cyanogen is the intermediate in the production of oxamide, a relatively new commercial fertiliser and a stabiliser in nitrocellulose preparation. Degussa have recently developed a method involving oxidation of hydrogen cyanide with hydrogen peroxide, followed by hydrolysis of the cyanogen to oxamide, $H_2N.CO.CO.NH_2$. An interesting one-step production of oxamide has been accomplished by Hoechst [45,46]. In their process, hydrogen cyanide and oxygen are contacted with a solution of copper nitrate and acetic acid to form either oxamide or cyanoform amide in high yield (99.5% and 80% respectively). Replacement of acetic acid with acetonitrile leads to cyanogen formation.

Cyanogen Chloride   The classical method of preparing cyanogen chloride is to react hydrogen cyanide with chlorine.

$$HCN + Cl_2 \longrightarrow ClCN + HCl$$

but the oxidation can also be effected by electrolysis of aqueous solutions of hydrogen cyanide and ammonium chloride [47] or by the gaseous oxidation of hydrogen cyanide with oxygen and hydrogen chloride in the presence of transition metal catalysts [48].

Cyanogen chloride, a gas at room temperature (b.pt. 13°) is usually directly converted to cyanuric chloride by passing through a bed of carbon catalyst. Cyanuric chloride is the trimer of cyanogen chloride and possesses a triazine structure.

$$3 \; ClCN \longrightarrow$$

Cl—C(=N—CCl=N—CCl=N) (triazine ring with three Cl substituents)

The chloride atoms are labile and can be replaced by a multitude or organic groups. The chemistry of triazine compounds has been reviewed [44].

In 1975, 15% of the hydrogen cyanide produced in the U.S.A. was consumed in cyanuric chloride manufacture, and the growth in triazine herbicides and its other uses as optical brighteners, reactive dyestuffs etc. is likely to continue at a fast rate.

TOXICITY OF CYANIDES

The toxicity of hydrogen cyanide and many of its derivatives is well known. An atmospheric concentration of 300 ppm HCN is rapidly fatal and the maximum limit for prolonged exposure is about 10 ppm. In solution, the recommended limit of free cyanide is 0.1 ppm.

Stringent regulations for waste discharges have been imposed on cyanide industries and have made hydrogen cyanide prohibitively expensive for small consumers. Cyanides can be rendered relatively harmless by addition of iron salts to form stable, complex ferrocyanides, but the usual method for disposal of wastes is by decomposition to ammonium compounds and ultimately carbon dioxide and nitrogen. Oxidation with chlorine or sodium hypochlorite, by electrolysis (e.g. the Cyanil Process [50]) and hydrogen peroxide [51] can lower levels of soluble cyanide down to 0.1 ppm.

References

1. H.E. Williams, Cyanogen Compounds, Edward Arnold & Co., London, 1948
2. V.K. Krieble and A.L. Peiker, J. Amer. Chem. Soc., 1933,55,2326
3. R.L. Webb, S. Frank and W.C. Schneider, J. Amer. Chem. Soc., 1955,77,3491
4. J.P. Ferris and L.E. Orgel, J. Amer. Chem. Soc., 1965,87,4976
5. J.J. Ritter and P.P. Minieri, J. Amer. Chem. Soc., 1948,70,4045
6. British Patent 1,178,950
7. Anon, Ind. Eng. Chem., 1959,51,1232
8. U.S. Patent 1,934,838
9. U.S. Patent 2,496,999
10. S. Serban and I. Tomescu, Rev. Chim. (Bucharest), 1967,18,197
11. Y.K. Bingham, J. Catal., 1971,21,27
12. U.S. Patent 2,797,148
13. U.S. Patent 2,590,146
14. F. Endter, Chem. Ing. Tech., 1958,30,305
15. U.S. Patent 3,305,661
16. Can. Patent 748,793
17. M.P. Freeman and C.C. Mentzer, Ind. Eng. Chem. Process Des. Develop. 1970, 9,39
18. U.S. Patent 3,376,211
19. U.S. Patent 3,324,166
20. Anon, Chem. Mkt. Rep. Nov. 15th, 1976
21. Belg. Patent 623,100
22. Brit. Patent 1,051,404
23. Ger. Patent 1,238,011
24. Ger. Patent 1,267,676
25. G.M. Strongin, A.I. Shishkina and R.P. Vakhutina, Tr. Kim. Tekhnol., 1968, 3,108
26. U.S. Patent 3,402,994
27. A.G. Sharpe, Chemistry of Cyano Complexes of the Transition Metals, Academic Press, 1976
28. Manox Ltd., private communication
29. J.F. Keggin and F.D. Miles, Nature, 1936,137,577
30. P. Ratcliffe (Manox Ltd.), private communication
31. J.A. Sistino, Pigment Handbook, 1973,1,401, Wiley N.Y.
32. Brit. Patent 889,673, Ital. Patent 612,601
33. Ger. Patent 1,188,232
34. H. Holtzman, Ind. Eng. Chem., 1945,37,855
35. Eurostat, Statistical Office of the European Communities, October 1976
36. M.L. Kastens and W.G. McBurney, Ind. Eng. Chem., 1951,43,1020
37. F. Kaess, Chem. Ing. Tech., 1959,31,80

38. Annual Fertiliser Review 1973, Food and Agric. Org. of the United Nations.

39. Anon, Jap. Chem. Week, Sept. 1976

40. Anon, Chem. Process. Ind. Management, Aug. 1973

41. S.A. Miller and B. Bann, J. Appl. Chem. 1956,6,89

42. Ger. Patent 1,249,875

43. W.F. Fierce and W.J. Sandner, Ind. Eng. Chem., 1961,53,985

44. British Patent 1,084,477

45. Ger. Patent 2,308,941

46. Anon, Chem. Eng., April 1976, 45

47. U.S. Patent 3,294,657

48. Jap. Patent 7036,204

49. E.M. Smolin and L. Rapport, The Chemistry of Heterocyclic Compounds
     "s-Triazines", Interscience, 1959

50. Anon, Can. Chem. Proc., Aug. 1975, 28

51. Anon, Chem. Week., Jan. 1972, 55

The Inorganic Chemistry of Nuclear Fuel Cycles

By   J.R.FINDLAY, K.M.GLOVER, I.L.JENKINS, N.R.LARGE,
J.A.C.MARPLES, P.E.POTTER and P.W.SUTCLIFFE

United Kingdom Atomic Energy Authority

Summary

The aim of this article is to introduce the reader to the many aspects of the inorganic chemistry of the nuclear energy industry.

The resources of thorium and uranium are firstly discussed together with the various thermal and fast breeder nuclear reactor systems in which these materials are being exploited.

Among the topics reviewed are the extraction of thorium and uranium from their ores, the enrichment of uranium, the chemistry relating to the heat transfer media for gas-, water- and liquid metal- cooled nuclear reactor systems, and the preparation and fabrication of the various nuclear fuels.

The complex chemical changes which occur during irradiation of nuclear fuels within a reactor due to the formation of over 30 elements from the fission of the actinide nuclei, the subsequent reprocessing of the fuel for the separation of uranium and plutonium from the fission product elements, and the inorganic chemical aspects of waste management are considered.

Finally some aspects of the actinide elements other than thorium, uranium and plutonium, with particular emphasis on their industrial application, are discussed.

The Modern Inorganic Chemicals Industry
1.   Introduction

This article sets out to introduce the reader to the inorganic chemistry
of the nuclear fuel cycles. An article of this limited length can only serve
as a guide to further reading into the many facets of the relevant chemistry;
of necessity, it also reflects the preferences and interests of the authors,
and must therefore be incomplete and superficial in its coverage.

2.   Raw Materials for the Fuel Cycles

The fuel cycles which we will discuss require the utilisation initially of
the 2 actinide elements which occur naturally, namely uranium and thorium.
These elements are the only actinides to occur to any appreciable extent in
nature. The other elements which are formed in the fuel cycles are produced by
nuclear transmutation reactions; the scale of these reactions varies greatly.
Plutonium (At. No. 94) is produced in tonne quantities in reactors in several
countries. Neptunium (At. No. 93), protoactinium (At. No. 91), and americium
(At. No. 95) are available in gram quantities, whilst curium (At. No. 96),
berkelium (At. No. 97) and californium (At. No. 98) are prepared in milligram
quantities; the higher elements einsteinium (At. No. 99), fermium (At. No. 100).
mendelevium (At. No. 101), nobelium (At. No. 102), lawrencium (At. No. 103),
are prepared almost on the atomic scale[1].

The scale of the separation techniques varies widely. Uranium is mined
throughout the world in large quantities and its separation together with that
of thorium employs familiar extraction techniques[2]. The purification of these
elements together with neptunium, plutonium and americium relies largely on
solvent extraction and ion exchange methods. The occurence and extraction of
thorium and uranium will be considered together with aspects of the formation
of neptunium, plutonium, americium, and the higher actinides within the various
nuclear reactor systems.

2.1   Thorium.

The most stable isotope of thorium is ^{232}Th which is an α-emitter
and has a half-life of 1.4×10^{10} years allowing it to be handled without undue
precaution. Although thorium is a fairly abundant element, only a few tonnes
of thorium are extracted per year, and mainly as a by-product of rare earth
element extraction. The earth's crust contains 0.001-0.002 wt % of thorium,
and it is approximately 3 times as common as uranium. Thorium was discovered
by Berzelius in 1828 in a Norwegian mineral, but it was not used industrially
until 1885 when Auer von Welsbach discovered that thoria (ThO_2) became incan-
descent when heated. The gas mantle which von Welsbach developed was widely
used and in 1973 still represented the largest single use of thorium.

The world consumption of thorium in 1970 has been estimated to be about
270 tonnes. Of this, 50% was used as thorium nitrate for the preparation of
gas mantles, 40% was used in the metallurgical industry, and the remaining 10%

was used for refractories, catalysts and in nuclear reactors. In the future, the utilisation of thorium in nuclear reactor systems could be very high. The estimates of the demand in the year 2000 by the US Bureau of Mines are: non-nuclear uses - 50-620t, and for nuclear energy - 700-6500t depending on the need for and the successful exploitation of thorium burning systems. Even if the demand for thorium were to reach these higher values, no supply problems would be anticipated for a considerable period. This is because thorium is obtained as a by-product of the extraction of both rare earth elements and uranium. The available thorium reserves are estimated to be 0.5 Mt, the total resources are in excess of 1 Mt. This may be compared with an estimated cumulative usage of thorium up to the year 2000 of 40,000-60,000t.

The only commercial source of thorium is the monazite sands. There are some rarer materials, however, containing larger quantities of thorium; thorianite contains up to 90% ThO_2 and has been worked in Madagascar. Thorite a silicate which contains up to 62% thorium is found in the western USA and in New Zealand. Monazite, the chief ore, is principally a phosphate of the rare earths which contains between one and 10% thoria. It is found in granites and other igneous rocks, but in concentrations too small to be commercially useful. On weathering of the thorium containing rocks, the heavy thorium containing monazite sands are formed; these sands are found in Australia, Brazil, India, South Africa, Sri Lanka and the USA. After mining, the monazite sands are first concentrated by sluicing to remove the lighter sands and then further concentrated magnetically to produce monazite of about 95% purity. The monazite is then opened by either acidic or basic digestion.

The potential for the utilisation of thorium in nuclear reactors depends on the fact that ^{232}Th (the only isotope found in nature) can absorb neutrons to produce fissile ^{233}U, allowing it to be used as a fertile material in a breeder reactor. The methods of utilising thorium for nuclear power production are well documented[3]. The major advantage of the ^{232}Th - ^{233}U cycle lies in the high net neutron yield of ^{233}U that can be achieved in thermal reactors, so that the fuel and fissile inventory required is small for a given power output. The greater neutron economy is, however, offset commercially by the difficulties in reprocessing caused by the hard γ-rays from the daughter products of the short-lived α-emitter ^{228}Th arising from the reaction chain:

$$^{232}\text{Th} \xrightarrow{\text{n,2n}} {}^{231}\text{Th} \xrightarrow{\beta^-} {}^{231}\text{Pa} \xrightarrow{\text{n,γ}} {}^{232}\text{Pa} \xrightarrow{\beta^-} {}^{232}\text{U} \xrightarrow{\alpha} {}^{228}\text{Th}$$

Thorium-containing nuclear fuels, fabricated into particles coated with carbon and silicon carbide, dispersed in a graphite matrix and cooled by helium are potential fuels for the High Temperature Reactor systems. The Dragon (OECD), Peach-Bottom (USA) and THTR (German Fed. Rep.) reactors have all demonstrated the potential of this type of fuel cycle and the coated particle fuels.

2.2 <u>Uranium</u>. Uranium was the first of the actinide elements to be recognised
as a new element in a pitchblende specimen from Saxony by Klaproth in 1789.
Uranium is the heaviest element to occur in nature in recoverable amounts, the
isotopes, all α-emitters, occur in the following proportions: ^{238}U, 99.28%
with a half-life of 4.5×10^9 years, ^{235}U, 0.71%, half life 7×10^8 years,
^{234}U, 0.005%, half life 2.35×10^5 years.

Uranium occurs widely in nature and the concentration in the earth's crust
is 4 ppm; the corresponding concentration for copper is 40 ppm. Sea water contains 0.002 ppm uranium. From the time of the discovery of uranium in pitchblende up to just after the second world war, pitchblende from similar deposits
supplied most of the world's requirements. In 1899 carnotite was described and
soon afterwards was found to occur in Jurassic sandstone in Colarado and Utah,
but it was not until 1948 that pitchblende was also recognised in this region.
Coffinite, a uranium silicate, was discovered in 1948 forcing a revision of the
ideas on the genesis.

Uranium deposits can be classified as follows[4]:

1.   Uranium in sandstone,

2.   Uranium in conglomerates,

3.   Vein and similar type deposits,

and 4.   Other uranium deposits.

Uranium deposits as sandstones are collectively the most important as they comprise 35% of the total reserves at a cost of < $66/kg U_3O_8. The major part of
this type of deposit occurs in Colorado and Wyoming. Uranium in conglomerates
is confined mainly to the well-known fields of the Witwatersrand and Elliot
Lake and Blind River, USA. These reserves form 15% of the total at the price
given above. Vein deposits were the first type from which substantial quantities of uranium were recovered and the reserves in such deposits are now about
20% of the world's total. The other deposits tend to be thorium ores or uranium
minerals which are more refractory than those in vein deposits; these contain
15% of the total reserves. Perhaps the best known example of this latter category is the Rössing deposit in South West Africa in which fine grained
uraninite occurs in quartz, alkali-feldspar pegmatites. Betafite and davidite
also occur in appreciable amounts and there is an abundance of secondary uranium
minerals in near surface rocks. The estimated reserves of uranium in non-
Communist countries is shown in Table 1. Reserves are defined as those deposits
in known locations which are of such grades, quantity and configuration that
they can be recovered within the given production cost range, with currently
proven mining and processing technology. Additional resources refer to uranium
assumed to occur in unexplored extensions of known deposits or in undiscovered
deposits in known uranium districts which could be produced at a given cost.

Table 1

Estimated Reserves of Uranium in Non-Communist

Countries at a Cost of < $66/kg U_3O_8*

Country	Reserves, 10^3tU	Estimated Additional Resources, 10^3tU
Algeria	28	-
Argentina	30	40
Australia	354	102
Brazil	10	10
Canada[+]	173	605
France	55	40
Gabon	20	10
India	32	25
Niger	50	30
South Africa, including South West Africa	276	75
Spain	103	105
Sweden	300	-
USA	490	-
Others	54	45
Totals (rounded)	1970	1100

* See reference 4

Official Canadian figures for recoverable
resources minable at up to $88/kg U_3O_8

The estimated availability figures for uranium production are given in Table 2 but the estimated demand for uranium over the next 25 years requires a more accurate assessment that at present available; the largest estimates of demand by USERDA[5] are lower than the lowest OECD/NEA/IAEA figures[6]. It seems likely that the demand figures will be revised downwards, as has been the case since 1968, up to 1990 and beyond. If the data shown in Table 1 are acceptable, there are substantial supplies for the future assuming that there would be normal additional yearly tonnages. If however, fast breeder reactor systems are not introduced within the next decade, there will be an ever increasing need to find new reserves and resources.

Table 2

Uranium Availability, Tonnes*

Country	Planned or produced 1975	Likely production under favourable circumstances 1985
Argentina	50	500
Australia	-	15000
Canada	3560	12000
Denmark	-	500
France	1700	3000
Gabon	800	1200
Germany	30	250
Japan	5	20
Niger	1200	5000
Portugal	115	150
South Africa	2700	14000
Spain	150	500
USA	9610	31000
Sweden	-	500
Others	-	6400
Totals (rounded)	20000	90000

* See reference 4

The extraction of uranium from the ore is complex, since the ore normally contains only a few hundred ppm uranium. The details of the treatments depend on the exact nature of the ore but a typical flow sheet is shown in Table 3. For more details the reader is referred to a number of review articles [7-15].

The main isotopes of uranium ^{235}U and ^{238}U can be separated by a number of techniques of which the main methods are gaseous diffusion[16-17] and the gas-centrifuge[18-19]; uranium hexaflouride (UF_6) is employed for both methods. Recent developments have included considerations of laser and gas jet or nozzle separation techniques[20]. The UF_6 is produced directly from the 'yellow cake' uranium dioxide (UO_2)[21-23]. Uranium tetrafluoride (UF_4), which is used for metal fuel production, is obtained in a similar manner.

Table 3

Extraction of Uranium Oxides from the Ore

ORE

　Crushing
　Ore dressing ....　　Physical concentration (electronic or
　　　　　　　　　　　　　　　　sedimentation). Roasting.

CONCENTRATE

　　　　　　　　　　　　　　　　⎧Sulphuric acid leaching (95-98% efficient)
　Wet extraction ....　　⎨or
　　　　　　　　　　　　　　　　⎪Sodium carbonate leaching (for alkaline-
　　　　　　　　　　　　　　　　⎪carbonate ores): after this process the
　　　　　　　　　　　　　　　　⎩next stage can usually be omitted.

CRUDE LIQUOR ($\sim$5g U/1)

　　　　　　　　　　　　　　　　⎧Ion exchange (absorb on resin from SO_4^{--} rich
　　　　　　　　　　　　　　　　⎪soln.: elute with strong HNO_3 or HCl)
　Intermediate purification　　⎨or chemical precipitation (with H_2O_2+NH_4OH:
　　　　　　　　　　　　　　　　⎪for conc. leach liquors)
　　　　　　　　　　　　　　　　⎪or solvent extraction (as in final
　　　　　　　　　　　　　　　　⎩purification stage).

INTERMEDIATE LIQUOR ($\sim$15 g/1)
　Precipitation ....　　With ammonia or sodium hydroxide
　　　　　　　　　　　　　　　　or magnesia.

HIGH GRADE CONCENTRATE
(50-80% U_3O_8)

　　　　　　　　　　　　　　　　⎧By solvent extraction from uranyl nitrate
　　　　　　　　　　　　　　　　⎪solution with 20% tributyl phosphate or
　　　　　　　　　　　　　　　　⎪alkyl amines or organophosphorous compounds
　Final purification ....　　⎨in kerosene or hexane using a counter-
　　　　　　　　　　　　　　　　⎪current process. Originally ether was used:
　　　　　　　　　　　　　　　　⎪the necessary double extraction gave higher
　　　　　　　　　　　　　　　　⎩purities than the single TBP one.

PURE URANYL NITRATE SOLUTION
　Denitration ....　　⎰Ammonium diuranate precipitation and
　　　　　　　　　　　　　　　　⎱ignition or thermal denitration at $300^\circ C$.

PURE UO_3 OR U_3O8
(600 ppm total impurities)

　　　　　　　　　　　　　　　　⎰Reduction with hydrogen at $700^\circ C$ (UO_3 gives
　Reduction ....　　⎨a more reactive oxide suitable for metal
　　　　　　　　　　　　　　　　⎱production).

PURE UO_2 ('Yellow Cake')

3.   The Transuranium Elements

3.1   Neptunium. ^{237}Np with a half-life of 2.14 x 10^6 years is the only isotope
of neptunium which is easily prepared and is sufficiently stable to be obtained
in massive quantities.   It is formed by neutron irradiation of uranium^{24} by the
2 reactions,

$$^{235}U \xrightarrow{n,\gamma} {}^{236}U \xrightarrow{n,\gamma} {}^{237}U \xrightarrow[6.7d]{\beta^-} {}^{237}Np$$

$$^{238}U \xrightarrow{n,2n}$$

The second reaction accounts for 70% of the neptunium production from the
irradiation of natural uranium.   The preparation and recovery of neptunium is
therefore best undertaken by extraction from irradiated fuel elements as a by-
product during the recovery of uranium and the separation of the larger quantities of plutonium also formed.   The 2 main extraction processes used are
solvent extraction and ion exchange.

3.2   Plutonium.   Plutonium is only found in nature in very small quantities
because of the short half-lives of its isotopes.   Only a little primordial ^{244}Pu
remains[25], although some ^{239}Pu is continuously produced by the natural neutron
flux irradiation of uranium ores; the amount produced varies with uranium content and environment.   For example, one sample of pitchblende was found to
contain one atom of ^{239}Pu in 3 x 10^{11} atoms of uranium[26].

There are several plutonium isotopes with half-lives long enough to permit
their handling in large quantities. ^{239}Pu is the most readily available isotope
being produced in reactors by the irradiation of uranium

$$^{238}U \xrightarrow{n,\gamma} {}^{239}U \xrightarrow[23.5m]{\beta^-} {}^{239}Np \xrightarrow[2.33d]{\beta^-} {}^{239}Pu$$

The higher isotopes are produced by further neutron capture and consequently
are most difficult to make, particularly as fission rather than capture is more
likely to occur in some cases and ^{243}Pu is likely to decay by β^- emission before
capture can occur.   The half-lives of the Pu isotopes are shown in Table 4.

The isotope content of plutonium recovered from power reactors depends
upon the type of reactor as well as the burn-up: some typical examples are
given in **Table** 5.   A very low burn-up is required if extensive formation of
the higher isotopes is to be avoided.

The radiological and health physics aspects of this element have been
reviewed in the Plutonium Handbook[27] and more recently in a Medical Research
Council pamphlet entitled 'The Toxicity of Plutonium'.   Although the maximum
permitted body burden is only 0.75 µg and the maximum permissible concentration
in laboratory air is 3.2 x 10^{-14} g/l, hundreds of grams may be safely handled
in glove boxes[28].   The criticality hazard, which is the chance of a spontaneous

Table 4

Plutonium isotopes: Half-lives

Isotope	Decay mode	Half-life	Method of Formation
^{238}Pu	α	89.6y	^{237}Np$(n,\gamma)^{238}$Np $\xrightarrow{\beta^-}$
^{239}Pu	α	24,400 y^a	^{238}U$(n,\gamma)^{239}$U $\xrightarrow{\beta^-}$ ^{239}Np $\xrightarrow{\beta^-}$
^{240}Pu	α	6600 y	^{239}Pu(n,γ)
^{241}Pu	β^-	13 y	^{240}Pu(n,γ) or ^{238}U(α,n)
^{242}Pu	α	3.8×10^5y	^{241}Pu(n,γ)
^{243}Pu	β^-	5.0 h	^{242}Pu(n,γ)
^{244}Pu	α	8.3×10^7y	^{243}Pu(n,γ)

^{a} F L Oetting, Plutonium 1970 and Other Actinides, W N Miner (Ed) Met. Soc. AIME, New York (1970), finds 24,065 years by a calorimetric method.

Table 5

Isotope Contents of Pu Produced in Different Nuclear Reactors

Reactor	Burn-up (MWd/tonne)	Percentage of each isotope				
		238	239	240	241	242
Production	600	0	95	4.6	0.4	
Pressurised water	5000	0	91	8.5	1.5	
	25000	0.8	63	24	11	2.7
Boiling water	5000	0	89	10.5	1.3	
	25000	0.3	51	32	13.6	5.6
Fast breeder	3 years exposure Core		61	33	4.7	1.2
	Blanket		96	3.5	0.2	

self sustaining neutron chain reaction occurring, depends markedly on the form of the plutonium. The minimum critical mass of a solid sphere is about 10 kg unreflected or 5.6 kg when the neutrons are fully reflected. The minimum amount which can become critical in aqueous solution is only 509g when homogeneously distributed and the neutrons are fully reflected. The maximum external dose for plutonium is generally less restrictive than the other hazards; the actual radiation levels increase with the amount of higher isotopes present and handling times have to be reduced for plutonium from power reactors.

3.3 Americium. There are 2 isotopes of americium, with half-lives sufficiently long to allow the handling of massive quantities. ^{241}Am has a half-life of 458 years and ^{243}Am 7400 years. Both are formed by multiple neutron capture:

$$^{239}\text{Pu} \xrightarrow{\text{n}} {}^{240}\text{Pu} \xrightarrow{\text{n}} {}^{241}\text{Pu} \xrightarrow{\text{n}} {}^{242}\text{Pu} \xrightarrow{\text{n}} {}^{243}\text{Pu}$$

$$13.2\text{y} \downarrow \beta^{-} \qquad\qquad 5\text{h} \downarrow \beta^{-}$$

$$^{241}\text{Am} \qquad\qquad\qquad {}^{243}\text{Am}$$

^{241}Am is more readily available although it has a shorter half-life than ^{243}Am, because few steps are required for its formation. During the process, two thirds of the ^{239}Pu atoms and approximately 0.7 of the ^{241}Pu are consumed by fission. A detailed review of americium chemistry has recently appeared[29].

3.4 Curium, Berkelium and Californium. Two isotopes of curium are currently available in gram quantities; these are both α-emitters with short half-lives: ^{242}Cm - 162 days and ^{244}Cm - 18 years. Some heavier isotopes have long half-lives but are not easily prepared by irradiation.

For the preparation of ^{242}Cm, cermets of americium oxide (AmO$_2$) in aluminium have been irradiated[30].

$$^{241}\text{Am} \xrightarrow{\text{n,}\gamma} {}^{242}\text{Am} \xrightarrow[16\text{h}]{\beta^{-}} {}^{242}\text{Cm}$$

^{244}Cm is one of the elements produced by prolonged irradiation of ^{239}Pu. This process is also used to make several of the other higher actinide isotopes at the Oak Ridge Laboratory in Tennessee[30-32]. The reactions which occur are:

$$100\% \; {}^{239}\text{Pu} \xrightarrow{\text{n}} 30\% \; {}^{240}\text{Pu} \xrightarrow{\text{n}} {}^{241}\text{Pu} \xrightarrow{\text{n}} 10\% \; {}^{242}\text{Pu} \xrightarrow{\text{n}} {}^{243}\text{Pu}$$

$$\downarrow \qquad\qquad\qquad\qquad \downarrow$$

70% fission   66% fission

$$\xrightarrow{\beta^{-}} {}^{243}\text{Am} \xrightarrow{\text{n}} {}^{244}\text{Am} \xrightarrow{\beta^{-}} {}^{244}\text{Cm}$$

$$^{244}\text{Cm} \xrightarrow{\text{n}} {}^{245}\text{Cm} \xrightarrow{\text{n}} {}^{246}\text{Cm} \xrightarrow{\text{n}} {}^{247}\text{Cm} \xrightarrow{\text{n}} 0.8\% \; {}^{248}\text{Cm} \xrightarrow{\text{n}} {}^{249}\text{Cm}$$

$$\downarrow \qquad\qquad\qquad \downarrow \qquad\qquad\qquad\qquad\qquad\qquad \downarrow \beta^{-}$$

85% fission   50% fission ^{249}Bk

$$^{249}\text{Bk} \xrightarrow{\text{n}} {}^{250}\text{Bk} \xrightarrow{\beta^{-}} {}^{250}\text{Cf} \xrightarrow{\text{n}} {}^{251}\text{Cf} \xrightarrow{\text{n}} 0.3\% \; {}^{252}\text{Cf}$$

$$\downarrow$$

50% fission

Some kilograms of ^{239}Pu are irradiated at a moderate neutron flux in a reactor (Savannah River) which has the capacity to remove the large amount of heat produced. The product (mostly ^{242}Pu with a little americium and curium) is then processed by conventional solvent extraction processes, refabricated into targets and irradiated again in the high flux isotope reactor (HFIR). On processing the target, the higher actinide elements are extracted. These include ^{249}Bk with a half-life of 314 days, ^{252}Cf which has a half-life of 2.6 years and is spontaneously fissile and ^{249}Cf - half-life of 360 years. ^{247}Bk with a longer half-life (1400 years) is not easily made.

4. Nuclear Reactors

The commercial development of nuclear power for electricity generation has led to a continual development of reactor design and a steady increase in the number of reactors planned or installed. The present scale of this industry can be assessed from the summaries prepared by the International Atomic Energy Agency (IAEA) and other bodies[33-34]. In the member states of the IAEA at July 1976, there were 187 operating power reactors with a generating capacity close to 80 GW (e); a further 348 reactors with a capacity of 328 GW (e) were either under construction or planned. Throughout the world there are also over 350 research reactors of varying design or application.

There are several differing designs of reactors which have arisen from the different compromise solutions that have been devised to optimise the complex factors that influence reactor design. Nuclear physics, economics and materials technology are just some of the topics that have to be considered. The broad classification into thermal and fast reactors, depending upon the speed of the neutrons maintaining the chain reaction, is well known. Beyond the fission reactor lies the fusion reactor which, by comparison, is in its early stages with considerable development still required before commercial exploitation is achieved. Fusion systems are not considered further, save to give reference to further reading[35]. After early work on differing systems, the development of thermal reactors has followed 2 distinct lines dictated by the choice of coolant and moderator. These are the water-cooled reactors where water is also the moderating material - both light water and heavy water are used - and the gas-cooled reactors where the moderator is graphite[36].

The water-cooled reactors are sub-divided again following the various design choices that can be made. There are the well-known pressurised water reactors (PWR) where the coolant remains in the liquid phase under pressure. Then there are the boiling water reactors (BWR) in which the coolant is allowed to boil to raise steam directly in the reactor core. Both systems use enriched uranium dioxide clad in zirconium alloy fuel tubes. For reasons of increased neutron utilisation and the ability to use natural uranium without enrichment, a boiling system using heavy water has been developed principally by Canada and

is known as the CANDU reactor system. The British Steam Generating Heavy Water
Reactor represents a further development of an intermediate nature using a low
enrichment uranium oxide fuel. The moderator is heavy water but light water is
used for cooling[37]. Both the BWR and PWR are pressure vessel reactors where the
core and coolant are contained in a large pressure vessel. The SGHWR and CANDU
are pressure tube systems where the coolant and moderator are separated; the
coolant is contained within tubes which pass through the reactor core and which
contain the fuel.

The early gas cooled reactors, of which the British Magnox reactors are
well known, use natural uranium in metallic form clad in a magnesium alloy
cooled by carbon dioxide[38]. The reactors are large for their heat output because
of their large number of fuel channels and the size of the graphite moderator
structure in which they are contained. The carbon dioxide coolant system has
been developed further in Britain as the Advanced Gas Cooled Reactor (AGR) in
which the coolant temperatures are increased above those at which magnesium-
clad uranium metal fuel may be used[39]. Stainless steel-clad fuel in the form
of enriched uranium dioxide is used in a graphite moderated system. The gas-
cooled system that has attracted world wide interest is the helium-cooled high
temperature reactor (HTR) which is graphite moderated and is fuelled by enriched
uranium oxide in particulate form embedded in a graphite matrix[40]. Each particle
is coated with layers of pyrolytic carbon and usually silicon carbide which act
as a containment and pressure vessel for each particle to retain the fission
products both solid and gaseous which are released from the fuel. The system
is of particular interest because of its high outlet temperatures ($\sim 800^\circ$C) that
could be used directly for process heat or to drive direct cycle gas turbines
and would result in a high station efficiency. The reactor is also very suitable for operation on the thorium - ^{233}U cycle as opposed to use of ^{235}U.
Homogeneous solution, fused salt, liquid metal, and gaseous systems have been
investigated also but none have achieved commercial exploitation. Some hybrid
systems, e.g. water-cooled graphite moderated reactors, exist but again commercial exploitation is limited. The present distribution amongst the various
systems described is given below based on reactors operating at July 1976.

	PWR	BWR	Heavy Water	Magnox Type	AGR	HTR	Other
Number	66	49	11	36	5	2	13
GW (e)	39	23	3	7	2.5	0.3	3

Fast reactor development is at a much less advanced state with the main
emphasis on the sodium-cooled system. Sodium is chosen because of its good heat
transfer properties which are essential to remove heat from the very compact
core that must be employed in an economic design. From the nuclear physics
aspects, the properties of sodium are superior to those of water but inferior

to those of a gas such as helium. Prototype sodium-cooled fast reactors for power generation are in operation in France, Britain and Russia and are under construction in other countries[41]. The fuel for the prototype reactors is either enriched uranium oxide or mixed uranium-plutonium oxide. The uranium-plutonium oxide fuel represents that which would be used in a commercial fast reactor and would have an outer blanket of uranium oxide or possibly carbide. A choice of design using a pool or loop concept is available for the reactor itself. In the pool concept, the whole of the primary circuit is contained in one large sodium tank. In the loop design, primary sodium from the core is fed to the intermediate heat exchangers, between the primary and secondary sodium circuits, mounted externally from the core vessel.

Although the sodium system is by far the most highly developed, other systems have been considered and offer considerable potential. Steam—cooled systems have been disregarded because of the poor nuclear properties of water for fast reactor application. A gas—cooled system is particularly attractive because of the near ideal physics conditions that can be achieved. Much of the technology developed for the high temperature thermal reactor can be translated to the fast reactor application using a particulate fuel possibly directly cooled and with a coating of silicon carbide and graphite. The neutron utilisation is particularly good and leads to a high breeding ratio provided the amount of moderating material (e.g. the graphite in the fuel coating) is limited. Fused salt systems have been considered, again because of their good neutron utilisation, but are not being followed extensively.

5. Reactor Components

The wide variety of conditions encountered in the designs of the many reactor types that are now developed has resulted in a series of in-depth technical studies of materials properties and behaviour in appropriate environments. In addition to the consideration of conventional engineering factors such as corrosion, strength and stability, other constraints are set by the inherent nuclear properties of the materials employed and by their behaviour under irradiation. Each system gives rise to its own particular range of problems often reducing to one or 2 key aspects that have to be examined extensively. Some of these topics are outlined in relation to the various reactor systems in which they appear.

In thermal reactors, the choice of materials in the primary circuit is limited to those with a low neutron absorption cross section. High cross section materials are excluded except for the purposes of reactivity control. Further it must be certain that such materials as are chosen are compatible with the environment in which they are placed. A 30 year life may be required for permanent in-core structural components and evidence that this endurance may be achieved is required by inspection or other means. The redistribution of

material arising from corrosion or other reactions around the reactor circuits
must be considered for reasons of mass and activity build-up, which have consequent effects upon heat transfer, component handling and waste disposal.

In the water reactor field, zirconium is highly favoured as a fuel cladding
material. It is normally associated with 1-3% of hafnium which has a high
cross-section; for reactor applications, therefore, it must be produced virtually
free from hafnium. Zirconium, however, is not particularly suitable for use
directly in water. Protection of the metal by an oxide film is essential and
there is a strong tendency for the film to spall while still comparatively thin.
Use is therefore made of various alloys containing a few per cent of other
metals. The commonest of these is the zircaloy range, which does not suffer
from spalling until the oxide films are quite thick, and which has corrosion
properties not very dependent on their treatment during manufacture. Various
refinements have been made within the range and a series of alloys of differing
properties and composition have been developed. Zircaloy-4 is an alloy offering
some advantages over the better known zircaloy-2; its lower nickel content leads
to a lower uptake of corrosion hydrogen, and therefore to a reduced risk of
hydrogen embrittlement. Where greater mechanical strength is required, a
zirconium-niobium alloy can be used, but this has the disadvantage that its
corrosion behaviour is more sensitive to its treatment during manufacture, and
its corrosion rate in water is affected by the presence of oxygen in the absence
of radiation. Zirconium alloys, and their corrosion behaviour, have recently
been reviewed[42].

Steels are commonly used for out-of-core structural materials and for
pressure vessels. Cobalt, which is a common impurity in steels, has a fairly
high activation cross-section, and ^{60}Co has a fairly long half-life (5.3 y).
The movement of corrosion products around the circuit can lead to considerable
contamination levels on out-of-core surfaces, with consequential high radiation
doses to maintenance personnel, unless steels with a very low cobalt level are
used. A useful survey of the transport and deposition of activated corrosion
products has been given[43]. For heat exchangers and condensers high nickel
alloys such as Inconel and Incoloy are often used, but because of its better
corrosion properties, titanium offers some advantages.

The principles which apply to the selection of structural materials also
influence the way in which the chemistry of the coolant is controlled[44]. For
pressurised water reactors, alkaline conditions are generally favoured for the
coolant, because this leads to conditions where the solubility of iron is low
and the dependence of solubility on temperature is favourable. The levels of
circulating corrosion products and the deposition on core components are
therefore minimised. The principal dosing agents are lithium hydroxide and
ammonia. Where lithium is used it must be virtually free from ^6Li which would

give rise to unacceptable levels of tritium by the $^6Li(n,\alpha)^3H$ reaction; it is necessary also to add hydrogen to maintain reducing conditions. Ammonia gives less satisfactory control of corrosion products, but, where boiling takes place in the core, it may be used in place of lithium hydroxide, which may concentrate under such conditions and give rise to stress corrosion cracking of steels and to very rapid corrosion of Zircaloy. An alternative approach in a boiling system is to use low conductivity water with carefully controlled oxygen levels under neutral conditions, and in many cases this is found to give the best performance.

It is often found desirable to exercise part of the control of reactivity by introducing neutron absorbing material into the coolant. For this purpose the usual reagent is boric acid. This can be used in conjunction with alkaline dosing with lithium hydroxide, since boric acid undergoes very little dissociation at high temperatures and hence has little effect on the coolant pH.

In gas-cooled thermal reactors, problems centre around the chemical reactions of the coolant with the reactor components. In the case of carbon dioxide radiolytically induced reactions may occur with the graphite moderator or with the metal surfaces that are deleterious over the 30 year lifetime of the reactor[45-46]. In the early reactors using magnox clad fuel elements, these reactions are not of particular consequence, because temperatures and radiation levels are low. In the advanced gas cooled reactor (AGR) designs, reactions are more rapid and, if allowed to proceed unchecked, could consume about 5% of the graphite moderator structure each year. The graphite is protected by adding to the coolant small concentrations of carbon monoxide, methane and water vapour which inhibit the corrosion process. However, in the radiation field, compositional changes in the gas occur and various quasi-equilibrium states or reaction rate competitions are set up which depend in complex ways upon temperature, gas flow rate over the reacting surface and radiation dose. These result in situations where oxidation or corrosion may be less inhibited in some places than others and also in tendencies to form carbon deposits on surfaces such as those of the fuel pins. A layer of carbon as thin as 50 μm on a fuel pin surface can significantly alter the heat transfer to the coolant and lead to an increase in temperature of the fuel. Both oxidation and deposition problems have been studied extensively to specify the optimum gas composition to minimise these effects.

The high temperature gas-cooled reactors are cooled by helium because of its inert nature. However, impurity gases are always present to some degree and effort has centred upon means of purifying the coolant principally to remove the oxidising species[47]. A maximum impurity concentration is specified that can be allowed without deterioration of the core and component materials. Typically concentrations of carbon monoxide and hydrogen must each be maintained below 10 volume parts per million with a corresponding oxygen potential appropriate to a 10:1 CO/CO_2 or H_2/H_2O composition. This is achieved by a very high standard of leak tightness on the gas circulating plant and by employing low temperature clean-up plant. This is essential also to contain the small but significant arisings of fission product contamination in the coolant gas from failed or broken coated particle fuel. The extent of such release must be kept typically below 10^{-5} of the total fission product activity and is treated together with impurities in the gas clean-up plant. A small by-pass flow is passed through a cyclone unit and cooled to $100^{\circ}C$ to deposit metallic and halogen fission products. A water-cooled charcoal bed is used to delay the noble gas fission products to eliminate most of their decay heat and activity. Thereafter the gas is passed through a combined oxidation reduction bed (Cu + CuO) oxidising impurities to carbon dioxide and water which may be removed as solids in a freezer bed. A similar series of problems would arise in a gas-cooled fast reactor which is again helium cooled and the thermal reactor technology would be transferred directly.

The use of liquid metal for the coolant in a fast reactor has given rise to many problems in the fields of purification, corrosion, material transfer and component endurance[48]. Radiation effects tend to be less important in these systems because of their highly ionised state and reactions are governed by conventional kinetic and equilibrium thermodynamic effects. However, compared with aqueous and gaseous systems, the supporting science is far less developed and uncertainty surrounds even relatively basic factors such as metal and nonmetal solubilities.

One of the possibly more surprising features is the relative inertness of liquid sodium in the pure condition principally because of its reducing nature. The presence of oxygen, which is one of the principal impurities of concern, alters this situation, raising the oxygen potential to the point where reaction with structural components is possible. Sodium is compatible with most steels and some refractory metals, and is adequately contained by stainless steel. Some corrosion does occur by a series of complex reactions which depend upon oxygen impurity levels and involves such reactions as the formation of sodium chromite films which have protective qualities. At low oxygen levels, some preferential extraction of manganese from the steel is observed; this causes concern because of the induced activity from ^{54}Mn which has a strong γ emission and a long half-life. Whilst the corrosion experienced is not too severe and can be restricted by control of the oxygen impurity, the behaviour of the corrosion products themselves is of concern because of their tendency to deposit around the circuit on surfaces such as those of the primary to secondary circuit heat exchangers with consequent loss of thermal performance. This aspect of behaviour together with that of the associated activity has required extensive study in loop experiments simulating the conditions in the reactor.

The behaviour of carbon in sodium is also of concern because sodium can transport carbon from one part of the circuit to another and carburise or decarburise the steels with which it is in contact and consequently impair their properties[49]. The overall process is controlled by carbon diffusion in the steels, and its behaviour in sodium which includes its rate of dissolution and thermodynamic activity at the different temperatures that exist in the circuit. Electrochemical meters to measure carbon activity have been developed and are being used to correlate activity measurements with the observed carburisation phenomena. Hydrogen is a further species of interest, because it is introduced into the secondary circuits by diffusion through the sodium/water heat exchangers of the steam generating plant and arises in the primary circuit as tritium which is generated as a fission product or as an activation product of lithium or boron present as an impurity or control rod material respectively. The distribution of hydrogen and tritium around the circuits of a reactor has been predicted from modelling studies based on the established physico-chemical relationships.

Clean-up and impurity control methods in sodium circuits based on cold trapping have been developed using the large changes in solubility that exist with temperature. Below $150^{\circ}C$, the solubility of oxygen and hydrogen has dropped to a few parts per million such that passage of a small by-pass flow through a trapping device cooled to this temperature reduces the impurity concentrations to a low level. Whilst the engineering of such devices to ensure a high efficiency and capacity is complicated and the absolute level that can be achieved for a given temperature is subject to some controversy, the cold trap provides an effective method of impurity control and a basis upon which the performance of differing circuits can be compared. Other purification methods of hot trapping using refractory metal getters or by using precipitation techniques are available to achieve very high purities or to remove specific impurities respectively.

Many other problems of circuit and materials behaviour exist in the various reactor designs that have been described; reference should be made to the extensive literature on each reactor system for further information.

6.   Fuel Fabrication for Nuclear Power Reactors

6.1   Introduction. Although compilations of data on the world's power reactors describe a large number of different designs of nuclear reactors as discussed in section 4, the main types of reactor presently used for civil power generation fall into a more restricted number of categories. The salient features of these major reactor types are summarised in Table 6. It can clearly be seen that in terms of fuel employed there are primarily only 3 types, namely metal, oxide or

TABLE 6

Characteristics of Major Reactor Types

Reactor	Magnox	AGR	LWR	CANDU	Water/Graphite	HTR	LMFBR
Country of origin	Britain France	Britain	USA	Canada	USSR	Several	Several
Fuel material	Metal	Oxide	Oxide	Oxide	Oxide	Carbide (or oxide)	Oxide
Fuel form	Rod	Pellet	Pellet	Pellet	Pellet	Spherical Particle	Pellet
Fuel Enrichment	Natural U 0.7% U235	2% U235	3-4% U235	Natural U	2% U235	93% U235	15-30% PuO_2
Cladding	Magnox	Stainless Steel	Zircaloy	Zircaloy	Zirconium Niobium alloy	Graphite/ Silicon carbide	Stainless steel
Moderator	Graphite	Graphite	H_2O	D_2O	Graphite	Graphite	-
Coolant	CO_2	CO_2	H_2O	D_2O	H_2O	Helium	Na

carbide, each being also distinguishable by its form, that is rod, pellet or
coated particle respectively.

Historically uranium metal or alloy was used as fuel in the early power
generation demonstration reactors of the fast, light water and magnox types;
but only the last named has continued to employ metallic fuel. Perhaps the
most significant technological advance leading to electricity generation from
nuclear power becoming competitive with that from conventional fossil fuel, was
the development of ceramic nuclear fuels. In terms of nuclear powered capacity
either installed, or being built, the oxide fuel form predominates in most
countries of the world mainly as slightly enriched uranium oxide in the form of
pellets.

Although metallic fuels were used extensively in the early generation of
power reactors, uranium and plutonium based metallic fuels possess properties
which place serious limitations on their development for the high power density
requirements of modern power reactors. Neither uranium nor plutonium are well-
behaved engineering materials, nor do they possess high melting points. They
exhibit a high degree of reactivity towards water and oxygen and have poor
irradiation stability.

By comparison, the good thermal performance, high irradiation stability
and chemical inertness in reactor coolants of uranium dioxide were more than
sufficient to override its lower fissionable atom density. The subsequent
search for other ceramic fuels particularly with higher fissile densities led to
the exploration of carbide and to a lesser extent nitride and other related
fuels. The reactivity of the carbides with water and carbon dioxide has limited
their possible use so far to high temperature gas-cooled reactors, and to sodium
cooled fast reactors, where improvement in breeding gain is desired.

The processes that have been used to fabricate these 3 types of fuel will
be outlined with direction to more detailed review articles. Some mention of
refinements that have taken place in the light of irradiation experience of
these fuels will also be made, together with brief indication of alternative
fuels that have been evaluated.

6.2 **Metallic Fuel Fabrication.** The main stages in the conversion of uranium
ore concentrate (UOC) to natural uranium metal fuel elements for magnox reactors[50] are

- nitric acid dissolution, filtration and purification by solvent extraction.
- evaporation of the purified uranyl nitrate solution, conversion
  successively to UO_3 by thermal denitration in air and thence to UO_2 by
  reduction in hydrogen.
- conversion to UF_4 by hydrofluorination with hydrogen fluoride (HF) gas.

- reduction of UF_4 with magnesium within graphite lined steel vessels,
  with molten uranium falling to the base of the vessel to form a billet
  on solidification.
- vacuum remelting and casting to rods followed by heat treatment and
  machining.

Adjustments to the composition of the uranium particularly with respect to
the presence of iron (200-500 ppm) and aluminium (500-1200 ppm) together with
control of grain structure by heat treatment have been shown to exert a marked
effect in decreasing the irradiation growth of the fuel[51-52].

The early series of French CO_2 - gas cooled reactors used natural uranium
metallic **fuel rods.** These employ the Sicral F1 alloy which contains very
similar iron and aluminium levels to the UK magnox fuel.

6.3 <u>Oxide Fuel Fabrication</u>. There are many similarities in the fabrication
processes used to provide fuel for several of the different types of thermal
reactor. Most require the uranium to be enriched in ^{235}U isotope and the
gaseous compound UF_6 from the enrichment process is the common starting material.

Both wet and dry routes[53] have been developed for converting UF_6 to
sinterable UO_2 powder. In the wet route, gaseous UF_6 is dissolved in water, and
reacted with either ammonia or ammonium carbonate. The precipitate either
ammonium diuranate (ADU) or ammonium uranylcarbonate (AUC) is separated, dried
and treated with a hydrogen-steam mixture for the reduction to UO_2. In the dry
processes, the UF_6 is reacted with superheated steam to form solid uranyl
fluoride (UO_2F_2) which is reduced by hydrogen to UO_2 powder, in either fluidised
beds or rotary kilns. Efficient removal of fluorine is imperative for both
routes.

Mass production of UO_2 pellets[54] is carried out by cold pressing the UO_2
powder to green pellets in mechanical or hydraulic presses and sintering at
about 1600-1700°C in a hydrogen containing atmosphere. This reduces the oxygen
level to near stoichiometric composition ($UO_{2.0}$-$UO_{2.02}$). Considerable (~50%)
shrinkage takes place on sintering which, though isotropic, is not uniform due
to pressed density variations in the green pellet. Various additives to act as
a pellet binder and die lubricant, and to achieve free flowing powders for uniform die filling, are often employed. These are removed prior to sintering in
a 'de-bonding' preheating step.

The required density of the sintered product differs between the reactor
types, being in excess of 95% theoretical for AGR pellets and rather less
(90-95%) in the case of water reactor fuel. Controlled microstructure in terms
of pore and grain structure has more recently emerged as important, in order to
achieve maximum resistance to in-reactor densification, swelling and gas
release[55-56].

The incorporation of plutonium to form a mixed uranium plutonium oxide fuel $((U,Pu)O_2)$ has on the one hand been evaluated as a means of recycling plutonium in thermal reactors. On the other hand it has been accepted as the universally favoured fuel for liquid metal-cooled fast reactors[57-58]. Although plutonium enrichment levels differ significantly (2 and ~20% PuO_2 respectively) the fabrication technology and processes have much in common with each other and with those for UO_2. The $(U,Pu)O_2$ powder is obtained either by mechanically mixing UO_2 and PuO_2 or by co-precipitation from solutions containing uranium and plutonium.

The preferred fuel form for $(U,Pu)O_2$ fuels is as sintered pellets but some attention has been given to the use of particulate fuels vibrationally compacted into fuel tubes. The particles are prepared by pellet crushing or by gel-precipitation techniques as described in section 6.4.

6.4 **Carbide Coated Fuel Fabrication.** The introduction of the coated fuel particle[59-60] for HTR systems is based on the requirement to contain both fuel and the fission products released within a high temperature low neutron-absorbing material. Graphite is an ideal material but suffers from strength limitations. The concept therefore is to use a small combustible kernel surrounded by a protective graphite coating. The first HTR cores largely used kernels of uranium or uranium thorium carbide (UC_2) or $(U,Th)C_2$ but oxides could be preferred in future HTR systems.

There are 2 main methods of manufacturing the kernels[61]. The dry method of powder agglomeration involves the mixing and homogenizing of the relevant oxide powders with binder and with or without carbon, depending on whether oxide or carbide is required. By the simple action of gyrating in a bowl, the powders agglomerate and continue to grow to the required diameter. After sieving, heat treatments are given to remove binders and to perform the carbothermic reduction of the oxide/carbon mixture to dicarbide where appropriate. A final sintering step is carried out for densification. The wet methods, using procedures such as 'sol-gel' or 'gel-precipitation' take advantage of the surface tension of a liquid to form drops which when sprayed into another phase (gas or liquid) solidify, maintaining a spherical form following a chemical reaction. The spheres thus formed are washed, dried, calcined and finally sintered. Incorporation of a dispersion of carbon within the colloidal liquid again allows the preparation of carbides as well as oxides by these methods.

The protective coatings employed have been almost exclusively of pyrolytic carbon, consisting frequently of multilayers of differing densities and structure. Inclusion of a layer of silicon carbide (SiC) increases the retention of fission products, particularly strontium, barium and caesium. The coatings of pyrolytic carbon are deposited onto kernels in fluidised beds by means of high temperature (1200-2000°C) decomposition of hydrocarbons, typically methane

and acetylene in an inert carrier gas. The choice of hydrocarbon, temperature
and deposition rate controls the variety of structural and physical properties
that can be achieved in the different layers. Methyl trichlorosilane
(CH_3SiCl_3) is used in the same process for the deposition of SiC.

7. The Chemistry of Irradiation

An irradiated nuclear fuel is one of the most complicated high temperature
systems found in industry today. The transformation of the actinide elements by
fission to produce more than 30 different fission product elements results in
many complex chemical changes. These chemical changes are very pronounced,
particularly within the fuel elements of a liquid metal-cooled fast reactor
(LMFBR) where 10-15% of the original actinide elements could be fissioned, and
within the fuel particles of the High Temperature Reactor. Because the changes
within the fast reactor fuel elements are so much greater than in most of the
thermal reactor systems, we shall confine the discussion of the 'chemical
changes' to those in the fast reactor systems, where the fissile element is
plutonium, and to a lesser extent to the HTR system. The quantity of fission
product elements formed depends both on the energy of the neutron initiating the
fission reaction and upon the particular actinide element. Typical yield curves
for the fission of ^{235}U, and ^{239}Pu, at thermal and fast neutron energies are
shown in Fig. 1.

In the sodium-cooled fast reactor, the fuel elements may consist typically
of a column of fuel pellets 5 mm in diameter, inside a stainless steel can 0.75 mm
thick. Most commonly, the fuel is in the form of a single phase solid solution
of uranium and plutonium oxides ($U_{1-x}Pu_xO_{2-y}$) where x = 0.15-0.40 and the optimum
value of y will depend on the operating conditions. The pellet densities will be
such as to obtain an overall smear density of 70-80% theoretical density. Recently attention has been given to gel-precipitated fuels, instead of pellet fuels.
These are in particulate form but have a similar overall smear density when
packed into the fuel element tubes and sinter to a pellet-like configuration
during operation. Fast reactor fuel elements are operating in steep temperature
gradients with the result that mass transport and some limited chemical reactions
between the fuel and stainless steel occur. In oxide fuel, a central void of
about one-third the diameter of the pin is formed and is surrounded by a nearly
100% dense region of columnar grains. There is then a region of equi-axial
structured grains where grain growth is still occurring slowly. Finally the
outer peripheral region contains essentially unrestructured fuel owing to the
relatively low temperatures during irradiation. The temperature of the fuel
centre can be 2300K in the hot centre and 1100K at the fuel surface in the coldest
region.

In the long term it may well be that the oxide fast reactor system will not

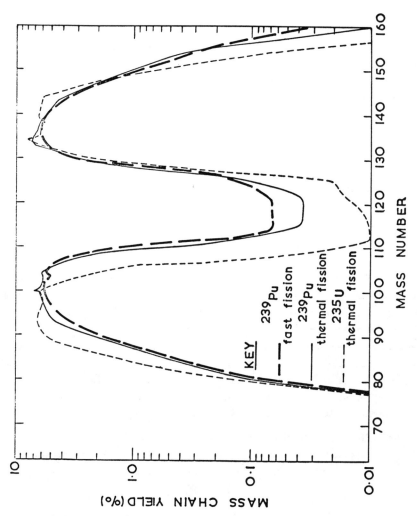

FIG. I. FISSION YIELD CURVES OF ^{235}U AND ^{239}Pu

breed plutonium for the fast reactor at a sufficiently rapid rate and a mixed uranium plutonium carbide fuel, $U_{1-x}Pu_xC_{1+y}$ will be required; y will be small and would represent the presence of a small amount of uranium-plutonium sesquicarbide in addition to the uranium-plutonium monocarbide solid solution. In addition to carbide, uranium and plutonium oxycarbide nitride, and carbonitrides as solid solutions may also be suitable fuels for a fast reactor giving a more efficient breeding than with an oxide fuel. Because of the higher thermal conductivity of these fuels, restructuring to give a central void would not be expected and indeed has not been observed in carbide fuel operating at normal power ratings, the centre temperature being typically 1650-2050K, depending on the pellet diameter.

For the HTR systems the fuel is in the form of a coated particle. The fuel kernels are dioxide or dicarbide of uranium or uranium and thorium of 100-800 μm diameter coated in pyrolytic carbon giving an overall particle diameter of 0.2-1.0 mm in diameter. The temperature gradient across a granule is usually only 600K/cm but sufficient to cause mass transport. We shall now consider some of the chemical features of reactor fuel systems in more detail.

7.1 **Mass Transport.** When a pure oxide solid solution $U_{1-x}Pu_xO_{2\pm y}(MO_{2\pm y})$ is subjected to radial and axial thermal gradients, as in a fast reactor fuel element, redistribution of the cations and anions will occur; as the concentration of fission product elements increase, these elements will also redistribute within the gradient. Transport can occur by solid state and gas phase transport[64-65]. The solid state transport of oxygen is considerably faster than that of the cations. For the gas phase transport, Pu- and U- containing molecules together with carbon and hydrogen impurities have been considered as playing an important role, the latter by the formation of H_2O/H_2 and CO/CO_2 mixtures. But it is now evident that other species should be considered binary and ternary oxides of the fission product elements, for example, molybdenum trioxide $((MoO_3)_n)$ and caesium molybdate (Cs_2MoO_4).

The diffusion of oxygen in hypostoichiometric oxide fuels is from the hotter to the colder regions. The relationship between the concentration gradient and the temperature is given by the relation:

$$\ln x = \frac{Q^*}{RT} + C$$

where x is the deviation from stoichiometry and Q^* is a heat of transport the value of which depends on whether transport is via the solid state or the gas, and varies with each mode of transport. It should be noted, however, that oxygen to metal ratios in irradiated oxide fuels are difficult to determine accurately. Indirect determinations of the oxygen content from oxygen potential measurements have been attempted by miniature solid state emf techniques for Zircaloy-clad UO_2 pins for PWR systems[66].

A further effect of mass transport in a temperature gradient is the so-called amoeba effect in HTR coated particle fuels. Here the oxide or carbide kernel is found to migrate through the coating especially in high temperature gradients. The same possibilities of solid state and gas phase transport are possible as for fast reactor fuels.

7.2 Transport of Uranium and Plutonium. Since the diffusion coefficients of uranium and plutonium in the mixed oxides are much smaller than for oxygen, vapour phase transport is the most likely predominant mode for the redistribution of these elements in a fast reactor fuel. Theoretical calculations[65] and experimental observations indicate that the U/Pu ratio in the gas phase is greater than in the solid for oxides close to the stoichiometric composition but that the situation is reversed for markedly hypostoichiometric compositions (e.g. $U_{0.8}Pu_{0.2}O_{1.92}$). This is because for MO_2, UO_3 gaseous molecules are the major gas phase species and at low oxygen to metal ratios, it is the PuO gaseous molecules that become the major species. Thus when a stoichiometric oxide fuel is heated in a temperature gradient the central portion of the fuel during pore migration and formation of the central void will become depleted in uranium due to the formation of UO_3 gas. The uranium collects mainly in the cold region of the columnar grains. The transport of uranium leads, of course, to enrichment of plutonium in the central region of the fuel pin with consequent increase of centre temperatures. Conversely for very hypostoichiometric fuels plutonium depletion in the centre is observed.

7.3 Transport of Fission Products. Gamma-ray spectroscopy has provided much information on the movement of fission products; possible mechanisms of migration have been recently reviewed[69]. For example caesium migrates, probably as Cs, CsI, and perhaps CsO_x to the colder clad regions of the fuel pins. Molybdenum also collects at the fuel-cladding interface and is most probably transported through the gas phase; the concentration of MoO_3 gaseous molecules is too low in oxide fuel to account for all the observed diffusion; transport may be by gaseous molybdates, such as Cs_2MoO_4.

7.4 Mass Transport in Carbide Fuels. Redistribution of the components of carbide fuels can occur in temperature gradients as for the oxide systems, although, owing their higher thermal conductivities, the gradients of temperature are not usually so steep, and the redistribution is less. If, however, the diameter of the carbide fuel pins is increased for a given power output, then temperatures of above $1800^\circ C$ have been estimated and considerable restructuring has been observed; grain growth occurs in the hotter regions, plutonium and fission product elements also move down the temperature gradient. The presence of impurity oxygen results in the transport of carbon down the gradient by means of CO gas[70].

7.5 <u>The Chemical Effects Due to Fission.</u> We shall first discuss the chemical
changes which occur in fast reactor fuels. A knowledge of the chemical changes
or constitution of a fuel material during irradiation or 'burn-up' of the
fissile atoms is of vital importance to its efficient performance in a reactor
core as well as to its later reprocessing. Clearly there is a great economic
incentive to achieve high 'burn-ups', that is to maintain the fuel pin integrity
up to values above 10% fission of the heavy metal atoms. Indeed experience with
the Dounreay Fast Reactor (DFR) has shown that burn-ups of at least this value
can be achieved. This is 4 times that used in the thermal reactor systems on
the basis of fissions per initial heavy metal atom[71].

The problem of considering the chemical state or constitution of any burnt
fuel can be reduced frequently to that of a multicomponent system within a
thermodynamically closed system (at least for a non-vented fuel pin) on which a
temperature gradient is imposed. The analyses have at first considered the fuel
clad to be inert, forming the boundary of the closed system; however, the
boundary clad is often stainless steel which undergoes chemical reactions with
the fuel, and thus disturbs the material balance of the system.

We shall now consider the fate of the individual fission product elements,
first within an oxide fuel; the burn-up of the plutonium atoms is accompanied
by an increase in the oxygen:metal ratio of the fuel matrix, that is an increase
in the oxygen thermodynamic potential during burn-up.

A typical fission product spectrum for the fast fission of ^{239}Pu is given
in Table 7. With a knowledge of the fission yields it is possible to allocate
the oxygen atoms according to the stabilities of the oxides[65]. The final
chemical state will depend on the initial oxygen to metal ratio as well as on
the magnitude of the temperature gradient. The stabilities of the majority of
the fission product oxides are quite well documented and an Ellingham plot of
the oxygen potentials as a function of temperature for the fission product
oxides and the oxide fuel enables a prediction to be made of those fission
product elements which would be present as oxides and those remaining in
elemental form[72].

The rare earth elements, Y, Zr, Nb, Sr and Ba will all be present as oxides
whilst Rh, Ru, Pd and Tc will usually be present as alloys formed from the
elements. The oxygen potential $(\Delta \overline{G}_{O_2})$ for the reaction $Mo + O_2 = MoO_2$ is very
close to that in some regions of the fuel matrix. Caesium and rubidium
dissolve both iodine and oxygen and are found in the cold regions of the fuel
element, that is in the region of the clad[70,73]. The actual **form of barium
and strontium is difficult to predict**; it is required to have information about
the formation of alkaline earth uranates, plutonates, and zirconates or solid
solutions of the form $Ba_{1-x}Sr_x(U_{1-y-z}Pu_yZr_z)O_3$ and their equilibrium with BaO
and SrO and also with the fuel matrix, and to have correlations between the

oxygen potential of the fuel matrix and the composition of the inclusion phases
containing Ba found in many examinations of irradiated fuels.

Table 7

Elemental Fast Fission Yields for ^{239}Pu

(365 days irradiation and 180 days cooling time)

Element	Fission Yield (%)	Element	Fission Yield (%)
Se	0.4	Sn	0.3
Br	0.3	Sb	0.2
Kr	1.9	Te	1.0
Rb	1.8	I	1.5
Sr	3.7	Xe	23.7
Y	1.8	Cs	21.0
Zr	19.1	Ba	5.0
Nb	v. low	La	5.3
Mo	22.9	Ce	11.9
Tc	6.0	Pr	~4
Ru	23.1	Nd	15.0
Rh	7.0	Pm	2.1
Pd	15.0	Sm	3.2
Ag	2.0	Eu	0.8
Cd	1.0	Gd	0.4
In	0.1		

The transition metals Mo, Tc, Ru, Rh and Pd are present in irradiated fuels
in the form óf a single phase alloy[70,73-77]. The actual concentration of the
elements in these inclusions depends on the temperature and position in the
irradiated fuel; the amount of molybdenum in the alloy, for example, decreases
with increase in the oxygen potential providing it is sufficiently high. This
is well illustrated in some studies of the inclusion in irradiated plutonium
oxide of varying stoichiometry[78]. The molybdenum plays the role of a buffer
for the oxygen potential in the fuel; above a certain potential the molybdenum
in the inclusions is oxidised at the same time as the fuel matrix, and in fact
holds the oxygen potential of the fuel almost constant. Some use of this $Mo-MoO_2$
reaction has been made as a redox indicator to determine the oxygen potential
as a function of radial position[79] to examine pins of $U_{0.8}Pu_{0.2}O_{2.00}$ irradiated
to 6.2% burn-up; the oxygen potential at the cold region most likely rose but
there was evidence of oxygen transfer to the stainless steel clad so that various

mixed oxide compounds, for example Cs_2MoO_4 are formed. Tellurium has been found combined with both paladium[78] and silver[73].

The swelling due to these solid fission products can easily be calculated from their chemical state, and is found to be equivalent to a volume change of approximately 0.5% $\Delta V/V$ per 1% burn-up. Similar estimates for carbides and nitrides suggest that the swelling will be about 1% $\Delta V/V$ per 1% burn-up.

For the carbide again a knowledge of the appropriate fission product phase diagrams[80] and the carbon potentials for the fission product carbide formation allows a prediction of the chemical state or constitution of a burnt-up fuel. Because of the much lower temperature gradients, material transport problems and restructuring will not occur to any significant extent although oxygen present as an impurity could enhance material transport by the formation of carbon monoxide gas[81]. If the radius of the pin is increased or the density lowered, the corresponding increase in the magnitude of the temperature gradient will result in more material transport effects[65]. The temperature of the fuel will be significant in determining the form of the fission products in the various regions of the fuel; such a variation of the chemical form of the fission products with temperature will influence the carbon valence; the carbon content of the fuel before irradiation is also an important factor in determining the final carbon balance.

It seems likely that the rare earths - La, Ce, Pr and Nd - will dissolve in the monocarbide matrix certainly at 10% burn-up, but the fate of yttrium and samarium is not known; they could dissolve in the matrix or form a separate phase, $(Y,Sm)C_x$, $x<1$. Zirconium and niobium will also be dissolved in the monocarbide matrix. In the hotter regions of the fuel it is likely that $U_{1-x}Pu_x(Ru_{1-y-z}Rh_yPd_z)_3$ compounds will form and indeed they have been found in irradiated carbide[82]; in the colder regions ternary compounds $(U_{1-x}Pu_x)_2(Ru_{1-y-z}Rh_yPd_z)C_2$ could form. The temperature is also important in deciding the fate of technetium and molybdenum, both ternary compounds $U_{1-x}Pu_x(Mo_{1-y}Tc_y)C_2$ and Tc-Mo alloys could form. In the cold region of a fuel, with initial composition $(UPu)C_{1.0}$, the carbon potential of the fuel will be buffered either by the reaction $(Sr,Ba) + 2C = (Sr,Ba)C_2$ or alternatively by a reaction $(U,Pu)C + (Mo,Tc) + C = (U,Pu)(Mo,Tc)C_2$, both of which have lower carbon potentials than that required for the formation of the $(UPu)_2C_3$ phase; in the hotter regions, however, sesquicarbide will form.

Little information is available on the nitride fuel materials in terms of the chemical constitution, although estimates of the constitution have been made[83]. Zr,Y and the rare earths should dissolve in the mononitride matrix. Ba_3N_2 and Sr_3N_2 will be present as separate phases as well as the elements from molybdenum to palladium either as metals or ternary nitrides. Caesium and rubidium should be found in the cold regions of the fuel element. During burn-up the sesquinitride phase $(UPu)_2N_3$ will form and the nitrogen potential of the fuel will increase as burn-up proceeds. For fuels operating with high temperature gradients, material transport via the gas phase should be considered; nitrogen may diffuse from the hot fuel centre towards the clad.

Analyses of the chemical changes which occur during the irradiation of HTR fuel particles have also been made[84]. Again metallic inclusions containing Mo-Tc-Ru-Rh and Te-Pd have been identified in oxide kernels; the oxide forming fission products are precipitated as compounds for example a mixed barium strontium zirconate $((BaSr)ZrO_2)$ or form solid solutions with the oxide kernel matrix. Caesium is found in fission gas pores. Fission product inclusions were not found in UC_2 kernels and probably form a supersaturated solution.

7.6 <u>Oxide Fuel-Clad Compatability.</u> The presence of a liquid phase of fission product elements, containing Cs, Mo, Te, I and O, gives an efficient means for the removal of the protective oxide layer on the stainless steel clad, resulting in further attack of the cladding wall. Two types of attack have been noted, a broad front attack and intergranular corrosion. The broad front attack is not too serious, since the maximum penetration observed is little more than 10% of the clad wall thickness. The intergranular attack is the more serious and can penetrate further into the clad wall; high local concentrations of Cs, Te, I and Mo are observed in the grain boundaries, with chromium (depletion of the steel). The deeper intergranular attack probably results from penetration and reaction of a caesium liquid phase containing Rb, Cr, Mo, U and Pu, aided by stress corrosion and the presence of $Cr_{23}C_6$ precipitates in the sensitized steel.

8. <u>Reprocessing Nuclear Fuels</u>

Reprocessing may be defined as the treatment of irradiated fuel to recover fertile or fissile materials (mainly uranium and plutonium) and to separate them from fission products and from one another. Special fuels which have been irradiated to produce the higher actinides for example curium and californium may also be processed. The particular feature that differentiates processing nuclear fuels from processing non-nuclear materials is that the substances to be dealt with are intensely radioactive and can be handled only be remote operation behind shielding. In addition, since fissile material is being processed, precautions must be taken to avoid an uncontrolled nuclear chain reaction (criticality accident) occurring. Criticality problems in reprocessing have been discussed[85].

General accounts of the chemical processing of nuclear fuels and the recovery of plutonium in the reprocessing of nuclear fuels have been written[86-87].

The most difficult stage in reprocessing is that in which the initial separation of unganted fission products from uranium and plutonium is carried out. The irradiated fuel is 'cooled' to allow short lived isotopes to decay and, after

100 days, the major fission product contribution to the radioactivity of the
fuel is from the nuclides shown in Table 8.

Table 8

Major Fission Products of Importance to Reprocessing (100 days cooling)

Periodic Group	0	1	2	3	4	5	6	7	8
	^{3}H	^{137}Cs	^{89}Sr	^{90}Y	^{95}Zr	^{95}Nb	^{129m}Te	-	^{103}Ru
	^{85}Kr		^{90}Sr	^{91}Y					^{106}Ru
			^{137m}Ba	^{140}La					^{106}Rh
			^{140}Ba	^{141}Ce					
				^{143}Pr					
				^{144}Ce					
				^{144}Pr					
				^{147}Nd					
				^{147}Pm					

Tritium and krypton are volatile and ^{131}I, with a half life of 8.05d, is
only a problem in fuels cooled for short periods e.g. 50d. The fission products
most difficult to separate from uranium or plutonium are ruthenium and the
elements zirconium and niobium. The degree of separation required between any
pair of elements is expressed as a decontamination factor (DF) which is defined
as:

$$DF = \frac{\text{Concentration of element A in original material}}{\text{Concentration of element A in product}}$$

The decontamination factors specified[88] for the second reprocessing plant at the
British Nuclear Fuels Works at Windscale were:

	DF required for		
	Fission Products	U	Pu
Uranium product	10^7	-	3×10^5
Plutonium product	3×10^8	10^7	-

These figures refer to reprocessing metallic uranium fuel irradiated to the
equivalent of 3000 MWd/t at a mean rating of 2.7 MW/t and cooled for 130d.
Higher DF's would be needed for materials of higher burn-up or for fast reactor
fuels where the initial amounts of fission products are greater and where the
uranium:plutonium ratio is much lower.

8.1 <u>The Choice of a Primary Separation Process.</u> All large scale reprocessing
plants use a solvent extraction process in a nitrate medium. This is for 2
main reasons, first nitric acid is relatively non-corrosive to stainless steel
and the fabrication of plant is simple and second the actinide nitrates are
readily extractable into many organic solvents but, in general, the fission
product nitrates are not. Small specialized processes such as that for the
production of the higher actinides may use other media, for example chloride,

but this would not be acceptable for large scale operations.

Separation methods based on precipitation, ion exchange and extraction into an organic solvent have all been examined. The first large scale production of plutonium was based on co-precipitating plutonium, as Pu(IV), with bismuth phosphate[89]. The method used repeated oxidation-reduction cycles (Pu(IV) → Pu(VI)) since Pu(IV) is co-precipitated with $BiPO_4$ but PU(VI) is not. In this way those elements tending to co-precipitate with Pu(IV) and $BiPO_4$ were removed by precipitating $BiPO_4$ while keeping the plutonium in the hexavalent state. This use of the different properties of the valency states of plutonium is common to all separation methods but the large scale use of a remotely operated precipitation plant posed enormous mechanical, engineering and safety problems. Precipitation is now only used for specialised parts of a reprocessing cycle.

Ion exchange is not used for the primary separation and only one large scale study (processing 45 kg of irradiated uranium per day) has been carried out[90]. There are 3 main objections to the use of ion exchange:

(1) Organic resins are susceptible to radiolytic damage due to the action of the intensely ionizing radiations from alpha and beta particle emitters[91]. Inorganic exchangers have been examined but have not found favour.

(2) The rates of absorption can be slow and it may be necessary to use heated columns to obtain rapid absorption/exchange rates[92].

(3) Continuous plant operation is desirable and, although possible, is more difficult than in a liquid-liquid system. The use of ion exchange in the atomic energy industry has been reviewed[93].

Solvents, and diluents, also undergo radiolytic attack[87] but this can be minimized by keeping the contact time between the solvent and the radioactive solutions to a minimum. Extraction equilibria are usually established in seconds and on the rare occasions when they are slow, use can be made of this in the separation process. It is also relatively simple to recycle used solvent through a clean-up stage to remove degradation products which may interfere with the extraction process. For these reasons and for those already mentioned, solvent extraction is the preferred process for large scale separations.

The usual mechanism by which solvent extraction is achieved is to form adducts. The most commonly used extractant is tri-n-butyl phosphate (TBP) which forms adducts of the type $Pu(NO_3)_4 \cdot 2TBP$. The TBP is used as a 5-40% v/v solution in an inert hydrocarbon diluent. The exact concentration of extractant is a compromise between the extractability of the adduct, the solubility of the adduct in the organic phase, the extractabilities of adducts formed by unwanted elements e.g. ruthenium or zirconium and niobium, and the physical

properties (densities, surface tensions, viscosities, etc.) of the organic
and aqueous phases.

8.2 Dissolution of Fuel and Preparation of Feed for Solvent Extraction.

Before the dissolution stage, the fuel element cladding must either be removed
or cut to expose the fuel. In the UK metallic uranium fuel from magnox reactors
is removed from its can mechanically. Stainless steel or zircaloy clad oxide
fuels are sheared into short lengths to expose the fuel. Special treatment may
be needed to prepare carbide or nitride fuels for dissolution; such fuels are
not yet in commercial use. Chemical decanning (selective dissolution of cladding) and electrolytic dissolution of both can and fuel have been used only on
a limited scale[94].

Irradiated fuel is usually dissolved in about 7M nitric acid (HNO_3) yielding a final solution of around 3M HNO_3. In a non complexing acid such as
perchloric acid, disproportionation of IV valent plutonium (Pu(IV)) at low
acidities as shown[95] occurs:

$$3Pu(IV) = 2Pu(III) + Pu(VI)$$

and the net reaction is:

$$3Pu(IV) + 2H_2O = 2Pu(III) + PuO_2^{2+} = 4H^+$$

This disproportionation is highly acid dependent and it is possible to stabilise
the Pu(IV) valency state by increasing the acidity. With complexing acids like
nitric, powerful complexes can be formed[96] and the situation is more complicated;
the stabilisation of, for example, the Pu(IV) valency state is then due to a
combination of hydrolytic and nitrate complexing reactions. All the plutonium
valency states undergo hydrolytic reactions at low acidities and Pu(IV) solutions are of particular importance in reprocessing. Hydrolysis of Pu(IV) solutions slowly yields a polymeric form of plutonium which is very difficult to
redissolve. For this reason it is bad practice to dilute plutonium solutions
with water as local regions of high pH can be produced with the subsequent
risk of polymer formation[97].

Much of the power of the solvent extraction technique arises from the
differing extractabilities of metals in various valency states. The distribution of a given species between an organic extractant and an aqueous phase
is given by the extraction coefficient $E_{o/a}$ defined as:

$$E_{o/a} = \frac{\text{Concentration of species in the organic phase}}{\text{Concentration of species in the aqueous phase}}$$

Values of $E_{o/a}$ for the different valency states of some actinides are
given in Table 9 and have been reviewed elsewhere[98].

Table 9

Extraction Coefficients for U, Np, Pu, Am System:

4M HNO_3 and 10% v/v TBP in kerosene

Valency State $\rightarrow$	III	IV	V	VI
U	-	10 (25% TBP)	-	20
Np	-	3.0	0.13	11.0
Pu	0.14	11.5	-	2.5
Am	0.08	-	-	-

8.3 **The Stabilisation of Valency States.** It has already been shown that the valency states of plutonium can be stabilised by the control of the hydrogen ion and the complexing ion concentration. The desired valency state can also be stabilised by the addition of oxidising or reducing agents. The electrode potential E is given by the relation:

$$E = E_o + \frac{0.0591}{n} \log_{10} \frac{\text{(oxidised)}}{\text{(reduced)}}$$

where E_o is the standard potential, n the number of electrons involved in the reaction, (oxidised) and (reduced) refer to the concentrations of the oxidised and reduced forms under consideration. An example is the reduction of Pu(IV) by ferrous ion:

$$Pu(IV) + Fe(II) = Pu(III) + Fe(III)$$

The completeness of the reduction depends on the Fe(II)/Fe(III) mole ratio which must be maintained at a high value to keep the equilibrium over to the right hand side.

8.4 **Principles of Solvent Extraction.** The distribution of a substance which is soluble in 2 liquid phases which are themselves immiscible may be defined by the ratios of activities a_A and a_B which that substance has in the 2 solvents A and B respectively. It is assumed that neither of the solvents is saturated with solute and that the molecular state of the dissolved substance is the same in both solvents (i.e. no association or dissociation). In practice the activities can be replaced by concentration and the 2 phases usually consist of an aqueous and an organic phase, the latter, for convenience being called the solvent phase. The amounts extracted are governed by the relative volumes of solvent and aqueous phase so that the effective extraction (distribution) coefficient E_f is given by[99]

$$E_f = \frac{L}{H} \cdot E_{o/a}$$

where L and H are the volume of organic (light) and heavy (aqueous) phase.

Where successive equilibrations (extractions) are to be carried out, the amount W_n of solute remaining in the aqueous phase is given by:

$$W_n = W_o \left(1 - \frac{v\,E_{o/a}}{V + v\,E_{o/a}}\right)^n = W_o \left(\frac{V}{V + v\,E_{o/a}}\right)^n$$

where W_o is the original weight of solute in a given volume (V) of initial aqueous phase and v is the volume of organic extractant used for each extraction. For W_n, the amount of solute remaining in the aqueous phase, to be minimised it is desirable that n, the number of extractions is large and that v, the volume of extractant is small. It is also desirable that $E_{o/a}$ is large.

In reprocessing irradiated fuel the aqueous phase contains a large number of solutes and the separations depend on the differences in extraction coefficient of the individual solute. Where the separation factor between species 1 and 2 (given by $E_{1o/a}/E_{2o/a}$) is small, as in the separation of adjacent rare earths, many extraction stages will be needed. The extractions are normally carried out in a counter-current fashion where, after an extraction stage, the organic and aqueous phases move in opposite directions to contact fresh aqueous or solvent phases. This is shown in Fig. 2 in the form of a vertical column; horizontal extractors (mixer-settlers) are also used. The various types of solvent extraction equipment that exist have been reviewed[99].

8.5 **Reprocessing in Practice.** A major reprocessing plant can process up to 7 t/d of irradiated fuel and most plants use TBP for the primary separation[87]. Many variants of the TBP-based separation process exist[100] and, in general, there are 2 modes of operation. In the widely used Purex process, the uranium and plutonium are co-extracted into TBP and are then separated after the reduction of Pu(IV) to the inextractable Pu(III). The uranium and plutonium are separated in the first cycle. In the Windscale-TBP process (Fig. 3) the uranium and plutonium are also co-extracted but they are then both backwashed into an aqueous phase. The uranium-plutonium separation is carried out after re-extracting both elements into TBP, the plutonium again being selectively backwashed into the aqueous phase as Pu(III). When processing oxide fuel in the Windscale plant, a preliminary solvent extraction cycle is carried out using dibutylcarbitol (Butex)[101]. In the TBP processes, unless special precautions are taken, most of the neptunium is found in the first TBP cycle with the highly active wastes and methods for its recovery have been described[102].

The uranium and plutonium streams from the primary separation go to further solvent extraction cycles where the remaining fission products and other impurities are removed. The purification cycles may use other solvents for example in France the tertiary amine trilaurylamine (TLA) in an inert diluent is used for plutonium purification[103].

Ion exchange may also be used for the purification of uranium, neptunium and plutonium[104].

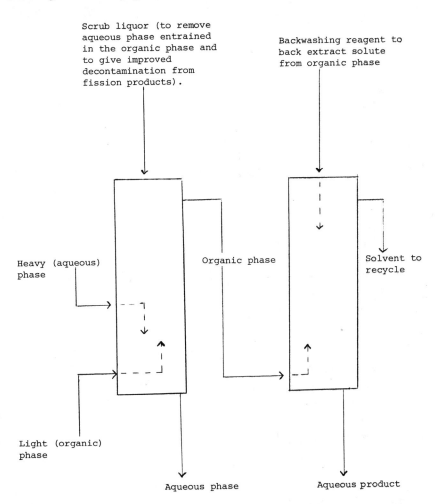

FIG.2   Counter current extraction system

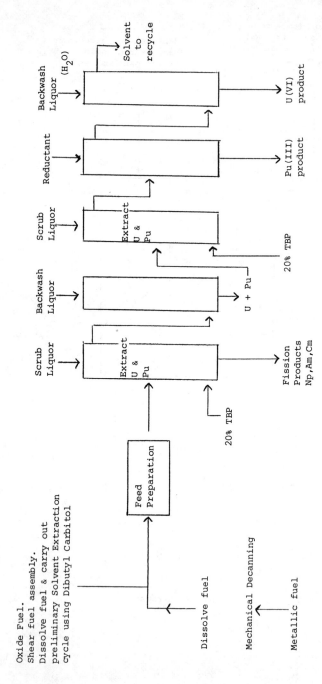

FIG.3   The Windscale-TBP process

8.6 **Specialised Processes.** ^{238}Pu is produced in the USA and the UK by the neutron irradiation of ^{237}Np:

$$^{237}Np \xrightarrow{n,\gamma} {}^{238}Np \xrightarrow{\beta^-} {}^{238}Pu$$
$$\downarrow$$
$$fission$$

It is used as a heat source for the production of electrical energy for example for satellite power sources and for heart pacers, and its production and purification have been described[105]. Some of the considerations in selecting the purification process have also been described[91].

The separation and purification of ^{243}Am (half-life 7,400y), ^{244}Cm (half-life 18.1y) and ^{252}Cf (half-life 2.6y) pose particularly difficult problems due to their very high specific activities and, in the case of ^{252}Cf to the presence of high neutron fluxes.

The uses of some of the higher actinides and their preparation by neutron irradiation of special targets in a nuclear reactor has been described[106]. The most highly developed separation process for the higher actinides is the TRAMEX[97] process (Tertiary Amine Extraction) in which the transplutonium elements are separated as a group. This group is then split into an americium-curium and a transcurium fraction by extraction into 2 Ethylhexyl phenyl phosphoric acid in Diethylbenzene.

Organophosphorous reagents are also used in the TALSPEAK (Trivalent Actinide Lanthanide Separation by Phosphorous Reagent Extraction) process[108] which separates the lanthanides from the trivalent actinides.

9. **Vitrification of the Highly Active Waste**

After the fuel has been dissolved in nitric acid and the valuable uranium and plutonium extracted, the remaining acid solution is a waste product. This Highly Active Waste (HAW) contains about 99.5% of the non-volatile fission products and also the higher actinides (Np, Am, Cm) produced by neutron capture reactions in the reactor. The solvent extraction process is very efficient and only a small fraction of the uranium and plutonium (typically much less than 0.5%) finds its way into this waste stream.

At present, this HAW is stored in double walled stainless steel tanks at the reprocessing plant - in the UK, at Windscale. The waste is extremely radioactive (6500 Ci/l at 1y cooling to 900 Ci/l at 10y) and the tanks are shielded with several feet of concrete. The waste is not a simple solution and, after it has been partly evaporated for storage, it contains a considerable amount of solid sludge. Consequently stirring is provided through air sparge pipes. The waste emits appreciable quantities of heat (20 watts/litre after 1 year and 2.5 watts/litre after 10 years) and 2 internal cooling coils are fitted. In

addition to the fission products, the waste also contains other elements. The
current composition (expressed as weight per cent of oxides) of the Windscale
waste is approximately:

Fission Products	39 wt %
Al_2O_3	20
Fe_2O_3	11
Cr_2O_3	2
NiO	1
MgO	25
ZnO	2

The aluminium comes from additions to the fuel, magnesium from the cladding
and iron chromium and nickel from corrosion of the process vessels. HAW from
oxide fuel reprocessing will not contain the aluminium and magnesium but will
contain the corrosion products. Some wastes may contain as much as 50% gad-
olinium oxide if the latter is added to the dissolver as a neutron poison to
prevent criticality incidents when high plutonium content fuels are being
processed. The UK and US methods have been described[109-110].

Tank storage has been proved to be satisfactory for storage periods of
decades, but needs an undesirably high level of surveillance for long term
storage. Accordingly, all countries with reprocessing plants have been developing processes for solidifying the waste[111]. Because the waste contains over 20
elements with a concentration of more than 0.5 wt %, it is thought that only a
glass or glass ceramic is capable of solidifying all of them satisfactorily -
merely calcining the waste produces an easily dispersable water-soluble product.

Some early work[112] used phosphate glasses for the fixation. This is not
now favoured because the leachability by water increases by 10-100 times when
the glasses crystallise or devitrify and because the phosphates corrode the
process vessels. At present, borosilicate glasses are preferred, the exact composition depending on the waste to be solidified and personal preferences. Some
selected compositions are given in Table 10 in wt % oxides.

Although there is some unanimity on the composition of glass to be used,
the process to be used for making it varies considerably from country to country.
In the UK, the original Fingal process[113] and the Harvest plant[114-116] at
present under development, carry out the whole process in the final container.
The liquid waste and a slurry of glass making chemicals is added at the top of
the heated vessel and evaporation, calcination and glass making occur in a

gradually rising series of zones. The process has the advantages of simplicity, of lack of moving parts and that the process vessel has to last for only a single run. On the debit side, it is not easy to optimise the conditions for any of the stages. In particular, fission product ruthenium is likely to be oxidised leading to the formation of gaseous RuO_4, and consequent loss of Ru.

Table 10

	UK	France	Germany	US	German Glass-Ceramic
Ref	(5)	(6)	(7)	(8)	(9)
Waste	25.7	24.5	20.0	26.1	20.0
Fission Products	10.0	20.2	16.3	24.0	16.3
SiO_2	50.9	49.0	40.4	27.3	26.0
B_2O_3	11.1	13.3	10.9	11.1	5.2
Na_2O	8.3	8.2	22.0	4.1	3.2
Li_2O	4.0			K_2O 4.1	1.2
Al_2O_3		5.0	1.1		10.8
CaO			2.2	1.5	6.0
TiO_2			3.4		6.4
MgO				1.5	1.2
SrO				1.5	
BaO				1.5	15.6
ZnO				21.3	4.0
ZrO_2					0.8

In France, the AVM plant (Atelier de Vitrification de Marcoule) is in an advanced stage of development[117] and will probably start up in 1978. Here the waste is first denitrated and calcined in a rotary calciner. The calcined glass frit is then fed into a melting furnace which is periodically emptied into the storage vessels. The advantages and disadvantages are almost exactly the converse of the British process. The German Vera process[118] is similar to the French process except that a spray calciner is used. In the US, various methods have been developed to the pilot plant scale[119]. Among other possible waste storage substances that have and are being considered are: (a) Dispersing phosphate glass spheroids in a metal matrix[120] (b) using a glass ceramic[117] and (c) 'tailoring' a ceramic composition to immobilise the more important fission products individually[121]. The chief problem with any process will be in dealing with the off-gases, ensuring that these do not generate large volumes of low active wastes.

Once the glass has been made, the containers will be stored in water filled ponds or in an air cooled repository for a few years at least. After that, the

possible disposal routes considered are in the deep ocean, beneath the ocean
floor or in some geological formation which is likely to remain isolated from
man for the forseeable future[122].

Various properties of the glasses have been studied so as to evaluate their
likely behaviour in storage or on disposal. Chief among these are their leaching characteristics and their radiation stability.

The leaching characteristics have been studied in many laboratories[118, 121,
123-127]. Typical rates would be 10^{-3} to 10^{-4} g. cm^{-2} d^{-1} at $100^{\circ}C$ and 10^{-6}
to 10^{-7} at $5^{\circ}C$. They can be markedly affected by the leachant: acid waters
for instance leach some glasses very quickly. An experiment at Chalk River[123]
where some glass blocks were buried in the soil below the water table gave
reassuringly low leach rates, decreasing with time. The leach rates of samples
deliberately or accidentally crystallised (devitrified) are in general not
much changed.

The stability of glasses to radiation damage has also been examined. Calculations show that in terms of atomic displacements, over 90% of the damage
will come from α-decay of the incorporated actinides. Experiments have been
carried out in which the equivalent thousands of years of storage is simulated
in the laboratory in a year or 2 by doping the sample with a few per cent of a
short half life α-emitter e.g. ^{244}Cm or ^{238}Pu. These experiments[118, 124-126]
have been uniformly reassuring, only small changes being observed in density
leach rate or structure and only small amounts of stored Wigner energy being
found.

10. Practical Applications of the Higher Actinide Elements[128-130]

The formation of higher actinide elements by neutron capture has been described earlier in section 3.4 with a schematic representation of the build-up
chains. These nuclides are playing an increasingly important role in industry
as energy and radiation sources. Their field of application extends from
nuclear power, through space travel, metereology, process control, analytical
techniques, biology and medicine to factories and offices.

10.1 Energy Applications. Plutonium 238 and curium 244 are used as energy
sources for terrestrial and extra terrestrial applications that call for the
conversion to electricity of the heat released during decay. These batteries
operate on thermoelectric conversion and have the advantage of taking up little
space and requiring no maintenance. SNAP 19 having a total weight of 12 kg
fuelled with plutonium 238 producing 60 watts electrical output has been cir-
cling the earth in the weather satellite NIMBUS B since May 1968. SNAP 27 also
fuelled with plutonium 238 was left on the moon by Apollo 12 astronauts. One
purely thermal application has been to use a unit containing 750 grams of plutonium 238, and producing 420 watts of thermal power to heat a fluid which

circulates through the veins of a deep sea diving suit.

10.2 Radiation Applications.

10.2.1 X-ray fluorescence. The radiation from americium 241 has a particularly important application in the determination of trace elements by X-ray fluorescence, in particular those elements having atomic numbers up to 69. The bombardment of a suitable low atomic number target material by high energy alpha particles produces a mono-energetic source of X-rays in the range 0.1 - 5keV which is used for low energy X-ray fluorescence analysis.

10.2.2 γ gauging techniques. Gauging is perhaps the most common industrial application of radioisotope sources. The thickness of sheet metal, glass, plastics and rubber moving on a conveyor belt at thicknesses up to 10 mm in steel is determined by transmission measurement of the 59.6 KeV americium 241 gamma. For substances of low atomic number for which the transmission measurement is not sensitive enough, gamma back scatter is used.

The thickness of coatings of plastics on metal are determined by X-ray back scatter fluorescence measurements using plutonium 238 or americium 241 low energy photons to excite X-rays from the substrate metal. The X-rays are attenuated by the coating and their intensity gives a measure of the coating thickness. For thin metal coatings on a metal substrate with a small difference in atomic number, the same technique is used, the intensity of the X-rays from the metal coating are measured.

Absorption of the americium 241 59.6 KeV gamma ray is used to measure the content of a high atomic number element in a low mean atomic number solution for example sulphur in oil, lead in petrol or ash in paper. In the filling of beer cans the intensity of the transmitted americium 241 gamma is used to activate switches when preset intensity levels are reached.

10.2.3 Neutron radiography. A very important application of the higher actinides is in their use as neutron sources in neutron radiography and activation analysis. Cf 252 emitting 2.31×10^{12} neutrons per second per gram is capable of supplementing reactors and accelerators as a source of neutrons and has the advantage of providing a portable point neutron source for use in remote and difficult locations. Alternatively the alpha activity of the higher actinides is used to produce neutrons by α-n reactions with elements of low atomic number such as beryllium. The most common source uses americium 241 and beryllium. Neutron radiography is used for the non-destructive examination of components incorporating light elements and metals such as inservice aircraft components, turbine blades and valve assemblies. Californium 252 is used as an intense implantable neutron source for cancer therapy. Neutron radiography compliments X-radiography in the medical and biological fields by giving an improved delineation of air filled structures in soft tissues.

10.2.4 <u>Neutron activation analysis.</u> Fast or thermal neutrons from americium 241
- beryllium or californium 252 sources are used to initiate characteristic reactions with specific elements; the intensity of the resulting prompt or delayed
gammas is a measure of the amount of element present. This technique is used
for the determination of fluorine in organic compounds without interference
from oxygen, for the process control of vanadium in oil at detection levels
between 0.1-100 ppm, for sulphur in coal, for down hole analysis of uranium and
copper at levels less than 0.01%, for silica, chromium and aluminium in ores,
for oil and water well logging, to determine the composition of river bed sedi-
ment and in forensic science for trace element analysis.

An important application of californium 252 is in nuclear safeguards and
the measurement of fuel uniformity in the nuclear industry. Neutrons from
californium 252 cause fissions in the nuclear material; the fission neutrons or
gamma intensity give a measure of the fissile content.

10.2.5 <u>Neutron scattering.</u> The thermalisation of fast neutrons from californium
252 by collisions with H_2 or other light elements is used to measure the moisture content of foodstuffs such as non fat milk, potato granules and powdered
orange juice in tins, soils, wood and concrete.

The attenuation of thermal neutrons is used to measure explosive charge
weights and fill heights in ordnance applications, and in process control to
measure the boron content of detergents and glass. The intensity of the transmitted beam as measured in a BF_3 or H_3 detector gives the percentage concentration of these elements.

10.2.6 <u>Alpha emission.</u> The alpha emission from americium 241 provides the
ionising radiation in smoke detectors which are extensively used in factories,
hotels and cinemas. Alpha particle scattering is used in surface analysis, a
100 mCi curium 242 source was used to perform the chemical analysis of the
lunar surface, einsteinium 254 alpha emission was used as the calibration source.

References

[1]G T Seaborg (eds. W Miller, H Blank) Heavy Element Properties Proceedings of 4th International Conference, Baden-Baden 1975 (North Holland,
American Elsevier, 1976), 3.

[2](ed. M J Jones) Geology, Mining and Extractive Processing of Uranium Proceedings of an International Symposium, London (The Institution of Mining and
Metallurgy, 1977).

[3]J A Lane, L G Alexander, L L Bennett, W L Carter, A M Perry, M W Rosenthal,
L Spewak Peaceful Uses of Atomic Energy Proceedings of 3rd International
Conference, Geneva 1964 (UN New York 1965), 6, 333.

[4]S H U Bowie, ref. 2, 76.

[5]E J Hanrahan, R E Williamson, Atomic Industrial Forum International Conference
on Uranium Proceedings of Conference, Geneva, Sept. 1976.

[6]Nuclear Energy Agency and International Atomic Energy Agency Uranium resources
production and demand, including other nuclear fuel cycle data (Paris: OECD
1976), 78.

[7]W D Wilkinson, Uranium Metallurgy, Vol. 1, Process Metallurgy Vol. II
Corrosion and Alloys, (Interscience - John Wiley, New York 1962).

[8]R C Merrett, The Extraction Metallurgy of Uranium (Colorado School of Mines
and US Atomic Energy Commission 1971).

[9]R G Bellamy, N A Hill, The Extraction and Metallurgy of Uranium, Thorium and
Beryllium (Oxford Pergamon Press, Oxford 1971).

[10]The Recovery of Uranium IAEA symposium, Sao Paulo, 1970 (IAEA, Vienna 1971).

[11]S E Smith, K H Garrett, The Chemical Engineer, 1972, 268, 440.

[12]P Mouret, P Vertes, J Sauteron, (ed. P Pascal), Preparation d'Uranium en
Nouveau Traite de Chimie Minerale (Masson et Cie, Paris 1960), XV(1), 95

[13]P Mouret, ibid (1967) XV(4), 12.

[14]H Huet, ibid (1967) XV(4), 19.

[15]J H Gittus, Uranium, (Butterworths, London, 1963).

[16]C Boorman (ed. H London) in Separation of Isotopes, (Newnes, London 1961), 332.

[17]J Shacter, E von Halle, R L Hoglund (eds. H F Mark, J F McKetta, D F Othmer)
in Encyclopaedia of Chemical Technology (Interscience, New York, 1965), 7, 91.

[18]W Growth, ref 16, 249.

[19]D G Avery, M Boggardt, P Jelinek-Fink, J V L Parry Peaceful Uses of Atomic
Energy Proceedings of 4th International Conference, Geneva 1974 (UN New York,
1972), 9, 53.

[20]Uranium Isotope Separation Proceedings of an International Conference, London,
March 1976 (British Nuclear Energy Society in conjunction with Kerntechnische
Gessellschaft 1976)

[21]F S Patton, J M Googin, W L Griffith, Enriched Uranium Processing, (Pergamon
Press, Oxford, 1963).

[22] C D Harrington, A D Rueble, Uranium Production Technology, (van Nostrand, Princeton 1959).

[23] S H Smiley, D C Brater, Peaceful Uses of Atomic Energy Proceedings of 2nd International Conference, Geneva, 1958 (UN New York, 1958), 4, 384.

[24] Production Technology of ^{237}Np and ^{239}Pu Proceedings of a symposium, Denver, 1964 (Ind. Eng. Chem. Proc. Des. Dev., 1964) 3, 289.

[25] D C Hoffman, F O Lawrence, J L Mewherter, F M Rourke, Nature, 1971, 234, 132.

[26] J A Leang, L J Mullins (eds. H F Mark, J J McKetta, D F Othmer), in Encyclopaedia of Chemical Technology, (Interscience, New York 1968), 15, 879.

[27] (ed. O J Wick), Plutonium Handbook, (US Atomic Energy Commission 1967), 2 volumes.

[28] See for example, (eds. G N Walton, A G Ashburn, J K Dawson, M B Waldron, D J O'Conner), Glove Boxes and Shielded Cells (Butterworths, London, 1958).

[29] W W Schulz, The Chemistry of Americium, ERDA Review TID 26971 (1976).

[30] R D Baybarz, Atomic Energy Res., 1970, 8, 327.

[31] D E Ferguson, J E Bigelow, Actinide Reviews, 1969, 1, 213.

[32] M Haissinsky, (ed. P Pascal), Noveau Traite de Chimie Minerale, (Masson et Cie, Paris, 1969), XV, 267.

[33] Power Reactors in Member States, 1976 Edition, (IAEA Vienna, 1976).

[34] Power Reactors Supplement, Nuclear Engineering International, 1977, 22.

[35] R S Pease, J. British Nucl. Energy Soc., 1976, 11, 2.

[36] A V Nero Jr., Guidebook to Nuclear Reactors, (California University, Berkeley, USA, 1976) LBL 5206.

[37] Steam Generating and Other Heavy Water Reactors, (British Nuclear Energy Society, London, 1968).

[38] Berkeley and Bradwell Nuclear Power Stations Proceedings of Symposium London, June 1963, (Institute of Mechanical Engineers, London, 1964).

[39] K Saddington, T N Marsham, J Moore, Peaceful Uses of Atomic Energy Proceedings of 3rd International Conference, Geneva, 1964 (UN New York, 1965), 5, 90.

[40] Gas-cooled Reactors with Emphasis on Advanced Systems, Proceedings of Symposium Julich, October 1975, (IAEA Vienna, 1976), 541.

[41] Fast Reactor Power Stations Proceedings of International Conference, London, March 1974, (British Nuclear Energy Society-Thomas Telford Ltd., London 1974).

[42] A B Johnson Jr., Reviews of Coatings and Corrosion, (Freund Publishing House Ltd., Israel, 1975).

[43] N K Taylor, UKAEA Report AERE-R 8164, 1976.

[44] P Cohen, Water Coolant Technology of Power Reactors, ANS-USAEC monograph (Gordon and Breach, New York, 1969).

[45] R Lind, J Wright, **Peaceful Uses of Atomic Energy** Proceedings of 3rd International Conference, Geneva, 1964, (UN New York, 1965), 9, 541.

[46] C J Wood, **CEGB Research**, Sept 1976, 3.

[47] See **Nuclear Engineering and Design**, 26, 1974.

[48] A W Thorley, J R Findlay, A Hooper, UKAEA Report TRG Report 2856(R) 1976.

[49] See **Liquid Alkali Metals** Proceedings of International conference, Nottingham, April 1973, (BNES - Thomas Telford Ltd., London, 1973).

[50] F Butler et al, **Peaceful Uses of Atomic Energy** Proceedings of 2nd International Conference, Geneva, 1958, (UN New York, 1958), 6, 317.

[51] G B Greenough, P Murray, **Nuclear Metallurgy**, 1962, 10, 83.

[52] J C C Stewart, et al, **Peaceful Uses of Atomic Energy** Proceedings of 3rd International Conference, Geneva, 1964, (UN New York, 1965), 10, 132.

[53] H Rogan, et al, **Peaceful Uses of Atomic Energy** Proceedings of 4th International Conference, Geneva, 1971, (UN New York, 1972), 8, 215.

[54] (ed. J Belle) **Uranium Dioxide: Properties and Nucl. Appls.** (USAEC 1961).

[55] H Stehle et al, **Nucl. Eng. Des.** 1975, 33, 230.

[56] H Rogan et al, **Trans. Amer. Nucl. Soc.** 1975, 20, 611.

[57] **Fuel and Fuel Elements for Fast Reactors** Proceedings of a symposium, Brussels, 1973, (IAEA Vienna, 1973) II.

[58] J M Horspool et al ibid, I, 3.

[59] W O Harms, (eds. J E Hove, W C Riley), **Modern Ceramics: Some Principles and Concepts** (Wiley, New York, 1965), 290.

[60] N Piccinni, **Adv. Nucl. Sci. Techn.** 1975, 8, 255.

[61] H Bairiot, J Vangeel, L Aerts, **Amer. Nucl. Soc.** Transactions of Winter Meeting Washington, November 1970, 13(2), 582.

[62] **Behaviour and Chemical State of Irradiated Ceramic Fuels**, Proceedings of a Panel, Vienna, 1972 (IAEA Vienna, 1974).

[63] **Thermodynamics of Nuclear Materials 1974**, Proceedings of a symposium, Vienna, 1974, (IAEA, Vienna, 1975), 2 volumes.

[64] E A Aitken, **J.Nucl. Mats.**, 1969, 30, 62.

[65] M H Rand, T L Markin, **Thermodynamics of Nucl. Mat.**, Proceedings of a symposium, Vienna, Sept. 1967, (IAEA, Vienna, 1968).

[66] M G Adamson, E A Aitken, S K Evans, J H Davies, **Thermodynamics of Nuclear Materials 1974** Proceedings of a symposium, Vienna, 1974, (IAEA, Vienna, 1975) I, 59.

[67] J E Battles, W A Shinn, P E Blackburn, R K Edwards, **Plutonium 1970** Proceedings of 4th International Conference, Santa Fe, 1970 (Am. Inst. Mining, Met. and Pet. Eng. and Inst. Met., 1970), 17 pt. II, 733.

[68] R W Ohse, W M Olsen ibid, 743.

[69] in ref. 62, 211.

[70] J M Horspool, N Parkinson, J R Findlay, P E Potter, M H Rand, L E Russell, W Batey, Fuels and Fuel Elements for Fast Reactors, Proceedings of International Symposium, Brussels, July 1973 (IAEA, Vienna, 1974), $\underline{I}$, 3.

[71] T N Marsham, UKAEA Publication Atom, 1973 $\underline{201}$, 150.

[72] O Kubaschewski, E L I Evand, C B Alcock, Metallurgical Thermochemistry, (Pergamon Press, Oxford, 1967) 4th edition.

[73] H Huber, H Kleykamp, GfK report KFK 1324, 1972.

[74] D R O'Boyle, F L Brown, J E Sanecke, J. Nucl. Mats., 1969, $\underline{29}$, 200.

[75] B T Bradbury, J T Demant, P M Martin, D M Poole, J. Nucl. Mats. 1965, $\underline{17}$, 227.

[76] B M Jeffery, J. Nucl. Mats., 1967, $\underline{22}$, 33.

[77] J L Bramman, T M Sharpe, D Thom, G Yates, J. Nucl. Mats. 1968, $\underline{25}$, 202.

[78] J H Davies, F T Ewart, J. Nucl. Mats., 1971, $\underline{41}$, 143.

[79] C E Johnson, I Johnson, P Blackburn, J E Battles, C E Crouthamel, in ref. 62, 1.

[80] H R Haines, P E Potter, in ref. 63, $\underline{2}$, 145.

[81] M G Adamson, UKAEA report AERE-R 6897 (1972).

[82] J L Bramman, R M Sharpe, R Dixon, J. Nucl. Mats., 1971, $\underline{38}$, 226.

[83] J M Leitnaker, R L Beatty, K E Spear, Am. Nucl. Soc. Trans., 1969, $\underline{12}$, 86.

[84] R Forthmann, H Grübmeier, H Kleykamp, A Naoumidis in ref 63, 147.

[85] E D Clayton, W A Reardon, Plutonium Handbook, (Gorden and Breach, Science Publishers, New York 1967), $\underline{2}$, Chapter 27.

[86] F S Martin, G L Miles, Chemical Processing of Nuclear Fuels, (Butterworths Scientific Publications, London, 1958).

[87] H A C McKay, D Scargill, A G Wain, Gmelin Handbuch Transurane, 1974, Band 7b, Teil A1, 177.

[88] B F Warner, W W Marshall, A Naylor, G D C Short, Peaceful Uses of Atomic Energy Proceedings of 3rd International Conference, Geneva, 1964 (UN New York, 1964) 10, 224.

[89] R L Stevenson, P E Smith, The Reactor Handbook, (Fuel Reprocessing, Interscience Publishers, 1961), vol 2, Chapter 4, 227.

[90] A M Aikin, Chem. Eng. Prog., 1957, $\underline{53}$, 82F.

[91] I L Jenkins, Actinides Reviews, 1969, $\underline{1}$, 187.

[92] J L Ryan and E J Wheelwright, Peaceful Uses of Atomic Energy Proceedings of 2nd International Conference, Geneva, 1958 (UN New York, 1958), $\underline{17}$, 137.

[93] I L Jenkins, A G Wain, Reports on the Progress of Applied Chemistry, 1974, $\underline{57}$, 308.

[94] R E Blanco, C D Watson, The Reactor Handbook, (Fuel Reprocessing, Interscience Publishers, 1961), Vol. 2, Chapter 3.

[95] J J Katz and G T Seaborg, The Chemistry of the Actinide Elements, (Methuen & Co. Ltd., London, 1957).

[96] J M Cleveland, Coordination Chemistry Reviews, 1970, 5, 101.

[97] J M Cleveland, Plutonium Handbook, (Gordon & Breach, Science Publishers, New York, 1967), Vol. 2, Chapter 13.

[98] H A C McKay, D Scargill, A G Wain, Gmelin Handbuch, Transurane, 1975, Band 21, Teil D2, 177.

[99] See for example, J F Flagg, Chemical Processing of Reactor Fuels, (Academic Press, New York and London, 1961).

[100] M A Awwal, D J Carswell, Chem. Rev., 1966, 66, 279.

[101] C M Nicholls, R Spence, Trans. Instn. Chem. Engrs., 1957, 35, 380.

[102] J S Nairn, D A Collins, Prog. in Nucl. Energy, Series III, 2, 518.

[103] A Chesne, G Koehly, A Bathellier, Nucl. Sci. Eng., 1963, 17, 557.

[104] F W Tober, Peaceful Uses of Atomic Energy Proceedings of 2nd International Conference, Geneva, 1958 (UN New York, 1958) 17, 574.

[105] Series of papers in Ind. Eng. Chem. Process Des. Dev. 1964, 3.

[106] G T Seaborg, Isotopes Radiat. Technol., 1968, 6, 1.

[107] See for example, references in R E Leize, R D Baybarz, B Weaver, Nucl. Sic. Eng. 1963, 17, 252.

[108] B Weaver, F A Kappelmann, J. Inorg. Nucl. Chem., 1968, 30, 263.

[109] D W Clelland, Treatment and Storage of High Level Radioactive Wastes, Symposium Vienna, 1962, (IAEA, Vienna, 1963), 63.

[110] B M Legler, G Bray, Chem. Eng. Prog., March 1976, 72(3), 52.

[111] Management of Radioactive Wastes from the Nuclear Fuel Cycle Proceedings of symposium, Vienna, Marcy 1976, (IAEA Vienna, 1976) 2 volumes.

[112] J R Grover, Management of Radioactive Wastes from Fuels Reprocessing Proceedings of symposium, Paris 1972, (OECD, Paris, 1973), 593.

[113] K D B Johnson, J R Grover, W H Hardwick, Peaceful Uses of Atomic Energy Proceedings of 3rd International Conference, Geneva, 1964, (UN New York 1965), 14, 244.

[114] A D W Corbett, G G Hall, G T Spiller in ref. 111, 1, 217.

[115] B F Warner, M J Larkin, R J Taylor, L Lowes, S N Johnson in ref. 111, 1, 231.

[116] J B Morris, B E Chidley in ref. 111, 1, 241.

[117] A Jouan, R Cartier, R Bonniaud, C Sombret in ref. 111, 1, 259.

[118] W Guler, W Hild, F Kaufmann, H Koschorke, H Krause, G Rudolph, J Saidl, K Scheffler in ref. 111, 1, 271.

[119] J L McElroy, W F Bonner, H T Blair, W J Bjorklund, C C Chapman, R D Kierks, L S Romero in ref. 111, 1, 283.

[120] J van Geel, H Eschrich, W Heimerl, P Grziwa in ref. 111, 1, 341.

[121] G J McCarthy, Nuclear Technology, 1977, 32, 92.

[122] P D Grimwood, G A M Webb, National Radiological Protection Board, Harwell, UK Report NRPB - R48, October 1976.

[123] W F Merritt in ref. 111, 11, 27.

[124] A R Hall, J T Dalton, B Hudson, J A C Marples, in ref 111, 11, 1.

[125] F Laude, R Bonniaud, C Sombret, G Rabot in ref. 111, 11, 37.

[126] J E Mendel, W A Ross, F P Roberts, R P Turcotte, Y B Katayama, J H Westsik in ref. 111, 11, 49.

[127] A K De, B Luckscheiter, W Lutze, G Mallow, E Schiewer and S Tymochowicz in ref. 111, 11, 63.

[128] E A Lorch, The Radiochemical Centre Ltd., Data available on request.

[129] C Keller, The Chemistry of the Transuranium Elements, Verlag Chemie, GmbH 1971.

[130] G T Seaborg Isotop. Radiat. Technol., 1968, 6, 1.